T0245253

CAMBRIDGE LIBRARY COLLECTION

Books of enduring scholarly value

Life Sciences

Until the nineteenth century, the various subjects now known as the life sciences were regarded either as arcane studies which had little impact on ordinary daily life, or as a genteel hobby for the leisured classes. The increasing academic rigour and systematisation brought to the study of botany, zoology and other disciplines, and their adoption in university curricula, are reflected in the books reissued in this series.

The Natural History of Birds

Georges-Louis Leclerc, Comte de Buffon (1707–88) was a French mathematician who was considered one of the leading naturalists of the Enlightenment. An acquaintance of Voltaire and other intellectuals, he work as Keeper at the Jardin du Roi from 1739, and this inspired him to research and publish a vast encyclopaedia and survey of natural history, the ground-breaking *Histoire Naturelle*, which he published in forty-four volumes between 1749 and 1804. These volumes, first published between 1770 and 1783 and translated into English in 1793, contain Buffon's survey and descriptions of birds from the *Histoire Naturelle*. Based on recorded observations of birds both in France and in other countries, these volumes provide detailed descriptions of various bird species, their habitats and behaviours and were the first publications to present a comprehensive account of eighteenth-century ornithology. Volume 9 covers water fowl and related birds.

Cambridge University Press has long been a pioneer in the reissuing of out-of-print titles from its own backlist, producing digital reprints of books that are still sought after by scholars and students but could not be reprinted economically using traditional technology. The Cambridge Library Collection extends this activity to a wider range of books which are still of importance to researchers and professionals, either for the source material they contain, or as landmarks in the history of their academic discipline.

Drawing from the world-renowned collections in the Cambridge University Library, and guided by the advice of experts in each subject area, Cambridge University Press is using state-of-the-art scanning machines in its own Printing House to capture the content of each book selected for inclusion. The files are processed to give a consistently clear, crisp image, and the books finished to the high quality standard for which the Press is recognised around the world. The latest print-on-demand technology ensures that the books will remain available indefinitely, and that orders for single or multiple copies can quickly be supplied.

The Cambridge Library Collection will bring back to life books of enduring scholarly value (including out-of-copyright works originally issued by other publishers) across a wide range of disciplines in the humanities and social sciences and in science and technology.

The Natural History of Birds

From the French of the Count de Buffon

VOLUME 9

COMTE DE BUFFON
WILLIAM SMELLIE

CAMBRIDGE UNIVERSITY PRESS

Cambridge, New York, Melbourne, Madrid, Cape Town, Singapore,
São Paolo, Delhi, Dubai, Tokyo, Mexico City

Published in the United States of America by Cambridge University Press, New York

www.cambridge.org
Information on this title: www.cambridge.org/9781108023061

This edition first published 1793
This digitally printed version 2010

ISBN 978-1-108-02306-1 Paperback

THE
NATURAL HISTORY
OF
BIRDS.

FROM THE FRENCH OF THE

COUNT DE BUFFON.

ILLUSTRATED WITH ENGRAVINGS;

AND A

PREFACE, NOTES, AND ADDITIONS,
BY THE TRANSLATOR.

IN NINE VOLUMES.

VOL. IX.

LONDON:
PRINTED FOR A. STRAHAN, AND T. CADELL IN THE STRAND;
AND J. MURRAY, N° 32, FLEET-STREET.
MDCCXCIII.

CONTENTS

OF THE

NINTH VOLUME.

Page

THE Swan — — — 1

The Goose — — — 25

2. The Magellanic Goose — — 57
3. The Goose of the Malouine, or Falkland Islands 58
4. The Guinea Goose — — 61
5. The Armed Goose — — 64
6. The Black-backed Goose — — 66
7. The Egyptian Goose — — 67
8. The Esquimaux Goose — — 69
9. The Laughing Goose — — 70
10. The Cravat Goose — — 71

The Brent — — — 76

The Barnacle — — — 81

The Eider — — — 90

The Duck — — — — 100

The Musk Duck — — — 138

The Wigeon — — — 143

The Crested Whistler — — — 153

The Whistler with Red Bill and Yellow Nostrils 154

The Black-billed Whistler — — 156

The

CONTENTS.

Page

The Gadwall — — — — 157
The Shoveler — — — 160
The Pintail — — — 166
The Long-tailed Duck from Newfoundland — 169
The Sheldrake — — — 171
The Pochard — — — 181
The Millouinan — — 185
The Golden Eye — — — 186
The Morillon — — — 191
The Little Morillon — — 194
The Scoter — — — 196
The Double Scoter — — — 204
The Broad-billed Scoter — — 205
The Beautiful Crested Duck — — 206
The Little Thick-headed Duck — — 209
The Collared Duck of Newfoundland — 210
The Brown Duck — — 212
The Gray-headed Duck — — 213
The White-faced Duck — — 214
The Marec and Mareca, Brazilian Ducks — 215
The Sarcelles — — — 217

1. The Common Sarcelle — — 218
2. The Little Sarcelle — — 222
3. The Summer Sarcelle — — 225
4. The Egyptian Sarcelle — — 229
5. The Madagascar Sarcelle — — 230
6. The Coromandel Sarcelle — — 231
7. The Java Sarcelle — — 232
8. The Chinese Sarcelle — — 233
9. The Feroe Sarcelle — — 235
10. The Soucrourou Sarcelle — — 236
11. The Soucrourette Sarcelle — 237
12. The Spinous-tailed Sarcelle — — 238
13. The Long-tailed Rufous Sarcelle — 239
14. The White and Black Sarcelle; or the Nun 240

15. The

CONTENTS.

Page
15. The Mexican Sarcelle — — 241
16. The Carolina Sarcelle — — 242
17. The Brown and White Sarcelle — — 243

SPECIES *which are related to the Ducks and Sarcelles* — — — 244
The Petrels — — — 252
1. The Cinereous Petrel — — — 256
2. The White and Black Petrel; or the Checker 258
3. The Antarctic Petrel; or Brown Checker 264
4. The White Petrel, or Snowy Petrel — 266
5. The Blue Petrel — — — 268
6. The Greatest Petrel; Quebrantahuessos of the Spaniards — — — 271
7. The Puffin-Petrel — — 273
8. The Fulmar; or White-gray Puffin-Petrel of the Island of St. Kilda — — 277
9. The Brown Puffin-Petrel — — 278
10. The Stormy Petrel — — 279

The Wandering Albatross — — 289
The Guillemot — — — 298
The Little Guillemot, improperly called the Greenland Dove — — — 301
The Puffin — — — 304
The Puffin of Kamtschatka — — 312
The Penguins and the Manchots; or the Birds without Wings — — — 314
1. The Penguin — — — 330
2. The Great Penguin — — 333

The Little Penguin; or the Sea-Diver of Belon 335
1. The Great Manchot — — 338
2. The Middle Manchot — — 341
3. The Hopping Manchot — — 346
4. The Manchot with a truncated Bill — 349

NOTES

CONTENTS.

Page

NOTES *and* HINTS *of certain Species of Birds that are uncertain or unknown* — — 354

APPENDIX, *by the* TRANSLATOR — 375

I. *Of Syſtems in Ornithology* — — 377

II. *Birds omitted by the Comte de Buffon, or ſince diſcovered* — — — 424

ADDENDA — — — 503

THE

No. 232

THE MUTE SWAN.

THE

NATURAL HISTORY

OF

BIRDS.

The SWAN.

LE CYGNE*. *Buff.*

1. *Anas-Olor.* Gmel.
   Anas-Cygnus manſuetus. Linn.
   Cygnus. Geſner, Johnſt. Charl. &c.
   Anſer Cygnus. Klein.
   Cygnus Manſuetus. Will. Ray, Sibb. &c.
   The Tame Swan. Edw. Penn. &c.
   The Mute Swan. Lath.
2. *Anas-Cygnus.* Gmel.
   Anas-Cygnus ferus. Linn.
   Cygnus Ferus. Briſſ. Ray, Will. Klein, &c.
   The Wild Swan, Elk, or Hooper. Will. Alb. Edw. and Penn.
   The Whiſtling Swan. Lath.

IN every ſociety, whether of men or of the lower animals, violence formed tyrants, mild authority conſtitutes kings. The lion and the tiger on the earth, the eagle and vulture in the air,

* In Greek Κυκνος: in Latin *Olor:* in Arabic *Baſkak.* Its name in Hebrew is uncertain. In Italian it is called *Cino* or *Cyno*; and at Venice *Ceſano*; at Ferrara *Ciſano:* in Spaniſh *Ciſne*; and in Catalonia

air, reign amidst the horrors of war, extend their domination by cruelty and the abuse of force. While the Swan upholds his stately empire on the water in gentleness and peace. Endowed with strength and vigour, and courage, but restrained by a sense of moderation and justice, he knows to fight and conquer, yet never urges an attack. Pacific king of the water-birds, he braves the tyrants of the air: he expects the eagle, without provoking and without fearing the rencounter. He repels his assaults, opposes to his talons the resistance of his feathers and the rapid strokes of a vigorous wing, which serves him as an *ægis* *; and often does victory crown his exertions †. This is his only formidable enemy; all the other ravenous birds respect him; and he is at peace with all nature ‡. He lives rather the friend than the monarch amidst the numerous tribes of aquatic birds, which all submit to his law. He is only the chief, the principal inhabitant of a peaceful republic §; nor have its citizens aught

Catalonia *Signe*: in German *Schwan*: in Saxony, and in Switzerland, *Oelb*, *Elbsch*, *Elbisch*, which Frisch derives from *Albus* (white): in Swedish *Swan*: in Illyrian *Labut*: in Polish *Labec*: in the Philippines, and particularly in the isle of Luçon, *Tagac*.

* Schwenckfeld and Aldrovandus.

† Aristotle, *Hist. Animal*, lib. ix. 2 and 16.

‡ *Illic innocui latè pascuntur olores.* Ovid. *Amor.* 2. *Eleg.* 6.

§ The ancients believed that the Swans spared not only the birds but even the fishes; which Hesiod indicates in his Shield of Hercules, by representing fishes swimming at ease beside the Swan.

to

to fear from a master who exacts no more than he grants, and whose sole wish is to enjoy tranquillity and freedom.

The graces of figure, the beauty of shape, correspond in the Swan to the mildness of his disposition: he pleases every eye; he decorates, embellishes every place that he frequents; he is beloved, extolled, admired *; and no species more deserves our applause. On none has nature ever diffused so much of those noble and gentle graces, which recal the image of her most charming productions: elegant fashion of body; roundness of form; softness of outline †; whiteness resplendent and pure ‡; motions full of

* "Interest," says M. Baillon, "which has disposed man to "subdue the quadrupeds and tame the birds, has had no part in the "domestication of the Swan. Its beauty, and the elegance of its "form, have engaged him to bring it to his habitation, merely to "decorate it. It has always had more attention paid it than its "fellow subjects; it has never been kept captive; it has been destined to adorn the pieces of water in his gardens, and there per-"mitted to enjoy all the sweets of liberty. . . . The abundance and "the choice of food have augmented the bulk of the tame swan; "but its form has lost none of its elegance; it has preserved the "same graces and the same freedom in all its motions; its majestic "port is ever admired; I doubt even whether all these qualities "are found to equal extent in the wild bird." *Note communicated by M. Baillon, king's councellor and bailiff of Waben, at Montreuil-sur-mer.*

† *Mollior & cygni plumis Galatea.* Ovid. *Metam.* 13.

‡ *White as a Swan.* This proverb has obtained in all nations; κυκνυ πολιωτερος was the expression of the Greeks, according to Suidas.—*Galatea candidior cygnis,* says Virgil.—In the language of the Syrians the name of white, and that of the Swan, were the same. *Guillem. Pastregius, Lib. de orig. rerum.*

 flexibility

flexibility and expreſſion; attitudes, ſometimes animated, ſometimes gently languiſhing:—All the features and actions of the Swan breathe the voluptuouſneſs, the enchantment which wrap our ſoul at the ſight of grace and beauty; all declare it, paint it, the bird of love *; all juſtify the ingenious and ſprightly mythology, that this delightful bird was the father of the moſt beautiful of women †.

The noble eaſe and freedom of its motions on the water beſpeak it not only the firſt of the winged ſailors, but the fineſt model preſented by nature for the art of navigation ‡. Its raiſed neck, and its round ſwelling breaſt exhibit the prow of a ſhip cleaving the waves; its broad ſtomach repreſents the keel; its body, preſſed down before, riſes behind into the ſtern; the tail is a genuine rudder; its feet are broad oars; and its wings, half opened to the wind, and

* Horace yokes the Swans to the car of Venus:

> ——— *quæ Gnidon*
> *Fulgenteſque tenet Cycladas, & Paphon,*
> *Junctis viſit oloribus.* Carm. lib. iii.

† Helen, born of Leda and the Swan, whoſe form Jupiter is ſaid by the ancients to have aſſumed: Euripides, to paint the beauty of Helen, and to allude, at the ſame time, to her birth, ſtiles her (*Oreſt.* act. v.) by the epithet ομμα κυκνοπτεϱον, *with aſpect lovely as the wings of the Swan.*

‡ No figure was more frequent on the ſhips of the ancients than that of the Swan; it appeared on the prow, and the mariners eſteemed it of good omen.

gently

gently inflated, are the ſails which impel the animated machine *.

Proud of his ſuperiority, and emulous of diſtinction, the Swan ſeems forward to unveil his beauties; ſeeks to charm the ſpectators, and to command their applauſe. And the ſight indeed captivates the eye, whether we behold the winged fleet at a diſtance gliding through the water, or view one, invited by ſignals †, approach the ſhore, and diſplay his elegance and grace by a thouſand ſoft, ſweet, undulating motions ‡.

To the endowments beſtowed by nature, the Swan joins the poſſeſſion of liberty. He is none of thoſe ſlaves which we can conſtrain or impriſon §. Even on our artificial lakes, he retains ſo much of the ſpirit of independence as to exclude every idea of ſervitude and capti-

* Finely deſcribed by our ſublime poet, Milton:

" The ſwan with arched neck
" Between her white wings mantling proudly, rows
" Her ſtate with oary feet." *Paradiſe Loſt*, Book vii.

† The Swan ſwims with much grace and rapidity when he chooſes; he comes to thoſe who call him. *Salerne*. The ſame author ſays, that this call is the word *godard*. According to Friſch, he anſwers to the name *Frank*.

‡ Aldrovandus.

§ A Swan confined in the court-yard is always melancholy; the gravel hurts his feet; he makes every effort to eſcape, which he would certainly effect, were not his wings clipt at each moult. " I " have ſeen one," ſays M. Baillon, " which lived in this condition " three years; it was reſtleſs and dejected, always lean and ſilent, " inſomuch that its voice was never heard; it was plentifully fed " however with bread, bran, oats, crabs, and fiſh; it flew away " when its wings were neglected to be clipt."

 vity.

vity*. He roves at will on the water, lands on the ſhore, wanders to a diſtance, or ſhelters himſelf under the brink; lurks among the ruſhes, or retires to the remoteſt inlets: then, leaving ſolitude, he returns to ſociety and to the enjoyment which he receives by approaching man; provided we are hoſpitable and friendly, not harſh and tyrannical.

Among our anceſtors, too ſimple or too wiſe to fill their gardens with the frigid beauties of art, inſtead of the lively beauties of nature, the Swans formed the ornament of every piece of water †. They cheared the gloomy ditches round caſtles ‡, they decorated moſt of the rivers §, and even that of the capital ||: and one of our moſt feeling and amiable princes ¶ took pleaſure in ſtocking the baſins of his royal manſions with theſe beautiful birds. We may at preſent enjoy the ſame ſpectacle on the fine waters of Chantilly, where the Swans are the chief ornament of that truly delicious place, of which every thing beſpeaks the noble taſte of its owner.

* The tame Swan likes freedom, and will not be confined. *Salerne.*

† This taſte was not unknown to the ancients; the tyrant Gelo conſtructed at Agrigentum a pool for feeding Swans. *Aldrovandus.*

‡ Aldrovandus.

§ According to Volaterran, there were no leſs than four thouſand on the Thames.

|| Salerne.

¶ Francis I.

The

The Swan ſwims ſo faſt, that a man, walking with haſty ſtrides along the banks, can hardly keep pace with it. What Albin ſays of this bird "that it ſwims well, walks ill, and flies in-"differently," is true only of the flight of the Swan degraded by domeſtication; for when free, and eſpecially when wild, it flies very loftily and vigorouſly. Heſiod gives it the epithet of *αεροσιποτης* *: Homer claſſes it with the great migratory birds, the cranes and the geeſe †; and Plutarch attributes to two Swans what Pindar ſung of two eagles, that Jupiter diſpatched them from the oppoſite extremities of the world, to diſcover the middle by their meeting.

The Swan, ſuperior in every reſpect to the gooſe, which lives only on herbs and grain, procures itſelf a rarer and more delicate food ‡. It continually practiſes wiles to enſnare and catch fiſh: it aſſumes a thouſand different attitudes, and draws every poſſible advantage from its dexterity and ſtrength. It evades or even reſiſts its enemies: an old Swan fears not in the water the ſtrongeſt dog: a ſtroke of its wing could break a man's leg, ſo violent it is, and ſo ſudden. Nor does the Swan dread any ambuſh or any

* *i. e.* that flies to the clouds.

† Iliad ii.

‡ The Swan lives on grain and fiſh, particularly eels; it ſwallows alſo frogs, leeches, and water ſlugs; it digeſts as quickly as the duck, and eats largely. *M. Baillon.*

foe; for its courage equals its addreſs and its force *.

The wild Swans fly in great flocks, and the tame Swans likewiſe walk and ſwim in company †. Every thing marks their ſocial inſtinct; and that inſtinct, the ſweeteſt in nature, beſpeaks innocent manners, peaceful habits, and that delicate and ſenſible diſpoſition which ſeems to beſtow on the actions that flow from it, the merit of moral qualities ‡. And the Swan prolongs its placid, joyous exiſtence to extreme age §. All obſervers aſcribe to it prodigious longevity: ſome repreſent it as even paſſing the term of three centuries; which muſt certainly be an exaggeration. But Willughby ſaw a gooſe, which was proved to have lived an hundred years; and he concludes from analogy that the period of the Swan muſt extend farther; both becauſe it is larger, and becauſe its eggs require longer time to hatch: for incubation in

* The Swan, ſays the ſame obſerver, is perpetually contriving to enſnare fiſh, which are its favourite food. . . . It can avoid the blows which its enemies aim at it. If a ravenous bird threatens the young, the parents will defend them intrepidly; they range round them, and the plunderer dares not to approach: if dogs aſſail them, they go before and make an attack. The Swan dives and eſcapes if the force of its enemy prove ſuperior to the reſiſtance which it can make. The Swans are ſeldom ever ſurprized by foxes and wolves, but in the darkneſs of night and during ſleep.

† Ariſtotle, *lib.* viii. 12.

‡ Ælian.—Ariſtotle.—Bartholin.

§ Ariſtotle.—Aldrovandus.

birds correfponds to geftation in quadrupeds, and bears fome relation, perhaps, to the body's growth, which is proportional to the duration of life. The Swan requires two years to attain its full fize, which is a very confiderable time; fince, in birds, the developement is much quicker than in quadrupeds.

The female Swan fits fix weeks at leaft *; fhe begins to lay in the month of February; and, as with the goofe, there is a day's interval between the dropping of each egg. She has from five to eight, and commonly fix or feven †; they are white and oblong, covered with a thick fhell, and are of a very confiderable fize. The neft is placed fometimes on a bed of dry herbs on the bank ‡; fometimes on a heap of broken reeds, heaped and even floating on the water ||. The amorous pair lavifh the fweeteft careffes, and feem in their pleafures to feek all the gradations of voluptuoufnefs. They begin by entwining their necks; thus they breathe the intoxication of a long embrace §; they communicate the fire

* Willughby.

† Five or fix. *Willughby.*—Seven at moft. *Schwenckfeld.*—Two or three; fometimes fix. *Salerne.*

‡ Schwenckfeld.

|| Frifch.

§ *Tempore libidinis blandientes inter fe mas & fœmina, alternatim cum fuis collis inflectunt, velut amplexandi gratiâ; nec mora, ubi coierint, mas confcius læfam a fe fœminam fugit; illa impatiens fugientem infequitur. Nec diutina noxa quin reconcilientur; fœmina tandem maris perfecutione relictâ, poft coitum frequenti caudæ motu & roftri, aquis fe mergens, purificat.* Johnfton.

which

which kindles in their veins; and after the male has fully indulged his appetite, the female ſtill burns; ſhe purſues, excites him anew, and then leaves him, with regret, to waſh in the water, and quench her remaining ardor *.

The fruits of theſe rapturous loves are tenderly cheriſhed and foſtered. The mother gathers, night and day, the young under her wings, and the father is ready to defend them with intrepidity againſt every aſſailant †. But his courage, on ſuch occaſions, bears no compariſon to the fury with which he attacks a rival that intrudes on the poſſeſſion of his beloved object ‡. He then forgets his mildneſs, becomes ferocious,

* Hence the opinion of its pretended modeſty, which, according to Albertus, is ſuch, that it will not eat after the moments of fruition till it has waſhed itſelf. Dr. Bartholin, improving on this idea, aſſerts, that, to cool its ardour, it eats nettles;—a receipt which would ſeem as proper for a doctor as for a Swan.

† Morin's *Diſſertation on the ſong of the Swan*—and Albertus.

‡ The Charente has its ſource in two ſprings, the one called *Charannat*, and the other, the wonderful abyſs, *Louvre*; which, joining their ſtreams, give exiſtence and name to the Charente: theſe afford a retreat to an innumerable multitude of Swans, the moſt amiable, the moſt beautiful, the moſt familiar of all the river birds: it is true, they are choleric and deſperate when provoked, which has been witneſſed in a houſe adjoining to the ſaid Louvre. Two Swans having attacked each other ſo furiouſly as almoſt to kill each other, four others of their companions haſtened to the ſpot, and, as if they had been human beings, endeavoured to ſeparate them, and to conciliate them into concord and mutual friendſhip; indeed, this deſerved more the name of prodigy than any other appellation. But if one treat them with gentleneſs, and coax and praiſe them a little, they will ſhew themſelves mild and peaceful, and take pleaſure in ſeeing the face of man. *Coſmographie du Levant, par André Thevet*; *Lyons*, 1554, *pp.* 189 and 190.

and

and fights with obſtinate rancour; and a whole day is often inſufficient to terminate the quarrel. They begin with ſtriking violently their wings, then join cloſe, and perſiſt till commonly one of them is killed; for they ſtrive to ſtifle each other by locking the neck, and forcibly holding the head under water *. It was probably theſe combats that made the ancients imagine that the Swans devoured one another †. Nothing is wider of the truth; only in this, as in other caſes, furious paſſions originate from a paſſion the moſt delicious; and it is love that begets war ‡.

At every other time, their habits are peaceful, and all their ſentiments are dictated by love. As attentive to neatneſs as they are addicted to pleaſure, they are aſſiduous each day in the care of their perſon: they arrange their plumage, they clean and ſmooth it; they take water in their bill, and ſprinkle it on their back and wings with an attention that implies the deſire of pleaſing,

* We certify all theſe facts as eye-witneſſes. *M. Morin.*

† Ariſtotle, *lib.* ix. 1 —Ælian was ſtill worſe informed, when he ſaid the Swan ſometimes kills its young. Theſe falſe ideas reſted leſs perhaps on facts in natural hiſtory, than on mythological traditions. Indeed, all the *Cycnuſes* in fable were exceeding wicked perſonages:—*Cycnus*, the ſon of Mars, was killed by Hercules, becauſe he was a robber: *Cycnus*, the ſon of Neptune, having ſtabbed Philonome his mother, was killed by Achilles; and laſtly, the beautiful *Cycnus*, friend of Phaeton and ſon of Apollo, was, like him, cruel and inhuman.

‡ Friſch aſſerts that the older Swans are the moſt vicious, and harraſs the younger; and that, to ſecure tranquillity in the hatches, the number of theſe old males ſhould be diminiſhed.

and

and which can only be repaid by the consciousness of being loved. The only time when the female neglects her attire, is that of incubation: her maternal solicitude then entirely occupies her thoughts, and hardly does she spare a few moments for the relief and support of nature.

The cygnets are hatched very ugly, and covered only with a gray or yellowish down, like goslings. Their feathers do not sprout till a few weeks after, and are still of the same colour. This unsightly plumage changes after the first moult, in the month of September: then they assume many white feathers, and others rather flaxen than gray, especially on the breast and the back. This laced plumage drops at the second moult, and it is not till eighteen months, or even two years, that these birds are invested with their robe of pure and spotless white; nor before that age can they have young.

The cygnets follow their mother the first summer, but they are compelled to leave her in the month of November, being chased away by the adult males, who wish to enjoy entirely the company of the females. These young birds, exiled from their family, unite in one body, and never separate till they pair.

As the Swan often eats marsh-plants, and particularly the *algæ*, it prefers rivers of a smooth and winding course, whose banks are well clothed with herbage. The ancients have cited

cited the *Meander* *, the *Mincio* †, the *Strymon* ‡, the *Cayster* ||, as ſtreams covered with Swans §. Paphos, the loved iſle of Venus, was filled with them ¶. Strabo ** ſpeaks of the Swans of Spain; and according to Ælian †† they were ſeen, at times, on the ſea of Africa. From this and other accounts ‡‡, we may conclude, that the ſpecies penetrates into the regions of the ſouth: yet the north ſeems the true country of the Swan, where it breeds and multiplies. In the provinces of France, wild Swans are ſcarce ſeen but in the hardeſt winters ||||. Geſner ſays, that

* Theocrit. *Idyl.* 19.

† *Et qualem infelix amiſit Mantua campum,*
Paſcentem niveos herboſo flumine cycnos.
Virg. *Georg.* ii. 198.
Mincius ingenti cycnos habet undâ natantes.
Bap. Mantuan.

‡ Belon remarks, that even at this day great numbers of Swans are ſeen on the Strymon.

|| Homer, *Iliad* ii.—Propertius, *Eleg.* 9.—Ovid. *Metam.* 25.

§ We muſt add the Po:——
. . . *piſcoſove amne Paduſæ*
Dant ſonitum rauci per ſtagna loquacia cycni.
Virg. *Æn.* xi. 457.
. . . *Eridani ripas diffugiens nudavit olor.*
Sil. Ital. *lib.* xiv.

¶ Schol. *in Lycophr.*

** Geogr. *lib.* iii.

†† Hiſt. Animal. *lib.* x. 36.

‡‡ According to Camel, the Swan occurs at Luçon, where it is called *tagac* (*Philoſ. Tranſ.* N° 285); but this author does not tell us whether it is the tame breed tranſported, or the natural wild kind, that occurs in this capital of the Philippines.

|||| Obſervations of Meſſrs. Lottinger, de Querhoënt, and de Piolenc.—“ In hard winters, they come upon the Loiret.” *Salerne.*—“ In 1709, the Swans, driven from the north by the extreme “ cold,

that in Switzerland, a long and ſevere winter is expected, when many Swans arrive on the lakes. In the ſame cold ſeaſon, they appear on the coaſts of France and of England, and on the Thames *. Many of our tame Swans would then join the wild ones, did we not clip the great feathers of their wings.

Yet ſome Swans neſtle and paſs the ſummer in the northern parts of Germany, in Pruſſia † and Poland ‡; and on nearly the ſame parallel of latitude, they are found on the mighty rivers about Azof and Aſtracan ||; in Siberia, among the Jakutes §; in Seleginſkoi, and as far as Kamtſchatka ¶. During the breeding ſeaſon, they are alſo found in immenſe numbers on the ſtreams and lakes of Lapland: there they feed on the *larvæ* of the gnat **, which cover the ſurface of the water. The Laplanders ſee them arrive

" cold, appeared numerous on the coaſts of Brittany and Nor-" mandy." *Friſch.*—" The intenſe cold, and the ſtorms of this " winter, have brought on the coaſt many ſea birds, and, among " others, many Swans."—*Letter dated from Montaudoin, 20th March* 1776.

* Britiſh Zoology.

† Klein and Schwenckfeld.

‡ Rzaczynſki.

|| Guldenſtaed.

§ Gmelin.

¶ The Swans are ſo common in Kamtſchatka, both in winter and ſummer, that every perſon eats of them: in the moulting ſeaſon, they are hunted with dogs, and felled with clubs: in winter they are caught on the rivers. *Kracheninicoff.*

** *Culex Pipiens.* Linn.

in the ſpring from the German ocean *: part ſtop in Sweden, and eſpecially in Scania †. Horrebow affirms, that they continue the whole year in Iceland, and inhabit the ſea when the freſh waters are frozen ‡. But if a few do remain, the bulk of them follow the common law of migration, and fly from a winter, which, as the ſhoals of ice are driven from Greenland, is attended with greater rigour in Iceland than in Lapland.

Theſe birds are as numerous in the northern parts of America as in thoſe of Europe. They inhabit Hudſon's Bay, and hence the name *Carry Swan's-neſt*, given by Captain Button to a point near the ſouthern extremity of the long barren iſland that ſtretches northwards in the bay. Ellis found Swans in *Marble Iſland*, which is only a pile of broken rocks round ſome pools of freſh water ||. They are likewiſe very numerous in Canada §, from whence they appear to migrate

* Obſervation of Samuel Rheen, paſtor at Pitha in Lapland, as cited by Klein.

† *Fauna Suecica.*

‡ He adds, that "during moult the Swans advance into the land, "and ſeek in flocks the waters on the mountains; it is at this "time that the inhabitants purſue them, and catch them, or kill "them eaſily, becauſe they cannot fly. Their fleſh is good, eſpe-"cially the breaſt of young ones, which makes a delicate diſh; "their feathers, and chiefly their down, form an important article "of trade."

|| Hiſt. Gen. des Voy. *tom.* xiv. 670.

§ The Swans, and other great river birds, ſwarm every where, except near dwellings, which they never approach. *Charlevoix.*— Among

migrate for winter quarters, into Virginia * and Louiſiana †. And theſe are found, on compariſon, to differ in no reſpect from our wild Swans. With reſpect to the black-headed Swans of the Falkland Iſlands, and of ſome of the coaſts of the South Sea, mentioned by travellers ‡, the ſpecies is ſo ill deſcribed, that we cannot decide whether it belongs to our Swan.

The differences which ſubſiſt between the wild and the tame Swan, have led to the opinion that they form two diſtinct and ſeparate ſpecies.

Among the Illinois there are plenty of Swans. *Lettres Edifiantes.*—Swans, which are called *horhey*, are ſeen principally near the Epicinys. *Theodat.*

* The Swans are numerous in Virginia during winter. *De Laët.*

† The Swans of Louiſiana are ſuch as thoſe in France, with this difference, that they are larger; however, notwithſtanding their bulk and their weight, they riſe ſo high in the air, that often they cannot be diſtinguiſhed but by their ſhrill cry: their fleſh is very good to eat, and their fat is a ſpecific for cold humours. The natives ſet great value on the feathers of Swans; they form them into diadems for their chiefs, and into caps, and twiſt the little feathers, as the wig-makers do hair, into cloaks for their women of rank. The young perſons of both ſexes make themſelves tippets with ſkin covered with its down. *Dupratz.*

‡ Among the birds with palmated feet, the Swan holds the firſt rank; it differs not from thoſe of Europe, except by its neck, which is of a velvet black, and makes an admirable contraſt with the whiteneſs of the reſt of its body; its legs are fleſh-coloured. This ſpecies of Swan, which we ſaw in the Malouine iſlands, occurs alſo on the river de la Plata, and at the Straits of Magellan, where I killed one in the bottom of Port Galant. *Bougainville.*—On the ſhore of the South Sea, we ſaw ſome Swans; they are not ſo large as ours; are white, except the head, half the neck, and the legs, which are black. *Coreal.*

ſpecies *. The wild Swan is ſmaller; its plumage is more inclined to gray than white †; it has no caruncle under the bill, which always is black at the point, and not yellow, except near the head. But the intenſity of colour, and even the caruncle or wattle on the front, are to be regarded as leſs the characters of nature, than the tokens and impreſſions of domeſtication. The Swans are ſubject to variations in the colours of their plumage and of their bill, as well as other tame birds; and Dr. Plott ‡ mentions a tame Swan, whoſe bill was red. Beſides, the difference in the colour of the plumage is not ſo great as would at firſt appear; we have ſeen tame Swans hatched gray, and continue long of that caſt. This colour ſubſiſts ſtill longer in the wild ones, and yet they grow white through age: for Edwards obſerved, that, in the hard winter of 1740, many wild Swans were ſeen in the neighbourhood

* Willughby and Ray.

† *N. B.* The Swan figured in the *Planches Enluminées* is the tame Swan; a wild one, preſerved in the king's cabinet, is entirely of a white gray, only deeper and almoſt brown on the back and the crown of the head.

‡ Britiſh Zoology.—*N. B.* We may alſo mention the Swan which Redi ſaw in the parks of the Grand Duke, which had the feathers of the head and neck tipt with a yellow or orange tinge; a peculiarity which may explain the epithet *purpurei,* applied by Horace to the Swans. (The expreſſion is "purpureis ales oloribus." *Carm. lib.* iv. *Od.* 1. But it is to be obſerved, that πορφυρεος, among the Greeks, and *purpureus* among the Latins, ſignified originally any pure virgin colour, and was afterwards appropriated to purple. Thus Virgil has "purpureo narciſſo, purpureo capillo, purpureum "ver, purpureum lumen, purpureum mare."—*T.*)

bourhood of London, entirely white. The tame Swan muſt therefore be regarded as a breed derived anciently and originally from the wild ſpecies. Klein, Friſch, and Linnæus have formed the ſame opinion; though Willughby and Ray pretend the contrary.

Belon reckons the Swan to be the largeſt of the aquatic birds; which is true, excepting, however, that the pelican has a much greater alar extent, that the albatroſs is as bulky, and that the flamingo is taller on its legs. The tame Swans are invariably ſomewhat thicker and larger than the wild ſort: ſome of them weigh twenty-five pounds, and meaſure, from the bill to the tail, four feet and a half; the breadth of the wings eight feet. The female is, in every dimenſion, rather ſmaller than the male.

The bill uſually exceeds two inches in length, and, in the tame kind, has above its baſe a fleſhy tubercle, inflated and prominent, which gives a ſort of expreſſion: this tubercle is covered with a black ſkin; and the ſides of the face, under the eyes, are covered with a ſkin of the ſame colour. In cygnets of the domeſtic breed, the bill is of a leaden caſt, and afterwards becomes yellow or white, with the point black. In the wild kind, the bill is entirely black, with a yellow membrane on the front: its form ſeems to be copied in the two moſt numerous families of the palmipede birds, the geeſe and ducks. In all of theſe the bill is flat, thick, indented at the edges, rounded into a blunt tip *, and terminated on the upper mandible by a nail of horny ſubſtance.

In all the ſpecies of this numerous tribe, there is, under the outer feathers, a thick down, which prevents the water from penetrating to the body. In the Swan this down is exceedingly fine and ſoft, and perfectly white: it is worked into muffs and furs, that are equally delicate and warm.

The fleſh of the Swan is black and hard; and the magnificence, rather than the excellence of the diſh, might recommend it in the Roman entertainments †. Our anceſtors affected the ſame oſtentation ‡. Some perſons have aſſured me that the cygnets are as good as geeſe of the ſame age.

Though the Swan is a ſilent bird, its vocal organs have the ſame ſtructure as in the moſt loquacious of the water fowl. The *trachea arteria* deſcends into the *ſternum*, makes a bend ||, riſes, reſts on the clavicles, and thence, by a ſe-

* *Tenet os ſine acumine roſtrum.* Ovid.

† See Athenæus.—The Romans fattened it as they did the gooſe, having put out its eyes, or ſhut it up in a dark priſon. See Plutarch, *De eſu carn.*

‡ Belon.—The grandees among the Muſcovites ſerve up Swans in their entertainments to ſtrangers. *Aldrovandus.*

|| According to Willughby, this conformation is peculiar to the wild Swan, and is not the ſame in the tame Swan; which ſeems to give occaſion to the difference which we ſhall remark between their cries. Yet this is inſufficient to conſtitute two diſtinct ſpecies; for the variation exceeds not the ſum of the impreſſions, both internal and external, which the domeſtic habits may in time produce.

cond inflexion, it reaches the lungs. At the entrance and above the bifurcation, is placed a true *larynx* furniſhed with its *os hyoides*, open in its membrane like the lip of a flute: below this *larynx*, the canal divides into two branches, which, after each forms an inflation, adhere to the lungs *. This ſtructure, at leaſt what regards the poſition of the *larynx*, is common to many aquatic birds; and even ſome of the waders have the ſame folds and inflexions of the *trachea arteria*, as we have remarked in the crane: and, in all probability, it gives the voice that ſonorous and raucous intonation, thoſe trumpet and clarionet ſounds which echo from the air and in the water.

Yet the ordinary voice of the tame Swan is rather low than canorous; it is a ſort of creaking, exactly like what is vulgarly called the ſwearing of a cat, and which the ancients denoted by the imitative word *drenſare* †. It would ſeem to be an accent of menace or anger, nor does love appear to have a ſofter ‡.

* Bartholin. *Cygni anatome ejuſque cantus.* Hafniæ, 1680.—Aldrovandus.

† *Grus gruit, inque glomis cygni prope flumina drenſant.* Ovid.

‡ Obſervations made at Chantilly, according to the directions of the Marquis Amezaga, and which M. Grouvelle, military ſecretary to his ſerene highneſs the prince of Conde, has been ſo obliging as to draw up.—"Their voice in the ſeaſon of love, and the accents "which they breathe in the ſofteſt moments, reſemble more a murmur than any ſort of ſong." See alſo, in the *Memoires de l'Academie des Inſcriptions*, a diſſertation of M. Morin, entitled, *Why Swans, which ſung ſo well formerly, ſing ſo ill now.*

Swans,

Swans, almoſt mute, like ours in the domeſtic ſtate, could not be thoſe melodious birds which they have celebrated and extolled. But the wild Swan appears to have better preſerved its prerogatives; and with the ſentiment of entire liberty, it has alſo the tones. The burſts of its voice form a ſort of modulated ſong *; yet the ſhrill and

* The Abbe Arnaud, whoſe genius is formed to revive the precious remains of elegant and learned antiquity, has obligingly concurred with us in verifying and appreciating what the ancients have ſaid on the ſong of the Swan. Two wild Swans which have ſettled on the magnificent pools of Chantilly, ſeem to have offered themſelves for this intereſting obſervation. The Abbe Arnaud has gone ſo far as to mark their ſong, or rather their harmonious cries; and he writes us in the following terms: " One can hardly ſay that the " Swans of Chantilly ſing, they cry; but their cries are truly and con- " ſtantly modulated; their voice is not ſweet; on the contrary, it is " ſhrill, piercing, and rather diſagreeable; I could compare it to no- " thing better than the ſound of a clarionet, winded by a perſon un- " acquainted with the inſtrument. Almoſt all the melodious birds " anſwer to the ſong of man, and eſpecially to the ſound of inſtru- " ments: I played long on the violin beſide our Swans, on all the " tones and chords; I even ſtruck uniſon to their own accents, " without their ſeeming to pay the ſmalleſt attention: but if a gooſe " be thrown into the baſon where they ſwim with their young, the " male, after emitting ſome hollow ſounds, ruſhes impetuouſly upon " the gooſe, and, ſeizing it by the neck, he plunges the head repeat- " edly under water, ſtriking it at the ſame time with his wings; " it would be all over with the gooſe, if it were not reſcued: " the Swan, with his wings expanded, his neck ſtretched, and his " head erect, comes to place himſelf oppoſite to his female, and ut- " ters a cry, to which the female replies by another, which is lower " by half a tone. The voice of the male paſſes from A *(la)* to B " flat *(ſi bemol)*; that of the female, from G ſharp *(ſol dièſe)* to A. " The firſt note is ſhort and tranſient, and has the effect of that which " our muſicians call *ſenſible*; ſo that it is not detached from the ſecond, " but ſeems to *ſlip* into it. Obſerve that, fortunately for the ear, " they

and ſcarce diverſified notes of its loud, clarion ſounds, differ widely from the tender melody, the ſweet and brilliant variety of our chanting birds.

But it was not enough that the Swan ſung admirably; the ancients aſcribed to it a prophetic ſpirit. It alone, of animated beings, which all ſhudder at the proſpect of deſtruction, chanted in the moment of its agony, and, with harmonious ſounds, prepared to breathe the laſt ſigh. When about to expire, they ſaid, and to bid a ſad and tender adieu to life, the Swan poured forth thoſe accents ſo ſweet, ſo affecting, and which, like a gentle and doleful murmur, with

" they do not both ſing at once; in fact, if while the male ſounded " B flat, the female ſtruck A, or if the male uttered A, while the female gave G ſharp, there would reſult the harſheſt and moſt inſupportable of diſcords. We may add, that this dialogue is ſubjected to a conſtant and regular rhythm, with the meaſure of two " times. The inſpector aſſured me that, during their amours, theſe " birds have a cry ſtill ſharper, but much more agreeable."—We ſhall add an intereſting obſervation which was communicated to us after the firſt pages of this article were printed. " There is a ſeaſon when the Swans aſſemble together, and form a ſort of commonwealth; it is during ſevere colds. When the froſt threatens to " uſurp their domain, they congregate and daſh the water with all " the extent of their wings, making a noiſe which is heard very far, " and which, whether in the day or the night, is louder in proportion as it freezes more intenſely. Their efforts are ſo effectual, " that there are few inſtances of a flock of Swans having quitted the " water in the longeſt froſts, though a ſingle Swan, which has ſtrayed " from the general body, has ſometimes been arreſted by the ice in " the middle of the canals." *Extract of a note drawn up by M. Grouvelle, military ſecretary to his ſerene highneſs the prince of Conde.*

a voice

a voice low *, plaintive, and melancholy †, formed its funereal ſong ‡. This tearful muſic was heard at the dawn of day, when the winds and the waves were ſtill ||: and they have been ſeen expiring with the notes of their dying hymn §. No fiction of natural hiſtory, no fable of antiquity, was ever more celebrated, oftener repeated, or better received. It occupied the ſoft and lively imagination of the Greeks: poets ¶, orators **, even philoſophers ††, adopted it, as a truth too pleaſing to be doubted. And well may we excuſe ſuch fables; they were amiable and affecting; they were worth many dull, inſipid truths; they were ſweet emblems to feeling minds. The Swan, doubtleſs, chants not its approaching end; but, in ſpeaking of the laſt flight, the expiring effort of a fine genius, we ſhall ever, with tender melancholy, recal the claſſical and pathetic expreſſion, *It is the ſong of the Swan!*

* *Parvus cycni canor.* Lucret. *lib.* iv.

† *Olorum morte narratur flebilis cantus.* Plin.

‡ According to Pythagoras, it was the ſong of exultation upon the immediate proſpect of paſſing into a happier ſtate.

|| Aldrovandus.

§ Ariſtotle, *lib.* ix. 12.

¶ Callimachus, Æſchylus, Theocritus, Lucretius, Ovid, Propertius, ſpeak of the ſong of the Swan, and draw compariſons from it.

** Cicero, Pauſanias, and others.

†† Socrates in Plato, and Ariſtotle himſelf, but from vulgar opinion and foreign report.

[A] Specific character of the Wild Swan, *Anas Cygnus:* "Its "bill is semicylindrical and deep black, its cere bright yellow, its "body white."—Specific character of the Tame or Mute Swan, *Anas Olor:* "Its bill is semicylindrical and deep black; its cere "black, its body white." Cygnets are even at present fattened at Norwich about Christmas, and sold for a guinea a-piece. In Edward the Fourth's time none was permitted to keep Swans, who possessed not a freehold of at least five marks yearly value, except the king's son: and by an act of Henry the Seventh, persons convicted of taking their eggs, were liable to a year's imprisonment and a fine at the will of the sovereign.

Nº 233

THE GOOSE.

The GOOSE*.

L'OIE. *Buff.*

1. *Anas Anser*, ferus. Linn. and Gmel.
Anser Sylvestris. Briss. and Frisch.
Anser Ferus. Gesner, Aldrov. Johnst. Will. Sibb. &c.
The Wild Goose. Albin and Will.
The Gray-lag Goose. Penn. and Lath.
2. *Anas Anser*, domesticus. Linn. and Gmel.
Anser Domesticus. Gesn. Aldrov. Johnst. Sibb. &c.
The Tame Goose. Penn. and Lath.

IN every genus, the primary species have borne off all the eulogies, and have left to the subordinate species only the scorn arising from the comparison. The Goose is in the same predicament with regard to the swan, as the ass when viewed beside the horse; neither of them is esti-

* In Greek Χην: in modern Greek Χινα: in Latin *Anser*: in Arabic *Uze*, *Avaz*, *Kaki*: in Italian *Oca*, *Papara*; the wild goose *Oca Salvatica*, the tame one *Oca Domestica*: in Spanish *Ganso*, *Pato*; the gander *Ansar*, *Ansarea* or *Bivar*; the gosling *Patico*, or *Hijo de Pato*; the wild goose *Ansar Bravo*: in Catalonian *Hoca*: in German *Ganz*, *Ganser*, *Ganserich*; the gosling *Ganselin*; the wild goose has the epithets *Wilde*, *Graue* (gray), and *Schnee* (snow). In Flanders the gander is called *Gans*, and the Goose *Goes*: in Switzerland *Ganss*: in Swedish *Goas*, and the wild kind *Wille Goas*: in Danish *Gaas*: in Polish *Ges*, *Gasior*, and the wild one *Ger Dzika*; which by the Greenlanders is named *Nerlech*; by the Hurons *Ahonque*; by the Mexicans *Tlalacatl*. The negroes on the gold coast call the tame sort *Apatta*.

mated

mated at its true value. The firſt ſtep of inferiority, appearing a real degradation, and recalling, at the ſame time, the idea of a more perfect model, exhibits, inſtead of the abſolute qualities of the ſecondary ſpecies, only an unfavourable contraſt with the primary. Laying aſide, then, for a moment, the too noble image of the ſwan, we ſhall find, that, among the inhabitants of our court-yards, the Gooſe holds a diſtinguiſhed rank. Its corpulence, its erect carriage, its grave demeanour, its clean gloſſy plumage, and its ſocial diſpoſition, which renders it ſuſceptible of a ſtrong attachment and a durable gratitude; finally, its vigilance, celebrated in high antiquity: all concur to recommend the Gooſe as one of the moſt engaging, and even of the moſt uſeful, of our domeſtic birds. For, beſides the excellence of its fleſh and of its fat, with which no bird is more abundantly provided, the Gooſe furniſhes the delicate down for the beds of the luxurious, and the quill, the inſtrument of our thoughts, which now writes its eulogy.

The Gooſe may be maintained at no great expence, and reared with moderate attention *. It is reconciled to the ordinary life of poultry, and ſuffers itſelf to be incloſed with them in the ſame court †; though that mode of exiſtence, and

* Schwenckfeld.

† Belon.

and eſpecially that conſtraint, is little ſuited to its nature: for to raiſe numerous flocks of large Geeſe, it is requiſite that they be kept near pools or ſtreams, ſurrounded with ſpacious margins, with graſſy patches or waſte grounds, where they may feed and ſport at liberty *. They are not permitted to enter meadows, becauſe their dung burns up the good herbage, and becauſe they dig into the ſoil with their bill. For the ſame reaſon, they are carefully removed from green corn, and are not permitted to range the fields till after harveſt.

Though the Geeſe can feed on graſs and moſt herbs, they prefer trefoil, fenugreek, vetches, ſuccory, and eſpecially lettuce, which is the greateſt regale of the little goſlings †. We ſhould carefully extirpate from their walk, henbane, hemlock, and nettles ‡, whoſe ſtings are very pernicious to the young birds. Pliny aſſerts, perhaps on ſlight foundation, that the Geeſe eat iron-wort for a purge.

The domeſtication of the Goose is neither ſo ancient nor ſo complete as that of the hen. The latter lays at all ſeaſons, more in ſummer, leſs in winter; but the former are unproductive in the winter, and ſeldom have eggs before the month of

* *Anſer nec ſine herbâ, nec ſine aquâ facile ſuſtinetur.* Pallad.

† *Lactuca molliſſimum olus libentiſſime ab illis appetitur & pullis utiliſſima eſca. Ceterum vicia, trifolium, fœnum græcum, & agreſtis intiba illis conſeratur.* Columella.

‡ Aldrovandus.

of March. Yet ſuch as are well fed begin to lay in February, and thoſe which are more ſparingly kept, often defer till April. The white, the gray, the yellow, and the black ſorts, follow that rule; only the white ones ſeem to be more delicate, and are really more difficult to rear. None of them ever makes its neſt in our court-yards *; they lay only every two days, but always in the ſame place. If their eggs be removed, they make a ſecond and a third depoſit, and even a fourth in warm countries †. It is, no doubt, by reaſon of theſe ſucceſſive layings, that Salerne ſays they continue till June. But if the eggs be conſtantly withdrawn, the Gooſe will ſtill perſiſt to lay, till at laſt ſhe waſtes away and dies. For the eggs, particularly thoſe of the firſt laying, amount to a large number; at leaſt ſeven, and commonly ten, twelve, or fifteen, and even ſixteen, according to Pliny ‡. Such may be the caſe in Italy; but in the inte-

* They ſink into ſtraw, there to lay, and the better to conceal the eggs; they have preſerved this habit of the wild Geeſe, which probably penetrate into the thickeſt ruſhes and marſh-plants, to hatch; and in places where the tame ones enjoy almoſt entire freedom, they gather ſome materials on which they depoſit their eggs. "In "the iſland of St. Domingo," ſays M. Baillon, "where many of "the inhabitants have tame Geeſe like ours, they lay in the ſavan-"nas, near the brooks and trenches; they form the bed with ſome "dry herbs, the ſtalks of maize or of millet: the females are leſs "prolific there than in France, their greateſt laying not exceeding "ſeven or eight eggs."

† Aldrovandus.

‡ *Lib.* v. 55.

rior

rior provinces of France, as in Burgundy and Champagne, the greateſt neſts contain but twelve eggs. Ariſtotle obſerves *, that often young Geeſe, like pullets, lay addle eggs before having intercourſe with the male. This fact is applicable to all birds.

But if the domeſtication of the Gooſe is more modern than that of the hen, it ſeems to have been more ancient than that of the duck, whoſe original features are leſs changed, ſo that the interval is greater between the wild and the tame Gooſe, than between thoſe two breeds of ducks. The tame Gooſe is larger than the wild, the parts of its body are more extended and more pliant, its wings are neither ſo ſtrong nor ſo ſtiff, the whole colour of its plumage is changed, and it retains ſcarce any trace of its primitive condition; it ſeems to have even forgotten the ſweets of its ancient freedom, at leaſt it ſeeks not, like the duck, to recover them. Servitude appears to have enfeebled it; it no longer has ſtrength to accompany or follow the flight of its ſavage brethren, who, proud of their force, neglect and even deſpiſe it †.

* *Lib.* vi. 12.

† "I have enquired," ſays M. Baillon, "of many fowlers who "kill wild Geeſe every year, but I could never meet with one who "had ſeen the tame birds among the wild, or who had killed hy-"brids. If tame Geeſe ſometimes eſcape, they do not become free; "they go to mingle in the neighbouring marſhes with others equally "tame, and thus only change maſters." *Note communicated by M. Baillon.*

That

That a flock of tame Geese prosper and increase by a quick multiplication, it is requisite, says Columella *, that the number of the females be triple that of the males. Aldrovandus allows six geese to one gander; and it is usual in our provinces to admit twelve, and even twenty. These birds prepare for the congress of love by first sporting in the water. They come out to copulate, and continue longer united and in closer embrace than most others; for the act is not a simple compression, but a real intromission, the male being provided with the proper organ †; and hence the ancients consecrated the Goose to Priapus.

The male shares with the female only the pleasures of love; he devolves on her the whole care of incubation ‡. She covers constantly and assiduously, and would even neglect to eat and drink, were not food placed near the nest §. Economists advise, however, to entrust the incubation and rearing of the goslings to a hen; so that the Goose may have a second and even a third hatch. The last one is left to the proper mother; and she can hatch ten or twelve eggs, whereas a hen cannot succeed with more than five. It would be curious to know whether, as

* *De Re Rust.* lib. viii. 13.

† Aristotle, *Hist. Animal. lib.* iii.

‡ *Id. Ibid.*

§ Aldrovandus.

Columella aſſerts, the Gooſe, wiſer than the hen, will cover no eggs but her own.

Thirty days are required for incubation, as in moſt of the large birds *; unleſs, as Pliny remarks, the weather be very hot, and then it ſucceeds in twenty-five days †. During the ſitting, a veſſel filled with grain, and another with water, are placed at ſome diſtance from the eggs, which the Gooſe never quits but to take a little food. It has been remarked, that ſhe ſeldom ever lays on two conſecutive days, and that there is always an interval of at leaſt twenty-four hours, and ſometimes of two or three days, between the excluſion of each egg.

The callow goſlings are fed firſt with the refuſe of the mill, or with rich bran kneaded with haſhed ſuccory or lettuce. This is the receipt of Columella, who recommends beſides to fill the young ones bellies before they are ſuffered to follow their mother to the paſture-ground; for otherwiſe, if they are tormented with hunger, they will ſet obſtinately on the ſtalks of herbs and little roots, and in ſtraining to tear them up, will diſlocate or break their neck ‡. Our common practice in Burgundy is to feed the newly hatched goſlings with haſhed chervil;

* Ariſtotle, *Hiſt. Animal. lib.* vi. 6.

† *Lib.* x. 59.

‡ *Saturetur pullus antequam ducatur in paſcuum; ſi enim fame premitur, cum pervenerit in paſcuum, fruticibus aut ſolidioribus herbis obluctatur ita pertinaciter, ut collum abrumpat.* Columella.

eight

eight days after we add a little bran, ſlightly moiſtened: and care is taken to ſeparate the parents when the proviſions are ſerved, for they would ſcarcely, it is ſaid, leave any thing to their brood. They afterwards have oats given them; and as ſoon as they can eaſily follow their mother, they are conducted to the green-ſward near the water.

Monſtrous births are perhaps more common in Geeſe than in other domeſtic birds. Aldrovandus has cauſed to be engraved two of theſe monſters; the one has two bodies joined to a ſingle head, the other has two heads and four legs proceeding from the ſame body. The exceſſive corpulence to which the Gooſe is naturally inclined, and which we ſeek to promote, muſt produce in its conſtitution alterations ſufficient to affect its generative powers. In general, very fat animals are little prolific; for the over proportion of adipoſe ſubſtance changes the quality of the ſeminal fluid, and even that of the blood. When the head of an extremely fat Gooſe is cut off, nothing but a white liquor flows, and, upon opening it, not a drop of red blood can be ſeen*. In ſuch caſes the liver, from the obſtruction occaſioned by the groſſneſs, ſwells to a prodigious ſize; nay, in a fatted Gooſe, the liver is often more bulky than all the other bowels together †. Theſe fat livers, on which

* Collect. Academ. *part. etrang. tom.* iv. *p.* 146.

† *Aſpice quam tumeat magno jecur anſere majus.* Martial.

our

our gluttons ſet ſo high value, agreed alſo with the taſte of the Roman Apiciuſes. Pliny deems it an important queſtion, to know what citizen invented that diſh *. They fed the Gooſe with figs, to make the fleſh more exquiſite †; and they had diſcovered that it fattens much quicker in a narrow, dark place ‡. But it was reſerved to our more than barbarous gluttony, to nail the feet, to put out or ſew up the eyes of theſe unhappy animals; cramming them at the ſame time with little balls, and denying them drink, that they might ſuffocate in their fat §. Uſually and more humanely we are contented with ſhutting them up for a month, and a buſhel of oats is ſufficient to make a Gooſe very fat. It is eaſy to know, by a very manifeſt external ſign, when we ſhould diſcontinue the feeding, and when the bird has received its due fat; for under each wing there is then a very diſtinct pellet of fat. It has been remarked, that the Geeſe bred near the margin of water are leſs expenſive to maintain, lay earlier, and fatten more eaſily, than others.

* *Noſtri ſapientiores anſeris jecoris bonitatem novere; fartilibus in magnam amplitudinem creſcit, exemptum quoque lacte augetur; nec ſine cauſâ in quæſtione eſt qui primus, tantum bonum invenerit, Scipio Metellus vir conſularis an M. Seſtius eâdem ætate eques Romanus.* Plin. *lib.* x. 22.

† *Pinguibus aut ficis paſtum jecur anſeris albi.* Horace.

‡ Columella.

§ J. B. Porta, refining on this cruelty, dares to give a horrible receipt, of roaſting the Gooſe alive, and eating it limb by limb, while its heart ſtill beats. See Aldrovandus, *lib.* iii. *p.* 133.

 Gooſe-

Goose-fat was much esteemed by the ancients for topical applications, and as a cosmetic. They recommended it for rendering firm women's breasts after delivery, and for preserving the skin fresh and sleek. That prepared at Comagene, with a mixture of aromatics, was boasted as a medicine. Aldrovandus gives a list of recipes, where this fat enters as a specific in all diseases of the *matrix*; and Willughby asserts that goose-dung is the most certain remedy for the jaundice. The flesh itself is not very salubrious; it is heavy, and difficult to digest *: yet was the Goose the chief dish at the suppers of our ancestors; and not till after the introduction of the turkey from America, did the Goose, in our court-yards and in our kitchens, hold only the second place †.

The most valuable article furnished by the Goose, is its down: this is plucked more than once a-year. As soon as the goslings are grown stout and well feathered, and the quills of the wing begin to cross on the tail, which happens at the age of seven weeks or two months, they are stript under the belly, under the wings, and on the neck. Their first feathers are therefore plucked in the end of May, or the beginning of June: and five or six weeks after, that is, in the course of July, there is a

* Galen.

† Salerne and Schwenckfeld.

second

ſecond plucking; and again a third in the beginning of September. During all that time, they are lean, their nouriſhment being diverted to the growth of the new feathers. But if they be left to recover their plumage early in autumn, or even at the cloſe of ſummer, they will ſoon gain fleſh, and afterwards grow fat, and againſt the middle of winter they will be very good for eating. The breeders are not plucked till a month or five weeks after incubation; but the ganders and geeſe which do not hatch may be ſtript twice or thrice annually. In cold countries, the down is richer and finer. The eſtimation in which the Romans held that brought from Germany, was more than once the cauſe of the ſoldiers neglecting their poſts in that country; for whole cohorts diſperſed in purſuit of Geeſe *.

It has been obſerved, that of tame Geeſe the great quills of the wings drop almoſt in a cluſter in one night. They ſeem then baſhful and timorous: they fly from a perſon's approach. Forty days are required for the protruſion of the new feathers; and at this time they continually eſſay their vigour and flap their wings.

Though the ſtep of the Gooſe is ſlow, oblique, and heavy, flocks may be led to a vaſt

* *Plumæ e Germaniâ laudatiſſimæ . . . pretium plumæ in libras denarii quini . . . & inde crimina plerumque auxiliorum præfectis, a vigili ſtatione ad hæc aucupia dimiſſis cohortibus totis.* Plin. *lib.* x. 22.

 diſtance,

diſtance, by ſhort journies *. Pliny ſays, that, in his time, they were conducted from the heart of Gaul to Rome, and that in theſe long marches, thoſe moſt fatigued took the front ranks, that they might be ſupported and puſhed forward by the body of the troop †. When they are collected cloſer together to paſs the night, the ſlighteſt noiſe wakes them, and they all ſcream at once. They alſo make a loud clamour when food is given them; whereas the dog is mute, if offered this boon ‡. Hence Columella is led to ſay, that Geeſe are the ſureſt guardians on a farm §; and Vegetius does not heſitate to aſſert, that they are the moſt vigilant ſentinels that can be planted in a beſieged city ‖. Every body knows, that, on the Capitol, they diſcovered to the Romans the aſſault attempted by the Gauls, and thus ſaved Rome. In memory of that important and ſalutary ſervice, the cenſor allowed each year a ſum of money for maintaining Geeſe; while, on the ſame day, dogs were whipt in public, as a puniſhment

* Salerne.

† *Mirum a Morinis uſque Romam pedibus venire: feſſi proferuntur ad primos, ita ceteri ſtipatione naturali propellunt eos.* Plin. *lib.* x. 59.

‡ Ælian, *lib.* xii. 33.

§ *Anſer ruſticis gratus, quod ſolertiorem curam præſtat quàm canis, nam clangore prodit inſidiantem.* De Re Ruſtica. — Ovid, deſcribing the hut of Philemon and Baucis, ſays, *Unicus anſer erat minimæ cuſtodia villæ.*

‖ De Re Milit. *lib.* iv. 26.

for

for their criminal ſilence in ſo critical a moment *.

The natural cry of the Gooſe is very noiſy, like the *clangor* of a trumpet or clarion; it is very frequent, and may be heard at a great diſtance. But the bird has alſo other ſhort notes, which it often repeats. If it is aſſailed or frighted, it ſtretches out its neck and gabbles with open mouth, and hiſſes like an adder. The Romans have expreſſed that odd ſort of noiſe by the imitative words *ſtrepit*, *gratitat*, *ſtridet* †.

Whether from fear or vigilance, the Gooſe repeats every minute its loud calls ‡: often the whole flock anſwer by a general acclamation; and of all the inhabitants of the court-yard, none is ſo vociferous or bluſtering. This great loquacity induced the ancients to give the name of Gooſe to indiſcreet prattlers, bad writers, and low informers; as its awkward pace and its uncouth geſtures make us apply the ſame appellation to ſilly and ſimple perſons. But beſides the marks of ſentiment and of underſtand-

* *Eſt et anſeri vigil cura, Capitolio teſtata defenſo, per id tempus canum ſilentio proditis rebus. Quamobrem cibaria anſerum cenſores imprimis locant. Eâdem de cauſâ ſupplicia annua canes pendunt inter ædem juventutis & ſummani, vivi in ſambucâ arbore fixi.* Lib. x. 22.

† —— *argutos inter ſtrepere anſere olores.*

Virg. Ec. ix. 36.

Cacabat hinc perdix; hinc gratitat improbus anſer.

Aut. Philomel.

‡ Ariſtotle, *Hiſt. Animal. lib.* i. 1.

ing which we discover in it *, the courage with which it protects its young, and defends itself against the ravenous birds †, and certain very singular instances of attachment and even gratitude, which the ancients have collected ‡, demonstrate that this contempt is ill-founded: and we can add an example of the firmest affection §. The fact was communicated to me by a man

* The sense which the Goose possesses in the highest perfection seems to be hearing; Lucretius thinks that it is smell:

Humanum longe præsentit odorem
Romulidarum arcis servator candidus anser. Nat. Rer. *lib.* iv.

† Aldrovandus.

‡ Pliny, *lib.* x. 22.

§ We give this note in the artless and animated style of the keeper of Ris, an estate belonging to M. Anisson Duperon, and the scene of this faithful and unshaken friendship. " Emmanuel was " asked how the white gander called *Jacquot* was tamed with him. " It is proper to observe that there were two ganders, a gray and " a white, with three females: these two males were perpetually " contending for the company of these three dames; when one or " the other prevailed, he assumed the direction of them, and hindered " the other from approaching. He who was master during the " night, would not yield in the morning; and the two gallants " fought so furiously, that it was necessary to run and part them. " It happened one day, that, being drawn tothe bottom of the gar- " den by their cries, I found them with their necks entwined, strik- " ing their wings with rapidity and astonishing force; the three fe- " males turned round, as wishing to separate them, but without " effect; at last the white gander was worsted, overthrown, and mal- " treated by the other: I parted them, happily for the white one, " which would have lost his life. Then the gray gander set a " screaming and gabbling and clapping his wings, and ran to join " his mistresses, giving each a noisy salute, to which the three dames " replied, ranging themselves at the same time round him. Mean- " while poor *Jacquot* was in a pitiable case, and retiring, sadly he " vented

a man of veracity and information, to whom I am partly indebted for the care and attention which

" vented at a diſtance his doleful cries: it was ſeveral days before " he recovered from his dejection, during which time I had occa- " ſion to paſs through the court where he ſtayed; I ſaw him always " thruſt out from ſociety, and each time I paſſed he came gabbling " to me, no doubt to thank me for the ſuccour which I had given " him on his defeat. One day he approached ſo near me, ſhowing " ſo much friendſhip, that I could not help careſſing him by ſtrok- " ing with my hand his back and neck, to which he ſeemed ſo ſen- " ſible as to follow me into the entrance of the court. Next day " as I again paſſed, he ran to me, and I gave him the ſame careſſes, " with which he could not be ſurfeited; but he ſeemed by his geſ- " tures to deſire that I ſhould lead him to his dear mates; I accord- " ingly did lead him to their quarter, and upon his arrival he be- " gan his vociferations, and directly addreſſed the three dames, " who failed not to anſwer him. Immediately the gray victor " ſprung upon *Jacquot*: I left them for a moment; he was always " the ſtronger; I took part with my *Jacquot*, who was under; I ſet " him over his rival, he was thrown under; I ſet him up again: in " this way they fought eleven minutes, and by the aſſiſtance which " I gave, he obtained the advantage over the gray gander, and got " poſſeſſion of the three dames. When my friend *Jacquot* ſaw him- " ſelf maſter, he would not venture to leave his females, and there- " fore no longer came to me when I paſſed; he only gave me at " a diſtance many tokens of friendſhip, ſhouting and clapping his " wings, but would not quit his prey, for fear that another ſhould " take poſſeſſion. Things went on in this way till the breeding " ſeaſon, and he never gabbled to me but at a diſtance: when his " females however began to ſit, he left them and redoubled his " friendſhip to me. One day having followed me as far as the ice- " houſe at the top of the park, the place where I muſt neceſſarily " part with him in purſuing my way to the Wood of Orangis, at " half a league's diſtance, I ſhut him in the park: he no ſooner ſaw " himſelf ſeparated from me than he vented ſtrange cries. How- " ever I went on my road, and I was about a third advanced, " when the noiſe of a heavy flight made me turn round my head; " I ſaw my *Jacquot* four paces from me: he followed me all the road,

which I have experienced at the royal press in printing my works. We have also received from St. Domingo an account pretty similar, and which shows that, in certain circumstances, the Goose appears capable of a very lively and strong personal attachment, and even of a sort of passionate friendship, which wastes and destroys it, when removed from the object of its affection.

As early as the time of Columella, the domestic Geese were distinguished into two kinds; that with the white, and that with the varie-

" partly on foot, partly on wing, getting before me and stopping " at the cross paths, to see what way I should take. Our expedi- " tion lasted from ten o'clock in the morning till eight in the even- " ing, and yet my companion followed me through all the wind- " ings of the wood, without seeming to be tired. After this he fol- " lowed and attended me every where, so as to become trouble- " some, I not being able to go to any place without his tracing my " steps, so that one day he came to find me in the church: another " time, as he was passing by the rector's window, he heard me talk- " ing in the room; and as he found the door open, he entered, " climbed up the stairs, and marching in, he gave a loud burst of " joy, to the no small affright of the rector.

" I am sorry, in relating such pleasing traits of my good and " faithful friend *Jacquot*, when I think that it was myself that first " dissolved the sweet friendship: but it was necessary that I should " separate him by force: poor *Jacquot* fancied himself as free in the " best apartments as in his own, and after several accidents of that " kind, he was shut up, and I saw him no more. His inquietude " lasted above a year, and he died from vexation; he was become " as dry as a bit of wood, as I am told; for I would not see him, " and his death was concealed from me more than two months after " the event. Were I to recount all the friendly incidents between " me and poor *Jacquot*, I should not, in four days, have done writ- " ing: he died in the third year of the reign of friendship, aged " seven years and two months."

 gated

gated plumage, the former more anciently domesticated than the latter. The freckled Geese, according to Varro, were not so prolific as the white ones *, which the farmer was advised by them to keep, as being also the largest †. Belon agrees entirely with the ancient writers on rural œconomy: but Gesner, who was almost his contemporary, asserts, that in Germany the gray sort are, for good reasons, preferred, being hardier and not less prolific; and Aldrovandus confirms the remark for Italy. It would seem as if the most ancient breed were emasculated by long domestication; and indeed the gray or variegated Geese are now inferior neither in size, nor in fecundity, to the white ones.

Aristotle, speaking of two breeds or species of Geese, a greater and a lesser, which are gregarious, seems by the latter to mean the wild Goose ‡. And Pliny treats particularly of this under its name *Anser ferus*. In fact, the Geese form two great tribes; of which the one, long since domesticated, is attached to our dwellings, and multiplies and varies in our hands; the other, much more numerous, has escaped from us, and remains wild and savage: for the whole difference results from the slavery of man on the one hand, and from the liberty of nature on the

* De Re Rusticâ, *lib.* viii. 13.

† Aldrovandus.

‡ *Lib.* viii. 15.

other.

other *. The wild Goofe is lean, and flenderer than the tame one: and the fame may be obferved of feveral breeds, according as they approach the primitive ftem, as between the common and the ftock pigeons. The wild Goofe has alfo its back brownifh-gray, its belly whitifh, and all its body clouded with rufty-white, and the tip of each feather fringed with the fame. In the domeftic Goofe, this rufty colour has varied, has affumed fhades of brown or of white, has even difappeared entirely in the white fort †. Some have a tuft on the head. But thefe changes are inconfiderable, if compared with thofe which the hen, the pigeon, and many other fpecies, have undergone in the domeftic ftate. The Goofe and the other water fowls which we have tamed, are much lefs removed from the wild ftate, and much lefs fubdued or enflaved, than the gallinaceous, which feem to be the native citizens of our court-yard. In countries where multitudes of Geefe are raifed, the whole attention needed, during the fummer months, confifts in calling them and conductiug them to the farm, where they have convenient and undifturbed retreats for neftling and educating their young; and thefe advantages, together with the afylum and food afforded them in winter, attach them to the abode, and reftrain them from deferting. The reft of their time is

* Belon.

† Ray.

fpent

ſpent beſide the brooks and pools, where they play and reſt on the banks. In a mode of life ſo nearly approaching to the liberty of nature, they reſume almoſt all its advantages, ſtrength of conſtitution, thickneſs and elegance of plumage, vigour and extent of flight *. In ſome regions even, where man, leſs civilized, that is leſs tyrannical, allows the animals ſtill to enjoy freedom, there are Geeſe really wild the whole ſummer, which become domeſtic in the winter. We have learnt this fact from Dr. Sanchez, and we ſhall here give the intereſting account which he communicated.

" I ſet out from Azof," ſays that learned phyſician, " in autumn 1736. Being ſick, and " afraid of falling into the hands of the Cu- " ban Tartars, I reſolved to walk, following " the courſe of the Don, and to ſleep every " night in the villages of the Coſſacs, who " are ſubject to the Ruſſian dominion. In " the firſt evenings of my journey, I remarked " a great number of Geeſe in the air, which " alighted and diſperſed through the hamlets. " The third day eſpecially, I ſaw ſuch a mul- " titude at ſun-ſet, that I enquired of the " Coſſacs, among whom I lodged that night, " whether they were tame Geeſe; and if they " came from a diſtance, as their lofty flight " ſeemed to indicate. Surprized at my igno-

* Scaliger.

" rance,

" rance, they replied that these birds came " from the remote northern lakes; and that " every year, on the breaking up of the ice, " in the months of March and April, six or " seven pairs of Geese leave each hut of the " village, which all take flight in a body, and " return not till the beginning of winter, as it " is reckoned in Russia, that is, at the first " snow; that these flocks arrive then, increased " sometimes an hundred-fold, and dividing " themselves, each little party seeks, with its " new progeny, the houses where they lived " the preceding winter. I had constantly that " spectacle every evening, for three weeks: " the air was filled with infinite multitudes " of Geese, which dispersed in bands: the girls " and women, at the doors of their huts, look-" ing at the flight, were calling out, 'There go " my Geese,' 'There go the Geese of such a " one:' and each of the bands alighted in the " court where they had spent the preceding " winter*. I continued to see these birds " till I reached Nova-Pauluska, where the " winter was already intense."

It is probably from some such relations that the wild Geese which visit us in winter are supposed

* The inhabitants make a slaughter among these Geese while their feathers are in down; they cut them in two and dry them; the down, famous for its goodness, is the subject of a great trade; the dry flesh is carried to the Ukraine, where the Cossacs barter it for spirituous liquors and some clothes. *Extract from the same narration of Dr. Sanchez.*

ſuppoſed to be domeſtic in other countries. But this notion is, as we learn from Belon, devoid of foundation; for the wild Geeſe are of all birds, perhaps, the moſt completely ſavage; and beſides, winter, the ſeaſon of their arrival, is the very time they ſhould be tame.

In France, the wild Geeſe paſs in October or the firſt days of November *. Winter, which then begins its reign in the north, determines their migration: and, what is remarkable, the domeſtic Geeſe, at this ſame time, ſhew by their inquietude, their frequent and long flights, a ſimilar deſire to journey †, the evident remains of original inſtinct.

The

* It is in the month of November, M. Hebert writes me, that the firſt wild Geeſe are ſeen in Brie, and they continue to paſs in that province till the hard froſts ſet in, ſo that their paſſage laſts nearly two months. The troops of theſe Geeſe are from ten or twelve to twenty or thirty, and never more than fifty: they alight in the plains ſown with corn, and do ſo much injury that attentive huſbandmen ſet children to watch their fields, and to frighten away the Geeſe by their ſhouts. It is in wet weather that they occaſion the moſt havoc, becauſe they tear up the wheat as they paſture on it; whereas in froſt they only crop it, and leave the reſt of the plant rooted in the ſoil.

† My neighbour at Mirande keeps a flock of Geeſe, which he every year reduces to fifteen, by ſelling a part of the old ones, and preſerving a part of the young. This is the third year that I have remarked that during the month of October theſe birds betray a ſort of reſtleſſneſs, which I look upon as the remnant of their diſpoſition to migrate. Every day, about four o'clock in the afternoon, theſe Geeſe take wing, paſs over my gardens, and make a circuit round the plain in their flight, and return not to their rooſt till night: they call each other by a cry, which I diſtinctly recognized to be that which the wild Geeſe repeat in their paſſage, to collect and unite their numbers. The month of October has been ſo mild

this

The flight of the wild Geese is always very elevated*; their motion is smooth, accompanied with no noise or rustling, and the play of the wings, in striking the air, seems never to exceed one or two inches. The regularity and conduct with which they are marshalled, implies a sort of intelligence superior to that of other birds, which migrate in confused and disorderly flocks. The arrangement observed by the Geese seems dictated by a geometrical instinct: it is at once calculated to preserve the ranks free and entire,

this year, that the grass has shot up in the pasture-grounds; independently of this abundance of food, the proprietor of this flock gives them grain every evening this season, lest he should lose a few of them. Last year one strayed away, and was more than two months after found at three leagues distance. After the end of October, or the first days of November, these Geese resume their tranquillity. —I conclude from this observation, that the most ancient domestication (since that of the Geese in this country, where there are no wild ones, must have taken place in remotest antiquity) never entirely effaces this character imprinted by nature, this innate desire to migrate. The tame Goose, degraded and incumbered, attempts a passage, exercises itself every day; and, though abundantly provided and wanting for nothing, I could warrant, that if wild ones passed at this season, they would always lead off some, and that nothing but example and a little courage are needed to make them desert: I doubt not, that if the same observations were made in the provinces where many Geese are fed, we should find that some are lost every year, and this in the month of October. I know not, however, if all the Geese reared in court-yards shew these marks of inquietude; but it must be considered that these are almost confined within walls, and never pasture or enjoy the view of the horizon; they are slaves which have lost every idea of their ancient liberty. *Observation communicated by M. Hebert.*

* "It is only in foggy weather that the wild Geese fly so near "the ground that they can be shot." *Idem.*

to

to break the reſiſtance of the air, and to leſſen the exertion and fatigue of the ſquadron. They form two oblique lines, like the letter V; or, if their number be ſmall, they form only one line: generally they amount to forty or fifty, and each keeps its rank with admirable exactneſs. The chief, who occupies the point of the angle, and firſt cleaves the air, retires, when he is fatigued, to the rear; and the reſt, by turns, aſſume the ſtation of the van. Pliny deſcribes the wonderful order and harmony that prevail in theſe flights *; and remarks that, unlike the cranes and the ſtorks, which journey in the obſcurity of the night, the Geeſe are ſeen purſuing their route in broad day.

Several ſtations have been noticed where the larger flocks divide, and diſperſe into different countries. The ancients mentioned Mount Taurus as the rendezvous of ſuch as ſpread through Aſia Minor †; and alſo Mount Stella, now called Coſſonoſſi (in Turkiſh *Fields of Geeſe*), whither prodigious flocks of theſe birds repair in the fall, and thence ſcatter through the whole of Europe.

* *Liburnicarum more roſtrato impetu feruntur, facilius ita findentes aëra, quàm ſi recta impellerent, a tergo ſenſim dilatante ſe cuneo, porrigitur agmen largèque impellenti præbetur auræ. Colla imponunt præcedentibus; feſſos duces ad terga recipiunt.* Lib. x. 23.

† Oppian ſays, that in paſſing Mount Taurus, the Geeſe take the precaution to ſtop their mouth with a pebble, that their natural diſpoſition to gabble may not betray them to the eagles; and the good Plutarch repeats the tale.

rope*. Several of theſe ſmall bodies, or ſecondary flocks, unite again, and form larger ſquadrons, amounting to four or five hundred; which we ſometimes ſee alight in our fields, where they are very deſtructive †, paſturing on the green corn, which they ſcrape from under the ſnow. Fortunately, the Geeſe are very unſteady and roving, remain a ſhort while in one place, and ſeldom return to the ſame diſtrict. They ſpend the whole day on the ground, among the cultivated fields or meadows; but retire every evening to the rivers or large pools. There they paſs the whole night, but arrive not till ſun-ſet, and ſome after twilight: each party is received by loud acclamations, to which it replies; ſo that, at eight or nine o'clock, and the darkeſt nights, they make ſuch noiſy and multiplied clamours, that we ſhould ſuppoſe them to be aſſembled by thouſands.

The wild Geeſe might, at this ſeaſon, be ſaid to be birds of the plain rather than birds of the water; ſince they never reſort to the ſtreams and pools but at night. Their habits are the reverſe of thoſe of the ducks, which leave the water at that time, and diſperſe to feed in the meadows, and do not return before the Geeſe repair to their diurnal haunts. On their arrival

* Rzaczynſki.

† Aldrovandus mentions Holland in particular as ſuffering by the viſits of wild Geeſe.

in

in the ſpring, the wild Geeſe ſcarce ſtop with us, and very few are then ſeen in the air: it is probable that they depart and return by different routes.

As the wild Geeſe ſo frequently ſhift their place, and as they have an acute ear, and are miſtruſtful and circumſpect, they are difficult to catch *, and elude moſt kinds of ſnares. That which Aldrovandus deſcribes, is perhaps the ſureſt and the beſt contrived. "When the "fields," ſays he, "are kept dry by the froſt, "a proper place is choſen for ſpreading a long "net, faſtened and ſtretched with cords, ſo "ſo that it may quickly drop: it is nearly like "a lark's net, but extends over a longer ſpace, "which muſt be covered with duſt. A few "tame Geeſe are ſet beſide it, to ſerve as calls. "It is requiſite that all theſe preparations be "made in the evening, and that the net be not "afterwards touched; for if in the morning the

* "It is almoſt impoſſible," ſays M. Hebert, "to ſhoot them "on their arrival, becauſe they fly too high, and begin not to de- "ſcend till they are over water. I have tried," he adds, with "little ſucceſs, to ſurprize them at day-break; I paſſed the "night in the fields; the boat was got ready in the evening, we "ſtepped into it long before day, and we advanced, concealed by "the duſk, a great way upon the water, and as far as the laſt of "the reeds: however, we were too far from the flock to fire upon "them; and theſe ſhy birds roſe all of them, and to ſuch a height, "that in paſſing over our heads they were beyond the reach of our "ſhot. All theſe Geeſe thus aſſembled had ſet off together, and "were waiting full day, had they not been diſturbed; then they "ſeparated and diſperſed in diviſions, and perhaps in the ſame or- "der in which they had collected in the preceding evening."

 "Geeſe

" Geeſe perceive the dew or rime bruſhed,
" they will grow ſuſpicious. They come
" to the cackling of the calls, and after long
" circuits, and many windings in the air,
" they alight: the fowler, concealed in a ditch
" at fifty paces diſtance, pulls the cord, and
" takes the whole flock, or part of it, under his
" net."

Our fowlers employ all their ſtratagems to ſurprize the wild Geeſe. If the ground be covered with ſnow, they throw a white ſhirt over their clothes. At other times, they diſguiſe themſelves with branches and leaves, ſo as to appear a walking buſh. They even cover themſelves with a cow's ſkin, and advance on all-four, holding their gun under them: and, with all theſe wiles, they often cannot approach the Geeſe, even during the night. It is ſaid, that one always ſtands ſentinel, with its neck extended and its head raiſed, and which, on the leaſt ſymptom of danger, ſounds alarm to the flock. But as they cannot ſuddenly mount, but run three or four paces clapping their wings, the fowler has time to fire on them.

The wild Geeſe do not remain with us the whole winter, unleſs the ſeaſon is mild; for in ſevere winters, when the rivers and pools are frozen, they advance farther ſouth, whence ſome return about the end of March, in their progreſs to the northern countries. They frequent, then, the hot and even the temperate climates in the time of their paſſages only; for we are not informed whether they breed in France*. A few breed in England, as well as in Sileſia and Bothnia †: a larger number breed in ſome cantons of Great Poland and Lithuania ‡: but the bulk of the ſpecies ſettle not till they have advanced farther north ||; nor do they ſtop on the coaſts of Iceland §, or on the extenſive ſhores of Norway ¶. They migrate in immenſe flocks as far as Spitzbergen **, Greenland ††, and the tracts adjoining to Hudſon's Bay ‡‡, where their fat and their dung

* Belon.

† Schwenckfeld.

‡ *Idem.*

|| Aldrovandus.

§ The wild Geeſe viſit Iceland only in the ſpring. It is uncertain whether theſe birds breed there, the more ſo as they are remarked not to halt, but to continue their flight towards the north: they are, properly ſpeaking, only birds of paſſage. *Horrebow.*

¶ There are only two kinds of wild Geeſe in Norway; the gray ones paſs in ſummer into the diſtrict of Nortland. The Norwegians believe that in winter they go to France... We know not where theſe Geeſe breed; however, ſome have been obſerved to multiply on the coaſt of Riefilde, in Norway. *Pontoppidan.*

** There is a great gulph (north-weſt of the iſland Baëren, between Spitzbergen and Greenland), and in the middle of it an iſland filled with wild Geeſe and their neſts. Heemſkerke and Barentz doubt not but theſe Geeſe are the ſame that are ſeen to come every year in great numbers into the United Provinces, particularly at Wieſingen, in the Zuyder-ſea, in North Holland and Frieſland, though hitherto it was unknown where they bred. *Recueil des Voyages de la Compagnie des Indes*; *Amſterdam,* 1702, *tom.* i. *p.* 35.

†† The wild gray Geeſe arrive at the opening of the ſummer in Greenland, to lay their eggs and rear their young. It is probable that they come from the neareſt coaſts of America; they return there to winter. *Crantz.*

‡‡ In the end of April, plenty of ducks and Geeſe arrive at Hudſon's

dung * prove resources to the miserable inhabitants of these frozen countries. There are also innumerable flocks on the lakes and rivers of Lapland †, as well as on the plains of Mangasea, along the Jenisea ‡; and in many other parts of Siberia, as far as Kamtschatka, whither they arrive in the month of May, and whence they depart in November, after having hatched. Steller saw them pass Bering's Island, flying in autumn towards the east, and in spring towards the west; and he thence infers that they come from America to Kamtschatka. Certain it is, the greatest part of these Geese, on the north-east of Asia, push southwards to Persia ‖, India §, and Japan, where their migrations are remarked as in Europe: we are assured even that in Japan they enjoy so much security, as to have forgotten their natural shyness ¶.

A fact

son's Bay. *Hist Gen. des Voy.*—On Nelson River there are many Geese, ducks, and swans. *Ellis.*—There are also numbers of Geese on Rupert River. *Lade.*

* The northern people season their meat with Goose-fat instead of butter. *Olaus Magnus.*—Dried Goose-dung serves the Esquimaux as a wick for their lamps instead of cotton; it is a poor shift, but much better than none at all. *Ellis.*

† Regnard. ‡ Gmelin.

‖ In Persia there are Geese, ducks, plovers, cranes, herons, divers, and woodcocks, every where; but most plentiful in the northern provinces. *Chardin.*

§ There are Geese, ducks, teals, herons, &c. in the kingdom of Guzaratte, in the East Indies. *Mandeslo.*—They are found also in Tonquin. *Dampier.*

¶ In Japan there are two sorts of Geese, which never intermix; the one white as snow, with the tips of the wings very black; the other

A fact which seems to corroborate the opinion that the Geese pass from America into Asia, is, that the same species which is seen in Europe and in Asia, occurs likewise in Louisiana *, in Canada †, in New Spain ‡, and on the west coast of North America. We know not whether the same species be found equally in the whole extent of South America. We learn, however, that the tame Goose, introduced from Europe into Brazil, is reckoned to have improved the delicacy and flavour of its flesh ||: on the contrary, it has degenerated in St. Domingo, where the Chevalier Lefebvre Deshayes has made several observations on the dispositions of these birds in the domestic state; and particularly on the tokens of joy which the ganders shew at the

other ash-gray; they are all so common and so familiar, as easily to permit a person to approach them. Though they are very pernicious in the fields, it is prohibited to kill them, under pain of death, in order to secure the privilege to those who purchase the right. The peasants are obliged to surround their fields with nets, to defend them from the ravages. *Kæmpfer.*

* Dupratz.

† The Geese, and all the large river birds, are every where abundant in Canada, except near habitations, which they are never seen to approach. *Hist. Gen. des Voy. tom.* xv. *p.* 227.—Among the Hurons there are wild Geese, which they call *ahonque*. *Theodat.*

‡ *Tlacalcatl* is a mountain Goose, like the tame, and either the same with our wild Goose, or akin to it. *Fernandez.*

|| It is said to be remarked that the ducks and Geese carried from Europe to Brazil, have there acquired a finer taste; on the contrary, the hens, which have there grown larger and stouter, have lost a part of their flavour. *Hist. Gen. des Voy. tom.* xiv. *p.* 305.

 birth

birth of the young *. He informs us alſo, that at St. Domingo is ſeen a migratory Gooſe, which, as in Europe, is ſomething ſmaller than the tame kind. And hence it would appear, that theſe birds of paſſage advance far into the ſouthern regions of the new world, as in thoſe of the old continent, where they have penetrated under the torrid zone †, and ſeem even to have traverſed its whole extent; for they are found in Senegal ‡, in Congo ||, in the vicinity of the Cape of Good

* Though the Gooſe bears here to be robbed thrice a year of its down, the ſpecies is however leſs valuable in a climate, where health forbids, in ſpite of effeminacy, to repoſe on the down, and where freſh ſtraw is the only bed on which ſleep can alight: nor is the fleſh of the Gooſe ſo good at St. Domingo as in France; it is never plump, it is ſtringy, and that of the Indian Gooſe is in every reſpect preferable. *Obſervation communicated by the Chevalier Lefebvre Deſhayes.*

Naturaliſts have not mentioned, I think, the ſingular expreſſions of joy which the gander gives his young the firſt times he ſees them eat. This animal ſhews its ſatisfaction by raiſing his head with a dignified air, and ſtamping with his feet, ſo that one ſhould imagine that he dances. Theſe ſigns of contentment are not equivocal, ſince they have place only in this circumſtance, and are repeated almoſt each time that the goſlings are fed in their tender age. The father neglects his own ſubſiſtence, to give vent to the joy of his heart: this dance is ſometimes of long duration, and if any incident occaſions an interruption, as when he chaſes the poultry to a diſtance from his young, he reſumes it with new ardour. *Idem.*

† All climates, M. Baillon writes me, ſuit the Gooſe and the duck, alike migratory and paſſing from the coldeſt countries into thoſe ſituated between the tropics. I have ſeen many arrive in the iſland of St. Domingo on the approach of the rainy ſeaſon, and they ſeem to ſuffer no ſenſible alteration in climates ſo oppoſite.

‡ On the coaſt of Senegal, the Geeſe and teals are well-flavoured. *Le Maire.*

|| Mandeſlo.

Hope,

Hope *, and perhaps in the lands of the ſouthern continent. In fact, we conceive the Geeſe which navigators have met with in the Magellanic lands, at Tierra del Fuego †, in New Holland ‡, &c. to be nearly a-kin to our ſpecies of Geeſe, ſince they have received no other name. Yet, beſides the common ſpecies, there exiſt in thoſe countries other ſpecies; which we now proceed to deſcribe.

* The country (at the bay of Saldana) is filled with oſtriches, herons, Geeſe, &c. *Gemelli Carreri.*—The ſize of the water Geeſe, which are found at the Cape of Good Hope, is the ſame with that of the tame Geeſe known in Europe; and with reſpect to colour, there is no other difference between them, except that the water Geeſe have on the back a brown ſtripe mixed with green. All theſe different kinds of Geeſe are excellent wholeſome food. *Kolben.*

† Geeſe are ſeen on the edge of the lagoons (in the bay of St. Julian) in the *Terra Magellanica. Quiroga.*—Wallis found Geeſe at Cape Forward, in the Straits of Magellan;—alſo in the bay of Cape Holland.—Cook found Geeſe and ducks at Chriſtmas Sound, in *Tierra del Fuego,* and called an iſland there, *Gooſe Iſland,* and a cove, *Gooſe Cove.*—Geeſe, ducks, teals, and other birds, occur at Port Egmont, latitude 51° S. in ſuch numbers, that our people were tired of eating them: it was uſual to ſee a canoe bring ſixty or ſeventy fine Geeſe without firing a ſhot; they were killed with ſtones. *Byron.*

‡ The water-fowl (at New South Holland) are the wild Geeſe, and the whiſtler ducks which perch. *Cook.*—Captain Cook left ſome pairs of tame Geeſe in New Holland, in hopes that they would multiply.

[A] Specific character of the wild Gooſe, *Anas Anſer*: "Its bill "is ſemi-cylindrical; its body cinereous above, and paler below; its "neck ſtriped." Great flocks of Geeſe are kept in the fens of Lincolnſhire, which are plucked about the neck, breaſt, and back, once or twice a year. The feathers form a conſiderable branch of trade; thoſe from Somerſetſhire are eſteemed the beſt, and thoſe from Ireland are reckoned the worſt.—The following is an extract from

from Mr. Pennant's firſt tour in Scotland: "The fens near Reveſby "Abby (in Lincolnſhire) eight miles beyond Horncaſtle, are of vaſt "extent; but ſerve for little other purpoſe than the rearing great "numbers of Geeſe, which are the wealth of the fenmen. During "the breeding ſeaſon, theſe birds are lodged in the ſame houſes "with the inhabitants, and even in their very bed-chambers: in "every apartment are three rows of coarſe wicker pens placed one "above another; each bird has its ſeparate lodge divided from the "other, which it keeps poſſeſſion of during the time of ſitting. A "perſon, called a *Gozzard*, (gooſe-herd) attends the flock, and "twice a-day drives the whole to water; then brings them back "to their habitations, helping thoſe that live in the upper ſtories "to their neſts, without ever miſplacing a ſingle bird.

"The Geeſe are plucked five times a-year; the firſt plucking "is at Lady-day, for feathers and quills; and the ſame is renewed "for feathers only, four times between that and Michaelmas. The "old Geeſe ſubmit quietly to the operation, but the young ones "are very noiſy and unruly. I once ſaw this performed, and ob-"ſerved that the goſlings of ſix weeks old were not ſpared; for "their tails were plucked, as I was told, to habituate them early to "what they were to come to. If the ſeaſon proves cold, numbers "of Geeſe die by this barbarous cuſtom.

"Vaſt numbers are drove annually to London, to ſupply the "markets; among them, all the ſuperannuated geeſe and ganders "(called here *Cagmags*) which ſerve to fatigue the jaws of the "good citizens, who are ſo unfortunate as to meet with them."

The MAGELLANIC GOOSE.

L'OIE DES TERRES MAGELLANIQUES.

Buff.

SECOND SPECIES.

Anas Magellanica. Gmel,

THIS large and beautiful Goofe, which feems peculiar to the country contiguous to the Straits of Magellan, has the lower half of its neck, its breaft, and the top of its back, richly enamelled with black feftoons on a rufous ground: the plumage of the belly is worked with the fame feftoons on a whitifh ground: the head and the top of the neck are of a purple red. There is a large white fpot on the wing: and the blackifh colour of the mantle is foftened by a purple glofs.

It would feem that thefe beautiful Geefe are what Commodore Byron ftiles the *painted Geefe* *, which are found at *Sandy Point,* in the Straits of Magellan. Perhaps this fpecies is the fame with that which Captain Cook calls *a new fpecies of Goofe,* and which he met with on the eaftern coafts of the Straits of Magellan, and of Tierra del Fuego, which are furrounded by immenfe floating beds of famphire.

* Anas Picta. *Gmel.*

The GOOSE of the MALOUINE, or FALKLAND ISLANDS.

THIRD SPECIES.

Anas Leucoptera. Gmel.
The White-winged Antarctic Goose. Brown.
The Sea Goose.
The Bustard Goose. Lath.

" OF ſeveral ſpecies of Geeſe," ſays M. de Bougainville, " on which we partly ſubſiſted in the Malouine iſlands, the firſt only " grazes. It is improperly called the *buſ- " tard.* Its tall legs are requiſite for wading " through the large herbs, and its long neck is " uſeful for deſcrying danger. Its pace is nim- " ble, as is its flight; and it has not the diſa- " greeable cackle of its family. The plumage " of the male is white, with a mixture of cine- " reous on the back and the wings: the female " is fulvous, and her wings decorated with " changing colours; ſhe uſually lays ſix eggs. " Their fleſh, which is ſalubrious, nutritive, and " well taſted, became our principal food, and " was ſeldom out of our reach. Beſides thoſe " bred on the iſland, the winds in autumn bring " large flocks, no doubt from ſome deſert " country,

" country, for ſportſmen eaſily diſtinguiſh theſe " new-comers by their indifference at the ſight " of men. Two or three other kinds of Geeſe, " which we found in theſe ſame iſlands, were " not ſo much ſought after, becauſe they con- " tract an oily taſte by feeding on fiſh *."

We term this ſpecies the *Gooſe of the Malouine iſlands*, becauſe in theſe iſlands it was firſt found by our French navigators; for the ſame Geeſe ſeem to be met with in Chriſtmas Sound, upon Tierra del Fuego, in Shag Iſland, and on other iſlands near Staten Land: at leaſt Captain Cook ſeems, on this head, to refer to Bougainville's deſcription, when he ſays, " The " Geeſe ſeem to be very well deſcribed un- " der the name of *buſtards*. They are much " ſmaller than our Engliſh tame geeſe, but eat " as well as any I ever taſted. They have ſhort " black bills, and yellow legs. The gander is " all white; the female is ſpotted black and " white, or gray, with a large white ſpot on " each wing." And a few pages before he gives a fuller deſcription, in the following terms: " Theſe birds appeared remarkable for the dif- " ference of colour between the male and the

* " The form of the latter," adds M. de Bougainville, " is leſs " elegant than that of the firſt ſpecies; there is one which riſes " with difficulty above the water; this is noiſy: the colours of " their plumage are ſeldom other than white, black, fulvous, and " cinereous. All theſe ſpecies, as well as the ſwans, have under " their feathers a very thick white or gray down."

" female.

" female. The male was ſomething leſs than " an ordinary tame gooſe, and perfectly white: " the female, on the contrary, was black, with " white bars acroſs, the head gray; ſome feathers " green, and others white. This difference ſeems " to be fortunate; for the female being obliged " to lead her young, the dark colour of her " plumage conceals her better from the falcons " and other birds of prey." Theſe three deſcriptions ſeem to belong to the ſame ſpecies, and differ not eſſentially from each other. Theſe Geeſe afforded Captain Cook's crew as acceptable repaſts as thoſe at the Falkland iſlands did the French *.

* " As ſoon as we got under the iſland, we found plenty of " ſhags in the cliffs; but without ſtaying to ſpend our time and ſhot " upon theſe, we proceeded on, and preſently found ſport enough: " for, on the ſouth ſide of the iſland, were abundance of geeſe. It " happened to be the moulting ſeaſon; and moſt of them were on " ſhore for that purpoſe, and could not fly. There being a great " ſurf, we found great difficulty in landing, and very bad climbing " over the rocks when we were landed; ſo that hundreds of the Geeſe " eſcaped us, ſome into the ſea, and others up into the iſland. We, " however, by one means or other, got ſixty-two, with which we " returned on board, all heartily tired; but the acquiſition we had " made overbalanced every other conſideration, and we ſat down " with a good appetite to ſupper on part of what the preceding day " had produced." *Cook's ſecond Voyage*, vol. ii. p. 182.

No. 234

THE CHINESE GOOSE

The GUINEA GOOSE.

FOURTH SPECIES.

Anas Cygnoides. Linn. and Gmel.
Anser Guineensis. Briss.
Anser Hispanicus, sive Cygnoides. Marsigl. Danub.
The Spanish Goose. Albin.
The Swan Goose. Ray and Will.
The Chinese Goose. Penn. and Lath.

THE appellation of *Swan-Goose*, given by Willughby to this large and beautiful bird, is very apt; but the Canada Goose, which is at least as beautiful, has an equal right to the name; and besides, all compounded epithets ought to be banished from natural history. The Guinea Goose exceeds all other geese in stature; its plumage is a brown gray on the back, and light gray on the fore side of the body, the whole equally clouded with rusty gray, and with a brown cast on the head and above the neck: it resembles therefore the wild goose in its colours; but its magnitude, and the prominent tubercle at the root of its bill, mark a small affinity to the swan; yet it differs from both by its inflated throat, which hangs down like a pouch or little dewlap: a very evident character, which has procured to these birds the denomination *Jabotieres*.

tieres *. Africa, and perhaps the other ſouthern countries of the old continent, ſeem to be their native abodes; and though Linnæus has termed them *Siberian Geeſe*, they are not indigenous in Siberia, but have been carried thither and multiplied in a ſtate of domeſtication, as in Sweden and Germany. Friſch relates that, having repeatedly ſhown to Ruſſians Geeſe of this kind, which were reared in his court-yard, they all, without heſitation, called them *Guinea Geeſe*, and not *Ruſſian or Siberian Geeſe*. Yet has the inaccurate denomination of Linnæus miſled Briſſon, who deſcribes this Gooſe under its true name of *Guinea Gooſe*, and again, a ſecond time, under that of *Muſcovy Gooſe*, without perceiving that his two deſcriptions refer preciſely to the ſame bird †.

Not only does this Gooſe, though a native of the hot countries, multiply when domeſticated

* From *jabot*, the craw.

† *Anas Cygnoides*, variety. Linn. and Gmel.
Anſer Muſcoviticus. Briſſ.
Anſer Ruſſicus. Klein.
The Crop Gooſe. Kolben.
The Muſcovy Gander. Albin. and Lath.

"It is ſomewhat larger," ſays Briſſon, "than the tame Gooſe: " . . . the head and the top of the neck are brown, deeper on the " upper ſide than on the under; . . . on the origin of the bill there " riſes a round and fleſhy tubercle; . . . under the throat alſo there hangs " a ſort of fleſhy membrane." Add, that Klein regards this Gooſe of Muſcovy or Ruſſia as a variety of the Siberian, which, we have ſeen, is the ſame with the Guinea Gooſe: "I ſaw," ſays he, "a variety of the Siberian Gooſe, its throat larger, its bill and legs " black, with a black depreſſed tubercle."

in

in the coldeſt climates; it alſo contracts an affinity with the common ſpecies, and the hybrids which are thus bred take the red bill and legs of our Gooſe, but retain of their foreign parent the head, the neck, and the ſtrong, hollow, and yet loud voice. The clangor of theſe large Geeſe is ſtill more noiſy than that of the ordinary kind, and they have many characters in common: the ſame vigilance ſeems natural to them. "Nothing," ſays Friſch, "can ſtir in the houſe "during the night, but the Guinea Geeſe will "ſound an alarm: and in the day-time they "give the ſame ſcreams if any perſon or animal "enter the court; and often will purſue, peck-"ing the legs." The bill, according to the remark of this naturaliſt, is armed at the edges with ſmall indentings, and the tongue is beſet with ſharp *papillæ*; the bill is black, and the tubercle which riſes upon it is vermillion. This bird carries its head high as it walks; and its fine carriage and its great bulk give it a noble air *. According to Friſch, the ſkin of the little dewlap or pouch under the throat is neither ſoft nor flexible, but firm and hard. This account, however, ſcarce agrees with the uſe which, Kolben tells us, the ſailors and ſoldiers at the Cape make of it †. I received a head and neck of

* Ray.

† The wild geeſe at the Cape have been called *crop geeſe (oies jabotieres.)* The ſoldiers, and the common people of the colonies, uſe theſe crops for tobacco-pouches; they will hold about two pounds. *Kolben.*

one

one of theſe Geeſe, and, at the root of the lower mandible, this pouch or dewlap was viſible: but as theſe parts were half burnt, we cannot deſcribe them exactly. I learn however from this packet, which was ſent from Dijon, that the Guinea Geeſe occur in France, as well as in Germany, Sweden, and Siberia.

[A] Specific character of the Guinea Gooſe, *Anas Cygnoides:* "Its bill is ſemi-cylindrical; its cere bunched; its eye-lids "ſwelled."

The ARMED GOOSE.

FIFTH SPECIES.

Anas Gambenſis. Linn. and Gmel.
Anſer Gambenſis. Briſſ. Will. and Ray.
Anſer Chilenſis. Klein.
The Spur-winged Gooſe. Lath.

THIS ſpecies is the only one, not only of the Geeſe, but of all the palmiped birds, which has ſpurs on the wings, like the kamichi, the jacanas, and ſome of the plovers and lapwings: a ſingular character, which nature has ſeldom repeated. With reſpect to ſize, this Gooſe may be compared to the Muſcovy duck; its legs are tall and red; its bill is of the ſame colour, and has, on the front, a little caruncle; the tail and the

the great quills of the wing are black; their great coverts are green, the ſmaller white, and croſſed by a narrow black ribband: the mantle is rufous, with reflections of dull purple; the ſpace round the eyes is of the ſame colour, which tinges alſo, though faintly, the head and the neck; the fore ſide of the body is finely fringed with ſmall gray zig-zags, on a yellowiſh white ground.

This Gooſe is ſtiled the *Egyptian* in our *Pl. Enl.* Briſſon has denominated it the *Gambian Gooſe*. It is indeed a native of Africa, and is found particularly about Senegal *.

* The wild geeſe are at Senegal of a colour very different from that of thoſe in Europe; their wings are armed with a hard, ſpiny, and pointed ſubſtance, two inches and a half in length. *Hiſt. Gen. des Voy. tom.* viii. *p.* 305.—*N. B.* This length ſeems to be exaggerated.—Another mentions that this Gooſe is called *bitt* at Senegal.

[A] Specific character of the *Anas Gambenſis*: "Its bill is ſemi-cylindrical; its cere bunchy; its ſhoulders ſpurred."

The BLACK-BACKED GOOSE.

L'Oie Bronzee. *Buff.*

SIXTH SPECIES.

Anser Melanotos. Gmel.

THIS alſo is a large and beautiful ſpecies, which is remarkable by a great fleſhy excreſcence of a comb-ſhape above the bill, and by the reflections of gold and bronze, gliſtening like burniſhed ſteel, with which its mantle ſhines on a black ground: the head, and the upper half of the neck, are ſpeckled with black amidſt the white, by means of little reflected feathers, that ſeem buckled on the back of the neck: all the fore ſide of the body is white, tinged with gray on the flanks. This Gooſe appears to have a thinner body and a ſlenderer neck than the common wild gooſe, though it is at leaſt as large. It was ſent to us from the coaſt of Coromandel: and perhaps the creſted gooſe of Madagaſcar, mentioned by the navigators Rennefort and Flaccourt, under the name of *raſſangue*, is only the ſame bird; which we recognize alſo with all its characters in the *ipecati-apoa* of the Brazilians,

Nº. 235

THE EGYPTIAN GOOSE.

Brazilians, of which Marcgrave has given a figure and description. Thus this aquatic species is one of those which nature has distributed in both continents.

The EGYPTIAN GOOSE.

SEVENTH SPECIES.

Anas Ægyptiaca. Linn. and Gmel.
Anser Ægyptius. Briss.
Anser Hispanicus parvus. Ray.
The Ganser. Albin.

THIS bird is probably what Granger, in his travels to Egypt, calls the *Nile Goose* *. It is not so large as our wild goose; its plumage is richly enamelled, and agreeably variegated; a broad spot of bright rufous is conspicuous on its breast; and all the fore side of the body is decorated, on a light gray ground, with a very delicate hatching of small zig-zags, cinereous, and tinged with rusty; the side of the back is worked in the same way, but with closer zig-zags, which produce a deeper rusty-gray; the throat, the cheeks, and the upper side of the head, are

* The birds of Egypt are the ibis, the Nile Goose, the horseman, the avoset, the heron, &c. *Granger.*

 white;

white; the reſt of the neck, and the ſpace round the eyes, are fine rufous or bay colour, which alſo tinges the quills of the wing next the body; the other quills are black; the great coverts are covered with a reflection of bronze-green on a black ground, and the ſmaller and middle ones are white; a little black ribband intercepts the extremity of the latter.

This Egyptian Gooſe journies or ſtrays in its excurſions, ſometimes to a vaſt diſtance from its native country: that repreſented in our *Planches Enluminées* was killed on a pool near Senlis; and from the appellation given by Ray to this Gooſe, it muſt alſo be ſometimes found in Spain *.

* *Anſer Hiſpanicus parvus.*

[A] Specific character of the *Anas Ægyptiaca*: "Its bill is "ſomewhat cylindrical; its body waved; its top white; a bright "white ſpangle, with a black bar, on its wing."

The ESQUIMAUX GOOSE.

EIGHTH SPECIES.

Anas Cærulescens. Linn. and Gmel.
Anser Sylvestris Freti Hudsonis. Briss.
The Blue-winged Goose. Edw. Penn. and Lath.

BESIDES the species of wild geese which migrate in such numbers during the summer to the north of our continent, it appears that there are also some kinds peculiar to the northern parts of the new world. The present frequents Hudson's Bay and the country of the Esquimaux: it is somewhat smaller than the common wild goose; its bill and legs are red; the rump, and the upper side of the wings, are pale blue; the tail is of the same colour, but duller; the belly is white, clouded with brown; the great quills of the wing, and those next the back, are blackish; the upper side of the back is brown, and also the lower part of the neck, of which the under side is speckled with brown on a white ground; the top of the head is of a burnt rufous.

[A] Specific character of the *Anas Cærulescens*: "It is brown, "below white; the coverts of its wings, and the hind part of its "back, are white."

The LAUGHING GOOSE.

NINTH SPECIES.

Anas Albifrons. Gmel.
Anser Septentrionalis Sylvestris. Briss.
Anas Erythropus. Muller, Kramer, and Browske.
The White-fronted Goose. Penn. and Lath.

EDWARDS gives the name of *Laughing Goose* to this species, which, like the preceding, occurs in the north of America. It is as large as our wild goose; its bill and legs are red; its front is white; all the plumage above the body is brown, more or less intense, and below white sprinkled with a few blackish spots. The one described by Edwards was sent to him from Hudson's Bay; but he says, that he has seen such birds in London during hard winters. Linnæus describes a goose found in Helsingia, which seems to be the same: whence it follows, that if this species be not entirely common to both continents, it passes, at least in certain circumstances, from the one to the other.

[A] Specific character of the *Anas Albifrons*: "It is brown; "below white spotted with black; its front and rump white; its "bill and legs flame-coloured."

The CRAVAT GOOSE

TENTH SPECIES.

Anas Canadenſis. Linn. and Gmel.
Anſer Canadenſis Sylveſtris. Briſſ. and Will.
The Canada Gooſe. Cateſby, Edw. Penn. and Lath.

A WHITE cravat, wrapped about its black neck, diſtinguiſhes ſufficiently this Gooſe, which is alſo one of thoſe peculiar to the northern parts of the new world, where at leaſt it derives its origin. It is ſomething larger than our domeſtic Gooſe, and has its neck and its body rather longer and more ſlender; its head and neck are black or blackiſh, which dark colour ſets off the white cravat that covers the throat. The prevailing caſt of its plumage is dull brown, and ſometimes gray. This Gooſe is known in France by the name of the *Canada Gooſe*: it has even multiplied under domeſtication, and occurs in ſeveral of our provinces. Within theſe few years, many hundreds inhabited the great canal at Verſailles, where they lived familiarly with the ſwans; they were oftener on the graſſy margins than in the water. There is at preſent a great mber of them on the magnificent pools that decorate the charming gardens of Chantilly. They have alſo multiplied in Germany and in England. This beautiful ſpecies may be viewed as forming the intermediate gradation between the ſwan and the Gooſe.

Theſe Cravated Geeſe migrate ſouthwards in America, for they appear during winter in Carolina; and Edwards relates, that in the ſpring they paſs in flocks to Canada, and thence return to Hudſon's Bay, and the other more northern parts of America.

[A] Specific character of the *Anas Canadenſis*: "It is cinereous; its head and neck are black; its cheeks and throat white.' It breeds in Hudſon's Bay, and lays ſix or ſeven eggs.—I muſt beg leave to ſubjoin the following extract from Mr. Pennant, to whoſe ingenious and accurate works I have ſo often been indebted.

"The Engliſh of Hudſon's Bay depend greatly on Geeſe, of "theſe and other kinds, for their ſupport; and, in favourable years, "kill three or four thouſand, which they ſalt and barrel. Their "arrival is impatiently attended; it is the harbinger of the ſpring, "and the month named by the Indians *the gooſe moon*. . . . They "prefer iſlands to the continent, as further from the haunts of men. ". . . The Engliſh ſend out their ſervants, as well as the Indians, "to ſhoot theſe birds on their paſſage. It is in vain to purſue them; "they therefore form a row of huts made of boughs, at muſquet-"ſhot diſtance from each other, and place them in a line acroſs the "vaſt marſhes of the country. Each hovel, or as they are called, "*ſtand*, is occupied by only a ſingle perſon: theſe attend the flight "of the birds, and on their approach mimic their cackle ſo well, "that the Geeſe will anſwer, and wheel and come nearer the ſtand. "The ſportſman keeps motionleſs, and on his knees, with his gun "cocked the whole time, and never fires till he has ſeen the eyes "of the Geeſe. He fires as they are going from him, then picks "up another gun that lies by him, and diſcharges that. The "Geeſe which he has killed he ſets up on ſticks as if alive, to de-"coy others; he alſo makes artificial birds for the ſame purpoſe. "In a good day (for they fly in very uncertain and unequal num-

"bers)

Besides these ten species of Geese, we find mention made by travellers of some others, which belong perhaps to the preceding.

1. The Icelandic Geese, of which Anderson speaks under the name of *Margee:* they are somewhat larger than a duck. In that island they appear in vast flocks.

2. The Goose, called *Helsinguer* by the same author; "which comes to settle on the east of "the island, and is so fatigued on its arrival, that "it may be knocked down with sticks *."

3. The Spitzbergen Goose, called *the red Goose* by the Dutch †.

4. The *Loohe* of the Ostiacs, a small Goose described by De l'Isle, from one killed on the banks of the Oby. "These Geese," says he,

"bers) a single Indian will kill two hundred. Notwithstanding "every species of Goose has a different call, yet the Indians are "admirable in their imitation of every one.

"The vernal flight of the Geese lasts from April to the middle "of May. Their first appearance coincides with the thawing of "the swamps, when they are very lean. The autumnal, or the "season of their return with their young, is from the middle of "August to the middle of October. Those which are taken in "the latter season, when the frosts usually begin, are preserved in "their feathers, and left to be frozen for the fresh provisions of the "winter stock. The feathers constitute an article of commerce, "and are sent to England." *Arctic Zoology*, vol. ii. 545.

* Anderson's Natural History of Iceland and Greenland.

† We saw (at Spitzbergen) a flock of *red Geese*; these geese have long feathers; there are numbers of them in Russia, Norway, and Jutland. *Recueil des Voyages du Nord*; *Rouen*, 1716, *tom.* ii. *p.* 110.

"have

"have their wings and their back of a deep "ſhining blue; their ſtomach is reddiſh, and "on the top of their head is a blue oval ſpot, "and a red ſpot on each ſide of the neck. "From the head to the ſtomach extends a ſil-"very ſtripe as broad as a quill, which pro-"duces a fine effect."

5. In Kamtſchatka are found, according to Kracheninikoff, five or ſix ſpecies of Geeſe, beſides the common wild Gooſe, viz. *the Gumeniſki, the Short-necked Gooſe, the Spotted Gray Gooſe, the White-necked Gooſe, the Little White Gooſe*, and *the Foreign Gooſe*. This traveller has mentioned no more than their names; and Steller ſays only that theſe Geeſe arrive in Kamtſchatka in the month of May, and retire in October.

6. The *Mountain Gooſe* of the Cape of Good Hope, of which Kolben gives a ſhort deſcription, diſtinguiſhing it from the *water* or common Gooſe and the *Crop* Gooſe *.

We ſhall not here ſpeak of the pretended *Black Geeſe of the Moluccas*, whoſe feet are ſaid *to reſemble thoſe of parrots* †: for ſuch incongruities

* Anas Montana. Gmel.—The Cape furniſhes three kinds of wild Geeſe; *the Mountain Geeſe, the Crop Geeſe, and the Water Geeſe*: Not but all of them are very fond of that element; yet they differ much both in colour and in ſize. The Mountain Gooſe is larger than the Gooſe reared in Europe; the feathers of its wings, and thoſe on the crown of its head, are of a very beautiful and ſhining green: this bird retires ofteneſt into the vallies, where it paſtures on herbs and plants. *Kolben.*

† *Hiſt. Gen. des Voy. tom.* viii. *p.* 377.

gruities can be imagined only by people ignorant of natural hiſtory.

To complete the numerous family of the Geeſe, we have only to add the ſpecies of the *Brent*, the *Bernacle*, and the *Eider*.

The BRENT.

LE CRAVANT. *Buff.*

Anas-Bernicla. Linn. and Gmel.
Brenta. Briſſ. and Klein.
Anas Torquata. Aldrov. Johnſt. and Will. *

THE name *Cravant* is, according to Geſner, no other than *Grau-ent*, which, in German, ſignifies *gray-duck*. In fact, the colour of this bird is brown-gray or blackiſh, and pretty uniform over its whole plumage; but its port and figure approach nearer to the gooſe than to the duck. It has the high head and all the other proportions of the gooſe, on a ſmaller model, and with a thinner body. Its bill is rather narrow, and pretty ſhort; its head is ſmall, and its neck long and ſlender: theſe two parts, and alſo the top of the breaſt, are of a blackiſh brown, except a very narrow white band, which forms a half-collar under the throat; a character which leads Belon to find a name relating to this bird in Ariſtophanes. All the quills of the wings and of the tail, and alſo the upper coverts of the latter, are likewiſe of a blackiſh

* In Italian *Ceſon*.

brown;

brown; but the lateral feathers, and all thoſe of the upper ſurface of the tail, are white. The plumage of the body is cinereous gray on the back, on the flanks, and above the wings; but it is dapple gray under the belly, where moſt of the feathers are edged with whitiſh; the iris of the eye is browniſh yellow; the legs and the membranes which connect the toes, are blackiſh, and alſo the bill, in which large noſtrils are perforated and expoſed.

The Brent has long been confounded with the bernacle. Willughby owns, that he once ſuppoſed they were only the male and female *, but afterwards perceived diſtinctly, from many characters, that theſe birds really formed two different ſpecies. Belon ſtiles the Brent (or *Cravant) the collared ſea duck*; and, in another place, he calls the bernacle, the *cravant* †; and the people on the French coaſts make the ſame miſtake ‡. The great reſemblance in the plu-

* Friſch ſays, that the Brent is called *baumgans*, or tree-gooſe, becauſe it builds its neſt in trees, which is altogether improbable: it is more likely that this name was borrowed from the bernacle, which was fabled to owe its birth to rotten wood.

† Aldrovandus is much more miſtaken when he takes the bird deſcribed by Geſner, under the name of *pica marina*, for the Brent or collared gooſe: this ſea-pie of Geſner is the guillemot; and this miſtake of ſo learned a naturaliſt as Aldrovandus, ſhows that deſcriptions in natural hiſtory, if ever ſo little faulty or confuſed, are of ſmall ſervice in giving a clear idea of the object meant to be repreſented.

‡ "The Brent or nun-gooſe is very common on this coaſt (of "Croiſic) where great flocks are ſeen; the people call it *bernacle*, "and I believed it to be the ſame till I ſaw one." *Note communicated by M. de Querhoënt.*

mage

mage and ſhape of the body, which obtains in theſe two birds, has given occaſion to the confuſion: yet the bernacle is jet black, while the Brent is only dark brown; and beſides, the latter frequents the coaſts of temperate countries only, while the bernacle appears only in the moſt northern countries. And that circumſtance alone convinces us that they are really two diſtinct and ſeparate ſpecies.

The cry of the Brent is a dull, hollow ſound, which we have often heard, and which we may denote by *ouan, ouan*; it is a ſort of hoarſe bark, which the bird frequently utters *. It has alſo, when purſued or even approached, a hiſs like that of the gooſe.

The Brent can live in the domeſtic ſtate †. We have kept one ſeveral months. Its food was grain, bran, or ſoaked bread. It conſtantly ſhowed a timid, ſhy diſpoſition, and avoided all familiarity; and though ſhut up in a garden with ſheldrakes, it always lived apart from them: it was even ſo cowardly, that a garganey, which had before lodged with it, made it run. It was obſerved to eat as much, perhaps even more, in the night than in the day. It was fond of bathing, and it ſhook its wings

* *Note communicated by M. de Querhoënt.*

† "A gentleman of this neighbourhood (at Croiſic) has preſerved one in his court-yard two years: the firſt ſpring it was very ſick at the time of laying; it died the ſecond, leaving one egg." *Idem.*

upon

upon coming out of the water. Yet fresh water is not its native element*; for all those which are seen on our coasts, arrive from the sea.—I shall here insert some observations on this bird, which were communicated to us by M. Baillon.

"The Brents were hardly known on the "coasts of Picardy before the winter of 1740. "The north wind then brought a prodigious "number of them; the sea was covered with "them: all the marshes being frozen, they "spread over the land, and committed great "destruction among the tender corn, which was "not sheltered by the snow; they devoured the "shoots to the very roots. The country peo-"ple, whose fields were exposed to this devasta-"tion, declared a general war against these "birds. They approached the Brents very "near the first days, and killed many with "sticks and stones. But they seemed as it "were to rise again; for new flocks continued "to pour in from the sea, and to destroy what "plants the frost had spared." . . . .

"Others appeared in 1765, and the sea-"shore was covered with them. But the north "wind, which had brought them, ceasing to "blow, they did not disperse in the fields, but "departed a few days after.

"Since that time, they have been seen every

* Belon.

"winter

" winter when the north winds blow conſtantly " twelve or fifteen days. Many appeared in " the beginning of 1776; but the ground being " covered with ſnow, moſt of them remained on " the ſea; and the reſt, which had entered the " rivers, or ſpread on the banks, a ſhort diſtance " from the coaſt, were compelled to return by " the ice floated in the ſtreams or heaped up by " the tide. As they were hunted, they grew " ſhy, and they are now ſprung at as great diſ- " tance as other game."

[A] Specific character of the Brent Gooſe, *Anas-Bernicla*: " It " is brown; its head, its neck, and its breaſt, black; a white col- " lar." The name *Brent* or *Brand* ſeems to be derived from the Saxon *Brennen*, to burn, and thence transferred to ſignify marking or ſtamping of any kind; in the preſent caſe it refers to the white ſpot imprinted on each ſide of the neck, which is black.

No. 236

THE BERNACLE GOOSE.

The BARNACLE.

LA BERNACHE. *Buff.*

Anas Erythropus. Linn. and Gmel.
Bernicla. Briff. Will. Ray, Sibb. &c.
Anfer Brenta. Klein.
In Scotland, Clakis. *

OF the marvellous productions which ignorance, ever credulous, has fo long fubftituted for the fimple and truely wonderful operations of nature, the moft abfurd perhaps, and yet the moft celebrated, is the growth of Barnacles and fcoters in certain fhells called the *conchæ anatiferæ* †, or on certain trees on the coafts of Scotland and the Orknies, or even on the rotten timbers of old fhips.

Some authors have written that fruits, whofe ftructure already exhibited the lineaments of a fowl, being dropt into the fea, turned into birds. Munfter ‡, Saxo Grammaticus and Scaliger affert

* It is fometimes ftiled the *Scotch Goofe*: in the Orknies it is called *Rod-Gans*: in Holland *Rot-Gans*: in Germany *Baum-Gans*: (tree goofe): in Norway *Raatne-Gans,* or *Goul*: in Iceland *Helfingen*: in Poland *Kaczka Drzewna.*

† Duck-bearing fhells.

‡ Geog. Univerf. *lib.* ii.

this *; Fulgoſus † even affirms, that the trees which bear theſe fruits reſemble willows, and produce at the end of their branches ſmall ſwelled balls containing the embryo of a duck, which hangs by the bill, and when ripe and formed, falls into the ſea, and takes to its wings. Vincent of Beauvais chuſes rather to attach it to the trunk and bark, whoſe ſap it ſucks, till, grown and completely feathered, it burſts from its impriſonment.

Biſhop Leſlie ‡, Majolus §, Odericus ||, Torquemada ¶, Chavaſſeur **, the biſhop Olaus Magnus ††, and a learned cardinal ‡‡, all atteſt this ſtrange generation. Hence, the bird has been called *tree-gooſe* §§; and one of the Orknies, the ſcene of the prodigy, has received the appellation of Pomona.

This ridiculous notion was judged not ſufficiently marvellous by Camden ||||, Hector

* In his Commentary on the firſt Book of Ariſtotle, *de Plantis*.

† *Lib*. i. 6.

‡ Chron. Scot.

§ *Dier. canicular. tract*.

|| In his voyage to Tartary.

¶ Hexameron.

** *De Gloria Mudi*.

†† Rer. Sept. *lib*. xix. 6 and 7.

‡‡ *Jacobus Aconenſis*.

§§ Pomona is the largeſt of the Orknies; and contains Kirkwall, the capital of thoſe iſlands. The origin of the name has given occaſion to many conjectures. The derivation hinted at in the text is as probable as the reſt; from *pomum* an apple, becauſe of the imaginary *animal-fruits*.—*T*.

|||| *Britannia*.

Boece,

Boece *, and Turnebius †; for, according to them, the old maſts and beams of ſhips, fallen to pieces and rotting in the water, became cruſted with embryos, in form of little muſhrooms, or big worms, which were covered by degrees with down or feathers, and at laſt compleated their metamorphoſis by changing into birds ‡. Peter Daniſi §, Dentatus ||, Wormius ¶, Duchêne **, talk much of this abſurd prodigy; which Rondelet, notwithſtanding his knowledge and good ſenſe, ſeems to credit.

Laſtly, according to Cardan ††, Gyraldus ‡‡, and Maier, who has written a treatiſe expreſsly on this bird, *without father or mother* §§, it originates neither from fruits nor worms, but from ſhells: and what is ſtill more wonderful than the prodigy itſelf, Maier opened an hundred of theſe *gooſe-bearing* ſhells, and found in all of them the rudiments of the bird completely

* *Hiſt. Scotiæ.*

† In Geſner.

‡ A grave doctor, in Aldrovandus, avers with an oath, that he had ſeen and kept the little Barnacles ſtill ſhapeleſs and as they dropt from the rotten timber.

§ Deſcription of Europe, article *Ireland.*

|| Apud Alex. ab Alex. *Genial dier.*

¶ Citing *the Epitome of the Scottiſh chronicles.*

** In his *Hiſtory of England.*

†† De variet. Rer. *lib.* vii. 3.

‡‡ See *Traite de l'Origine des Macreuſes.*

§§ *Tractatus de volucri arboreâ, abſque patre & matre, in inſulis Orcadum, formâ anſerculorum proveniente.* Aut. Mich. Maiero, Archiatro, Comite Imperiali, &c. *Francfurti*, 1629, in 12mo.

formed *.—Such wild whimsies and chimeras have been retailed concerning the origin of the Barnacles †: but as these fables once enjoyed great celebrity, and were admitted by many authors ‡, we have thought proper to relate them, in order to show how contagious are the errors of science, and how prone are men to the fascinations of the marvellous ||.

But

* Count Maier has stuffed his treatise with so many absurdities and puerilities, that they are alone sufficient to destroy his evidence. He proves the possibility of the miraculous generation of the Barnacles by the existence of hobgoblins, and that of sorcerers; he derives it from the immediate influence of the stars; and, if his simplicity were not excessive, we might accuse him of irreverence in the chapter which he entitles, *Quod finis proprius hujus volucris generationis sit, ut referat duplici suâ naturâ, vegetabili & animali, Christum Deum & hominem, qui quoque sine patre & matre, ut illa, existit.*

† In the northern languages *baum-gans*, and in Latin *anser arboreus.*

‡ Besides those which we have already cited, see *Traité de l'Origine des Macreuses*, by M. Graindorge, doctor of the faculty of Medicine, at Montpellier, and published by M. Th. Malouin, &c. *at Caen*, 1680, *in small duodecimo.*—*Deusingii fasciculus dissert. selectarum, inter quas una de anseribus Scoticis*; *Groningæ*, 1664, *in* 12mo.—*Ejusdem dissert. de Mandragoræ pomis, ubi, pag.* 38; *de anseribus Scoticis*; *Groningæ*, 1659, *in* 12mo.—*Jo. Ernesthus Hering dissert. de ortu avis Britannicæ*; *Wittembergæ*, 1665, *in* 4to.—Tancred Robinson's Observations on the Macreuse, and the Scotch Bernacle, *Philos. Trans.* vol. xv. N° 172.—Relation concerning Bernacles, by Sir Robert Moray, *Phil. Trans.* N° 137, &c.

|| I shall transcribe, for the entertainment of my reader, an account of this wonderful transformation, from our old botanist Gerard:

" But what our eyes have seene, and hands have touched, we shall " declare. There is a small island in Lancashire called the *Pile of* " *Foulders*, wherein are found broken pieces of old and bruised

" ships,

But even of our ancient naturaliſts many rejected theſe fables: Belon, always ſober and judicious, laughs at them; nor have Cluſius, Deuſingius, Albertus Magnus believed report. Bartholin diſcovered that theſe gooſe-bearing conchs contained only a ſhell-fiſh of a particular kind: and from the deſcriptions given of them by

" ſhips, ſome whereof have been caſt thither by ſhipwracke, and " alſo the trunks and bodies with the branches of old and rotten " trees, caſt up there likewiſe; whereon is found a certain ſpume " or froth that in time breedeth unto certain ſhels, in ſhape like " thoſe of the Muſkle, but ſharper pointed, and of a whitiſh colour, " wherein is contained a thing in form like a lace of ſilke finely " woven as it were together, of a whitiſh colour; one end whereof " is faſtened unto the inſide of the ſhell, even as the fiſh of Oiſters " and Muſkles are; the other end is made faſt unto the belly of a " rude maſſe or lumpe, which in time commeth to the ſhape and " form of a bird: when it is perfectly formed, the ſhell gapeth open, " and the firſt thing that appeareth is the foreſaid lace or ſtring; " next come the legs of the bird hanging out, and as it groweth " greater it openeth the ſhell by degrees, till at length it has all " come forth, and hangeth only by the bill: in ſhort ſpace after it " cometh to full maturitie, and falleth into the ſea, where it gathereth feathers, and groweth to a fowle bigger than a Mallard and " leſſer than a Gooſe, having black legs, and bill or beake, and feathers black and white, ſpotted in ſuch manner as our Mag-Pie, " called in ſome places *Pie-Annet*, which the people of Lancaſhire " call by no other name than tree-gooſe; which place aforeſaid, " and all thoſe places adjoyning, do ſo much abound therewith, that " one of the beſt is bought for three-pence. For the truth hereof, " if any doubt, may it pleaſe them to repaire to me, and I ſhall ſatisfy them by the teſtimonie of good witneſſes."

Few miracles are related more circumſtantially, or reſt on better evidence. So natural to man is credulity! which paſſes all bounds, when the prodigy of an event takes firm hold of the imagination, and lays the underſtanding aſleep —*T*.

Wormius *, Lobel †, and others, as well as from the figures publiſhed by Aldrovandus and Geſner, it is eaſy to perceive that they are the *pouſſe-pieds* ‡ of the coaſts of Brittany, which are affixed to a common pedicle, and ſend off a bundle of feathery filaments, that to a prejudiced imagination might appear the cluſtered lineaments of birds hanging from the branches. We need not remark the abſurdity of ſuch a notion: Æneas Sylvius relates, that chancing to be in Scotland, he inquired particularly for the place of the wonderful metamorphoſis of the Barnacle, but was referred to the remote Hebrides and Orknies; and he adds pleaſantly, that, as he ſought to advance, the miracle retired from him.

* The gooſe-bearing ſhell is triangular, ſmall, externally white-blue, gliſtening, light, compreſſed, an inch in length and in breadth; when ripe, it conſiſts of four valves, ſometimes more, of which the two anterior are thrice as large as the two poſterior, which adhere to them as appendices, very thin round a thicker part, by which they cling concealed to the ſea-weed; when opened, they ſhow rudiments of a little bird, and the feathers pretty diſtinct. *Wormius in Muſæo, lib.* iii. 7.

† We had ſhells with a rough thickiſh pedicle broken off from the bottom of an old ſhip; they are ſmall, whitiſh without, gliſtening, light, have the thinneſs of egg-ſhells, fragile, and bivalve. They are of the ſize of a compreſſed walnut, hang like fungous excreſcences from the bottoms of ſhips, where they ſeem to extract life for a young bird, whoſe rudiment is ſeen from the extreme part of the opened ſhell. *Lobel.*

‡ So called on account of the fibres which branch from it. It is the ſame with the *Barnacle* (the name alſo of the bird) a ſpecies of multivalve, the *Lepas Anatifera* of Linnæus.—*T.*

As the Barnacles breed in the diſtant parts of the north, no perſon for a long time had obſerved their birth, or even ſeen their neſts; and the Dutch, in a voyage which extended to the eightieth degree of latitude, were the firſt who diſcovered theſe *. Yet the Barnacles muſt neſtle in Norway, if it be true, as Pontoppidan relates, that they are ſeen the whole ſummer †. They appear in autumn and winter on the coaſts of Yorkſhire ‡ and Lancaſhire in England ||, where they are eaſily caught with nets, and ſhew none of the ſhyneſs and cunning natural to birds of their kind §. They occur alſo in

* " On the weſt ſide of Greenland, was a great winding and a " flat ſhore reſembling an iſland; we there found many eggs of " *Barnicles* (which the Dutch call *rotganſen)*; we found alſo ſome " of them hatching, which, on being driven away, cried *rot, rot,* " *rot,* (hence their name); one we killed with a ſtone, we cooked " it, and ate it with ſixty eggs which we had carried to the ſhip.

" Theſe geeſe or Barnicles were real geeſe, called *rotganſen,* " which come every year in great numbers about Wierengen in " Holland, and it was hitherto unknown where they laid their eggs " and reared their young; and hence it has happened that no au- " thors have ſcrupled to write that they are bred on the trees in " Scotland . . . Nor need we wonder that hitherto the retreats where " theſe birds hatch, were unknown, ſince no perſon has ever reached " the eightieth degree of latitude, much leſs ſeen the birds ſitting " on their eggs." *Trois navigations faites par les Hollandois au Septentrion, par Gerard de Vora; Paris,* 1599, *pp.* 112 and 113.

† Journal Etrangere, *Fevrier,* 1777.

‡ Liſter's Letter to Ray, *Philoſ. Tranſ.* N° 175.

|| Willughby.

§ Johnſon. He ſays this of the little Barnacle, which we ſhall find to be only a variety.

Ireland, particularly in Lough-foyl, near Londonderry, where they are obſerved diving inceſſantly to crop the roots of the large reeds, whoſe ſweet pith nouriſhes them, and, it is ſaid, makes their fleſh well-taſted *. Seldom they viſit France; yet one has been killed in Burgundy. whither it had been driven by the ſtormy winds of a boiſterous winter †.

The Barnacle is certainly of the family of the geeſe; and Aldrovandus juſtly blames Geſner for ranging it with the ducks. In fact, it is rather ſmaller and lighter, it has a ſlenderer neck, a ſhorter bill, and legs proportionally taller than the Gooſe. But it has its figure, its port, and all its ſhapes; its plumage is agreeably broken with large white and black ſpaces; and hence Belon has ſtiled it the *nun (nonnette*, ou *religieuſe)*. Its face is white, and two ſmall black ſtreaks join the eyes with the noſtrils; a black *domino* covers the neck, and falls with a round edge on the top of the back, and of the breaſt; all the mantle is richly waved with gray and black, with a white fringe, and all the under ſide of the body is of a fine clouded white.

Some authors ſpeak of a ſecond ſpecies of

* Nat. Hiſt. of Ireland, p. 192. [They appear in great numbers on the north of Ireland in Auguſt, and retire in March. They are caught in their paſſages, by nets ſtretched acroſs the rivers.—*T.*]

† It was brought from Dijon to M. Hebert, who communicated this fact.

Barnacle,

Barnacle *, which they reprefent as exactly like the other, only fomewhat fmaller. But this difference of fize is too inconfiderable to conftitute two fpecies; and we are of the fame opinion with Klein on this fubject, who, after a comparifon of thefe two Barnacles, concluded that they were only varieties.

* *Anas Erythropus*, var. Linn. and Gmel.
Bernicla Minor. Briff.
Brenthus. Gefner, Johnft. and Will.
Anfer Brenta. Klein.
The Pat, or Road Goofe. Will.

Thus defcribed by Briffon: "Above it is dull cinereous, the "margins of its feathers whitifh; below white; its top, and the "upper part of its neck, blackifh; the fore part of its head and "its throat fulvous; the lower part of its neck and its breaft "brown; its rump bright white; its middle tail quills black; the "outermoft white on both fides."

[A] Specific character of the Barnacle, *Anas Erythropus*: "It "is cinereous, its front white." Its length twenty-five inches, its alar extent fifty-three, and its weight five pounds. It is frequent likewife on the coafts of Hudfon's Bay.

The EIDER*.

Anas Mollissima. Linn. Gmel, Muller, and Klein.
Anser Lanuginosus. Briss.
Anas Sancti Cutberti, seu Farnensis. Will. and Ray.
Eider Anas. Sibbald.
The Colk. Martin's West. Isl.
The Eider, or Soft-feathered Duck:—The Cuthbert Duck. Will.
The Great Black and White Duck. Edw.
The Eider, or Cuthbert Duck. Penn. and Lath.

IT is this bird that furnishes the soft, warm, light down which bears its name. The Eider is a species of goose, which inhabits the northern seas, and descends no lower than the coasts of Scotland.

It is nearly as large as a goose: the principal colours of the male are white and black; and, different

* Sometimes the *Eider* is reckoned a species of goose. Thus, in German *Eider Gans*, and in Danish *Edder Gaasen:* sometimes a duck; as in German *Eider Ente*, and in Danish *Edder Anden:* at other times it has general names;—in German *Eidor Vogel:* in Swedish *Ad, Ada, Aed, Aeda, Eider, Gudunge:* in Danish *Edder Fuglen, Aer Fugl, Aer Bolte:* in Icelandic *Aedar Fugl, Adar, Aedder, Edder Fugl:* in Norwegian *Edder, Edder Fugl.* On the isle of Feroe it is called *Eider, Eder Fugl*, and *Eiderblicke* or *Aerblick* when its plumage has become white: at Bornholm *Aer Boer:* in Greenland *Mittek* or *Merkit*, according to Anderson, and the female *Arnaviak:* in Lapland *Likka.*

In French it is sometimes stiled the *down goose*, or the *down duck (oie à duvet*, or *canard à duvet).* It is said, in the text, that the name *Eider down*, which the French seem to have adopted from us, was corrupted into *aigle don*, and the bird which yields it supposed to be a kind of eagle.

No. 237

THE EIDER GOOSE.

different from the ufual difpofition, the former covers the back, and the latter the belly; and the fame obfcure black appears on the top of the head, and on the quills of the tail and of the wings, except the feathers next the body, which are white. Below the nape of the neck there is a broad greenifh plate; and the white of the breaft is wafhed with a brick or wine tint. The female is not fo large as the male, and all its plumage is uniformly tinged with rufty and blackifh, in tranfverfe and waving lines, on a brown gray ground. In both fexes, we perceive fcallops traced by little clofe feathers like velvet, which extend from the front on both fides of the bill, and almoft under the noftrils.

The Eider down is highly efteemed; and even on the fpot, in Norway and Iceland, it fells very dear *. This fubftance is fo elaftic and fo light, that two or three pounds of it, though preffed into a ball that may be held in the hand, will fo fwell as to fill and diftend the foot-covering of a large bed.

The beft down, which is called *live down (duvet vif)* is what the Eider pulls to line her neft, and which is gathered in the neft itfelf: for, befides the reluctance to kill fo ufeful a bird †, the

* Pontoppidan.

† Pontoppidan fays even, that in Norway it is prohibited to kill it for the down: "With the more reafon," he adds, "fince the feathers of the dead bird are fat, fubject to rot, and far from being "fo light as what the female plucks, to form a bed for its young."

the down taken from the dead body is inferior; either becauſe the down is in full perfection at the breeding ſeaſon, or becauſe the bird plucks only the fineſt and moſt delicate, that which covers the ſtomach and belly.

Care muſt be taken not to ſeek and gather the down in the neſts, till after ſome days of dry weather; nor muſt the birds be driven haſtily from their neſts, for in the fright they drop their excrements, with which the down is often fouled *. To clear it of the dung, the feathers are ſpread upon a ſieve of ſtretched cords, which are beat with a ſtick; ſo that the heavy clots fall through, and the light down jumps off.

The eggs are five or ſix in number †, of a deep green, and very good to eat ‡. If they be ſtolen, the female ſtrips her plumage again to make a ſecond hatch, which is ſmaller than the firſt. If the neſt be again plundered, as the female can furniſh no more down, the male lends his aid, and plucks the feathers from his

* Natural Hiſtory of the Eider, by Martin Thrane Brunnich, *art.* 41.

† "It is not uncommon," ſays Van Troil, "to find more, even "ten and upwards, in the ſame neſt occupied by two females, which "live together in perfect concord." *Letters on Iceland.*

‡ Anderſon pretends, that to have a number of theſe, a ſtick of a foot in height is planted in the neſt, and that the bird continues to lay till the heap of eggs riſes to the point of this ſtick, in order that ſhe may ſit to cover them. But were it as true as it is improbable, that the Icelanders employed this barbarous artifice, they would ill underſtand their true intereſt, to deſtroy a bird ſo precious to them, ſince, worn out with exceſſive laying, it generally expires.

breaſt;

breaſt; which is the reaſon that the lining of the third neſt is whiter than that of the firſt. But before we ſeize the ſpoils, we muſt now wait till the mother has hatched her eggs, which at moſt are only two or three, perhaps but one: for if her hopes of progeny are daſhed a third time, ſhe will for ever abandon the place; but if ſhe be permitted to rear her family, ſhe will return the following year, and bring with her the young Eiders.

In Norway and Iceland, the diſtricts to which the Eiders habitually reſort to build their neſt, are a ſpecies of property which is carefully preſerved, and tranſmitted by inheritance. There are ſpots that contain many hundred of theſe neſts; and we may judge, from the high price of down, what profit the proprietor muſt draw *. The Icelanders are at the utmoſt pains to invite the Eiders, each into his own eſtate; and when they perceive that theſe birds begin to haunt ſome of the iſlets which maintain herds, they ſoon remove the cattle and dogs to the main land, and procure the Eiders an undiſturbed retreat †. Theſe people have even formed, by art and perſevering labour, many ſmall iſlands by disjoining from the continent ſeveral promontories that ſtretch into the ſea. It is in theſe retreats of ſolitude and tranquillity

* To take an Eider's neſt on another's lands, is reputed theft in Iceland. *Van Troil.*

† Brunnich, § 48.

that

that the Eiders love to ſettle; though they are not averſe to neſtle near habitations, if nothing moleſt them, and if the dogs and cattle be removed. "A perſon," ſays Horrebow, "as "I myſelf have witneſſed, may walk among "theſe birds while they are ſitting, and not "ſcare them; he may even take eggs, and yet "they will renew their laying as often as three "times."

All the down that can be collected is ſold annually to Daniſh or Dutch merchants *, who come to buy it at Drontheim, and other parts of Norway and Iceland. Little or none of it is left in the country †. In that rough climate, the robuſt hunter, covered with bear's ſkin, enjoys, in his ſolitary hut, a peaceful, perhaps a profound ſleep; while, in poliſhed nations, the man of ambition, ſtretched on a bed of Eider down, and under a gilded roof, idly ſeeks to procure the ſweets of repoſe.

We ſhall here add ſome facts relating to the Eider, extracted from a ſmall work of M. Brunnich, written in Daniſh, and tranſlated into German, from which we directed a French verſion to be made.

* "A female in her neſt gives commonly half a pound of down, "which is reduced to one half in cleaning . . . The cleaned down "is valued by the Icelanders at forty-five *fiſh* (of which forty-eight "make a rix-dollar) the pound; the raw down at ſixteen fiſh . . . "The Icelandic company ſold, in 1750, down amounting in value "to 3757 rix-dollars (about 850 l. ſterling), beſides what was ſent "directly to Gluckſtad." *Van Troil.*

† Hiſt. des Voy. *tom.* xviii. *p.* 21.

In the breeding ſeaſon, ſome male Eiders are ſeen flying ſingle: The Norwegians call them *gield-fugl, gield-aee* *; they are ſuch as have not obtained mates, and have been worſted in the ſtruggles for the poſſeſſion of the females, which are fewer in this ſpecies than the males. Yet they ſooner arrive at maturity, ſo that the old males and the young females pair together, and hence their firſt hatch is ſmaller than the ſubſequent.

At the time of pairing, the male continually ſcreams *ha, ho,* with a raucous and moaning voice; that of the female reſembles the cry of a common duck. The firſt object of theſe birds is to place their neſt under ſhelter of ſome ſtones or buſhes, and particularly of junipers. Both male and female labour in concert; and the latter pulls the down from her breaſt, and heaps it, ſo as to form quite round the neſt a thick puffed roll, which ſhe preſſes on the eggs, when ſhe goes in queſt of food: for the male aſſiſts not in covering, but keeps watch in the vicinity, and gives notice if an enemy appears; the female then conceals her head, and if the danger is urgent, ſhe flies to her mate, who treats her harſhly, it is ſaid, if any accident happen to the brood. The ravens ſuck the eggs, and kill the young; the mother therefore haſtens to remove them from the neſt, and a few hours after they are hatched,

* i. e. *Free bird; free Eider.—T.*

ſhe

ſhe takes them on her back, and, with an eaſy flight, tranſports them to the ſea.

The male now leaves her, and neither of them returns more to land. Several hatches unite at ſea, and form flocks of twenty or thirty with their mothers, which lead them, and continually daſh the water, to bring up, with the mud and ſediment, inſects and ſmall ſhell-fiſh for ſuch of the young as are too weak to dive themſelves. This happens from the month of July, or even June; and the Greenlanders reckon the time of ſummer by the age of the young Eiders.

It is not until the third year that the male acquires regular and diſtinct colours: thoſe of the female are much ſooner unfolded; and in every reſpect the growth of the male is more rapid than that of the female. Both of them are at firſt covered or clothed with a blackiſh down.

The Eider dives very deep after fiſh, and feeds alſo on muſcles and other ſhell-fiſh, and ſeems very keen upon the garbage which the fiſhermen throw out of their barks. Theſe birds remain on the ſea the whole winter, even near Greenland, ſeeking the parts of the coaſt moſt clear of ice, and returning to land only in the evening, or previous to a ſtorm, which their flight to the coaſt during the day, it is ſaid, infallibly forebodes.

Though the Eiders journey, and not only ſhift from one place to another, but venture ſo far on ſea, that they have been ſuppoſed to paſs from

from Greenland to America *; yet they cannot properly be ſaid to be birds of paſſage, ſince they never leave the frozen climates, which their cloſe down ſo well fits them to bear. They can procure ſubſiſtence wherever the ſea is open: they advance from the coaſt of Greenland to the iſland of Diſco, but no farther; becauſe, beyond it, the ſea is covered with ice †; it appears even that they reſort thither leſs than formerly ‡. Yet they are found at Spitzbergen; for the Eider is the ſame with the *mountain duck* of Martens, though he himſelf miſtook it ||. From

* Brunnich.

† Anderſon.

‡ The Greenlanders ſay, that formerly they filled in a very little time a boat with Eider's eggs, in the iſlands round Ball River, and that they could not walk without treading on the eggs; but this plenty begins to diminiſh, though ſtill aſtoniſhing. *Idem.*

|| The mountain duck is a kind of a wild duck, or rather of a wild gooſe, as large as a middling gooſe; its plumage is mottled with different colours, and very beautiful; that of the male is marked with black and white, and the female has its feathers of the ſame colour with that of a partridge . . . They make their neſts in low places with their own feathers, which they pluck from under their belly, and which they mix with moſs; but theſe are not the ſame with what is called the *Eider down* (in this Martens is miſtaken, ſince every circumſtance of his deſcription characterizes the Eider.) We found in their neſts ſometimes two, ſometimes three, and even four eggs, of a pale green, and ſomewhat larger than thoſe of our ducks. Our ſailors, boring both ends, took out the white and yolk, and threaded them. The veſſels which had arrived before us at Spitzbergen, had taken numbers of theſe birds. The firſt days they were not at all ſhy, but in time they grew ſo cautious, that one could hardly approach ſo near them as to take a proper aim. It was in the ſouth haven, and on the 18th of June, that we firſt killed one. *Recueil des Voyages du Nord, tom.* ii. *p.* 98.

 the

the note of Steller, cited below, we may alſo gather that the Eider frequents Bering's Iſland, and the point of the Kuriles *. In our ſeas, the moſt ſouthern parts which theſe birds viſit, are the iſlands Kerago and Kona, near the coaſts of Scotland; Bornholm, Chriſtianſoë, and the province of Gothland in Sweden †.

* Steller ſaw, in the month of July, in Bering's Iſland, an eighth ſpecies of gooſe, about the ſize of the white ſpotted one: the wings were black; the ears of a greeniſh white; eyes black, edged with yellow; the bill red, with a black ray quite round it, an excreſcence as in the Muſcovy or the Chineſe gooſe; this excreſcence is bare and yellowiſh, except that it is ſtriped from one end to the other with ſmall feathers of a bluiſh black. The natives of the country report, that this gooſe is found in the firſt iſland Kurilſki, but is never ſeen on the continent. *Kraſcheninicoff.*

† Brunnich.

[A] Specific character of the Eider, *Anas Molliſſima*: "Its bill "is cylindrical; its cere cleſt behind, and wrinkled." The male is twice as large as a common tame duck; the female weighs three pounds and an half. The Eiders occur in the northern parts of both continents: in Greenland they build their neſts among the graſs, and in Sweden among the juniper buſhes. They dive to great depths for their food, which conſiſts of various ſorts of ſhell-fiſh: the Greenlanders purſue them, and dart them as they riſe fatigued. Their fleſh is good, and their ſkin is eſteemed an excellent inner garment. The moſt ſouthern retreats of theſe birds are the weſtern iſles of Scotland, Inch-colm in the Firth of Forth, and the Farn iſles on the Northumbrian coaſts. On the latter Mr. Pennant landed, 15th July 1769; and we ſhall borrow the following extract from his narrative.—"We found the female Eider ducks at that time ſitting: "the lower part of their neſts was made of ſea plants; the upper "part was formed of the down which they pull off their own breaſts, "in which the eggs were ſurrounded, and warmly bedded: in ſome "were three, in others five eggs, of a large ſize, and pale olive co- "lour, as ſmooth and gloſſy as if varniſhed over. The neſts were "built on the beach, among the looſe pebbles not far from the wa- "ter. The ducks ſit very cloſe, nor will they riſe till you almoſt "tread

" tread on them. The drakes ſeparate themſelves during the breed-
" ing ſeaſon. We robbed a few of their neſts of the down, and after
" carefully ſeparating it from the tang, found that the down weigh-
" ed only three quarters of an ounce, but was ſo elaſtic as to fill the
" crown of the largeſt hat. The people of this country call theſe St.
" Cuthbert's ducks, from the ſaint of the iſlands."

A Tour in Scotland, 8vo. *pp.* 35 and 36.

It appears from this extract, that the quantity of down which lines the Eider's neſts, is much ſmaller on the Farn iſles than in Iceland; a proof that theſe birds accommodate themſelves according to ſituation and climate.

The DUCK*.

LE CANARD. *Buff.*

1. *Anas Boschas.* Linn. and Gmel.
Anas Fera. Aldrov. Charleton, and Briss.
Anas Sylvestris. Klein.
Boschas Major. Will. Johnst. and Sibb.
The Common Wild Duck and Mallard. Will. Ray, Penn. and Lath.
2. *Anas Domestica.* Linn. Gmel. Aldrov. Johnst. Briss. &c.
Anas Cicur. Gesner.
The Common Tame Duck. Will. Ray, Penn. Lath. &c.

MAN made a double conquest when he subdued inhabitants at once of the air and of the water. Free in both these vast elements, equally fitted to roam in the regions of the atmosphere,

* The Greek name of the Duck, Νησσα, is derived from νεω, to swim; and the Latin *Anas* has the same origin. In Italian it is called *Anitra, Anatre, Anadra*; the wild kind *Anitra Salvatica, Cesone*: in Spanish *Anande*: in Portuguese *Aden*: in Catalonian *Anech*: in Genoese *Ania*: in Parmese *Sassa*: in German *Ente*, formerly *Ante*; the male *Racha, Racktscha*, words imitative of his hoarse voice, and corrupted into *Entrach* or *Entrich*; the wild sort *Wilde Ente, Mertz Ente, Gros Ente, Hag Ente*: in Silesian *Hatsche*, and the wild *Raetsch Endte*: in Flemish *Aente* or *Aende*: in Dutch the Drake is called *Woordt* or *Waerdt*, and the Duck *Eendt*: in Swedish the wild Duck is named *Graes End*, or *Blaonacke*; the tame *Ancka*: in Russian *Outha*: in Greenlandic *Kachletong*: in Polish *Raczka*; the wild kind *Kaczka Dzika*: in Illyrian *Kaczier*. The modern Greeks call the Ducks *Pappi*, or, according to some, *Papitra* or *Chena*: the people of India *Bebe*, according to Aldrovandus: the inhabitants of the isle of Luçon *Balivis*: the natives of Barbary *Brack*: those of the Society Islands *Mora*: the Mexicans *Metzcanauhtli*.

N°. 239

THE FEMALE DUCK.

No. 238

THE DRAKE.

mosphere, to glide through the ocean or plunge under its billows, the aquatic birds seemed destined by nature to live for ever remote from our society, and from the limits of our dominion.

Their only tie to the land is the necessity of depositing the fruit of their loves. By availing ourselves of that necessity, and of the feeling which so powerfully animates all creatures, we have enslaved them without imposing constraint; and by their fondness to their offspring we have attached them to our abodes.

Eggs, taken from the reeds and rushes amidst water, and set under an adopted mother, first produced, in our farm-yards, wild, shy, fugitive birds, perpetually roving and unsettled, and impatient to regain the abodes of liberty. But after they had tasted the pleasures of love in the domestic asylum, the same fowls, and more especially their descendants, grew gentler and more tractable, and, under our care and protection, bred the tame sorts: for it is a general remark, that, till animals propagate in the domestic state, some individuals may be enslaved, but the species will preserve their independence. If, in spite of their irksome bondage, the passion, which unites the sexes, kindles and dilates, it will sweeten their condition, and impart all the charms of freedom: they forget, they relinquish the prerogatives of the savage state: and the scene of their first pleasures, of their early loves, that scene, so dear to every feeling creature, becomes their favourite abode. The education of the family farther augments this attachment, and, at the same time, communicates it to the young, which, being citizens by birth of the residence adopted by their parents, never seek to change it. They know not other situations, and they contract a warm predilection for the place of their nativity; a passion felt even by slaves.

Yet have we subjugated only a small portion of the whole species, particularly in those birds which nature, bestowing a double privilege, has destined to rove in the air and on the sea. Some, indeed, have become our captives, but the bulk of them have eluded our attempts, and will for ever preserve their independence.

The species of the Duck and that of the goose, are thus divided into two great tribes; of which the one, long since tamed, propagates in our court-yards, forming one of the most useful and most numerous families of our poultry; and the other, no doubt still more extensive, constantly avoids us, and lives on the water, only visiting us in winter, and retiring in the spring, to breed in the distant, sequestered regions of the north.

It is about the 15th of October that the Ducks begin to appear in France *. At first, their flocks are small and unfrequent; but these are

* At least in our northern provinces; their appearance is later in the southern countries: at Malta, for example, as we are assured by the Commander Desmazy, they are not seen till November.

are ſucceeded in November by more numerous bodies. Theſe birds are diſtinguiſhed by the oblique lines and regular triangles which they form in the air. After they have all arrived from the northern countries, they are ſeen continually flying from one pool or river to another. Now is the time when the fowlers make great captures, by watching in the day, by lying in ambuſh at night, or by employing different ſnares or large nets. But all theſe methods of ſurprizing or decoying, muſt be dextrouſly managed, ſince Ducks are exceedingly miſtruſtful. They never alight till after making ſeveral wheels round the ſpot; as if their intention were to ſurvey it, and diſcover whether an enemy lurked in it. And when they ſettle, they take every precaution: they bend their flight, and dart obliquely on the ſurface of the water, which they raze and ſkim; then they ſwim at large, keeping always at a diſtance from the banks. At the ſame time, ſome of them watch for the public ſafety, and give alarm when they apprehend danger; inſomuch that the ſportſman is often deceived, and ſees them riſe before he can fire. Yet if he judges himſelf ſufficiently near, he need not be precipitate; for as the wild Duck ſprings vertically *, it does not get ſo ſoon out of reach as a bird that ſhoots directly onwards,

* Belon.

and it allows as much time for taking aim when flushed at the distance of sixty paces, as a partridge at that of thirty.

It is in the evening, *about night-fall*, by the edge of water into which female domestic Ducks are turned to attract them *, that the fowler lies in his hut, or covered and concealed any other way †, and fires on them with advantage. He knows the arrival of these birds by the rustling of their wings ‡, and he makes

* This manner of decoying the birds is ancient, since Alciatus cites the experiment in one of his epigrams:

Altilis allectator anas . . .
Congeneres cernens volitare per aëra turmas,
Garrit, in illarum se recipitque gregem.
Incautas donec prætensa in retia ducat.

† In time of snow I went a-ducking entirely covered with a large white sheet, having a white paper mask on my face, and a white ribband lapped about the barrel of my gun: they suffered me to approach without suspicion, and the white ribband enabled me to see half an hour longer; I shot even by the glimmering of the moon, and lost very few birds on the snow. *Note communicated by M. Hebert.*

‡ I shall here describe a method of fowling, of which I was both a spectator and an actor: it was in a plain between Laon and Rheims that a man, and we may easily judge that he was not the richest in the country, had taken his station in the middle of a meadow; there, wrapped in an old mantle, with no other shelter than a hurdle of hazel branches, which screened him from the wind, he waited patiently till some flock of wild Ducks should pass within his reach; he was sitting on a cage of ozier, divided into three compartments, and filled with tame drakes; his post was in the neighbourhood of a river, which winded in this meadow, and at a place where its banks rose seven or eight feet; to one of the banks of this river he had built a hut of reeds, like a sentry-box, perforated with loop-holes, which he could open or shut at pleasure, to spy his prey and

makes haſte to kill the firſt comers; for in this late ſeaſon the night creeps faſt on, and as the Ducks alight only in the duſk, the time is precious. But a greater capture may be made by ſpreading a net over the ſurface of the water, and leading the drag into the hut; in this way the whole flock of wild ducks decoyed by the domeſtic calls will be taken. This ſport requires a ſtock of patience; and the fowler, motionleſs and half-frozen, is more likely to catch cold than game. But the pleaſure uſually predominates, hope urges him to renew his application;—and the ſame night, that blowing his fingers, he ſwears never to return to his frozen

and take his aim: if he ſaw a flock of wild Ducks in the air (and they often paſſed, becauſe at this ſeaſon of ſport they were fired at on all ſides in the marſhes) he let looſe two or three of his tame drakes, which took flight and alighted within thirty paces of the ſentry-box, where he had ſcattered ſome grains of oats, which theſe drakes gathered greedily, for they were kept hungry; there were alſo ſome female Ducks faſtened to ſome poles ſtuck into the banks, and laid cloſe on the ſurface of the water, ſo that theſe Ducks could not come to the brink, but were obliged to call upon the tame drakes. The wild ones, after ſeveral turns in the air, ſtooped downwards and followed the tame drakes, or, if they lingered too long, the perſon diſpatched a ſecond flight of drakes, and even a third, and then ran from his obſervatory to his hut without being perceived; all the banks were ſtrewed with branches of trees and with reeds; he opened that loop-hole which anſwered beſt, obſerved the favourable moment when he could fire without killing his calls, and as he pointed on the ſurface of the water, almoſt horizontally, and ſaw the Ducks heads, he killed ſometimes five or ſix at a ſhot. *Extract of a Memoir of M. Hebert.*

poſt, he lays projects for the ſucceeding evening *.

In

* We owe to M. Baillon the idea and the detail of this ſort of ſport; for which we thank him, and which we ſhall give in his own words:

" A conſiderable number of wild Ducks is taken every winter " in our marſhes near the ſea; the contrivance employed to decoy " them into the nets is very ingenious; it manifeſtly proves the diſ- " poſition of theſe birds to ſociety. It is this:—

" They chooſe in the marſhes a flat covered with about two " feet of water, which they confine with a ſlight bank; the larg- " eſt and remoteſt hedges and trees are the beſt: on the edge they " form an earthen hut well lined with clay at the bottom, and co- " vered with ſods laid on plaſhed branches; there the fowler ſits, " and his head overtops the hut.

" They ſtretch in the water, nets like thoſe for larks, furniſhed " with two ſtrong iron bars, which hold them down on the mud; " the extending cords are fixed in the hut.

" The fowler faſtens ſeveral Ducks before the nets; and thoſe " of the wild breed, and procured from eggs gathered in the ſpring, " are the beſt: the drakes, with which they are paired in October, " are ſhut up in a corner of the lodge.

" The attentive fowler ſurveys the horizon on all ſides, eſpecially " towards the north; as ſoon as he perceives a flock of wild Ducks, " he takes one of the drakes, and throws it into the air: this bird " flies inſtantly to the reſt and joins them; the females, over which " it paſſes, ſcream and call; if it delays too long to return, a ſecond " is diſpatched, and often a third; the redoubled cries of the fe- " males bring them back, the wild ones follow, and alight with " them; the form of the hut ſometimes diſquiets them, but they in- " ſtantly gain confidence when they ſee their betrayers ſwim ſe- " curely to the females, which are between the hut and the nets; " they continue to advance, and the fowler attends the favourable " inſtant, and ſometimes takes a dozen or more at a ſingle draw.

" I have always remarked, that the Ducks trained to this ſport " ſeldom came within the incloſure of the net, but flew over it, and " knew the ſpot, though nothing appeared out of the water.

" All the marſh birds, ſuch as the whiſtlers, the ſhovelers, the

" teals,

In Lorraine, on the pools which border on the Sarre, Ducks are caught with a net ſtretched vertically, and like the draw-net uſed for woodcocks *. In many other places, the fowler ſitting in a boat, covered with boughs and reeds, approaches ſlowly the Ducks that are diſperſed on the water, which he collects together by

" teals, the pochards, &c. come to the call of the Ducks, or follow " the betrayers.

" This ſport is practiſed only in moon-light; the moſt favour-" able time is the riſing of that luminary, and an hour before day-" break. It is unprofitable, except in northerly or north-eaſterly " winds, becauſe the Ducks then journey, or are in motion to con-" gregate. I have ſeen to the amount of an hundred taken by the " ſame nets in one night. A man of weak conſtitution, or ſenſible " to cold, could not ſupport the hardſhips inſeparable from this " ſort of fowling: he muſt remain motionleſs, and often drenched " the whole night in the middle of the marſhes.

" I have often ſeen the wild Ducks deſcend to the call of the fe-" males of their own kind, how elevated ſoever they might be in " the air; the betrayers flew ſometimes with them more than a " quarter of an hour; each of the fowlers over whom the flock " paſſes, diſpatches others to them; they diſperſe, and each band of " traitors leads off a detachment; that of the fowlers which have " wild females is conſtantly the largeſt.

" In general ducking is a ſeducing but laborious ſport: a per-" ſon muſt brave the rigour of the weather, which, at that ſeaſon, " is often ſevere, his feet ſoaked in the water, and his toes chilled " with the froſt: he muſt patiently wait at night in the hut, or " walk out before day on the brooks and the rivulets. I remember " to have gone a-ducking every day for a month together, when the " weather was exceſſively cold, yet reſolving with myſelf that each " excurſion ſhould be my laſt; and to crown my hardſhips, I had " the mortification to ſee my excellent dog drowned, which was " caught among the ice. I ſpeak as an old ſportſman, recounting " my feats." *Extract from the excellent Memoir which M. Hebert has obligingly written for us on Ducks.*

* M. Lottinger.

ſetting

ſetting a little dog after them: the fear of an enemy prompts them to aſſemble, and they gradually join. They may be fired at, one by one, as they come near; and to prevent noiſe, a ſort of trunk-guns are uſed, or a diſcharge may be made on the whole flock with a large blunderbuſs, which ſcatters the ſhot, and which will kill or wound a good number; but no more than one fire can be given, for thoſe which eſcape know ever after the boat, and carefully avoid it *. This very amuſing ſport is called *the frolic (badinage)*.

The wild Ducks are alſo caught with hooks baited with *calves lights*, and faſtened to floating hoops. Indeed the fowling for Ducks is every where † one of the chief ſports of autumn ‡ and the beginning of winter.

Of

* Ducks have a ſort of memory, which recognizes the ſnare that they have once eſcaped. At Nantua, on the edge of a lake, a hut was conſtructed with branches of fir and with ſnow; and to make the Ducks to approach it, they are purſued at a diſtance by two boats; this plan ſucceeded eight or ten days, after which it was impoſſible to make them return. *M. Hebert.*

† Navarette makes the Chineſe practiſe the ſame ſtratagem for catching Ducks, that Peter Martyr deſcribes as an invention of the Indians at Cuba, who ſwimming on their lakes, he ſays, with their head only out of the water, and covered by a calibaſh, catch the geeſe by the feet.

‡ The method of ducking uſed by the Kamtſchadales is thus deſcribed: "Autumn is the ſeaſon of the great ducking at Kamtſchatka; " they go to the places covered with lakes, or full of rivers and inter- " ſected by woods; they clear the avenues acroſs theſe woods from " one lake to another, they ſtretch between the two, nets ſupported " by high poles, and which can be let down by ſlipping cords, of " which

Of all the provinces in France, Picardy is that wherein the breeding of tame Ducks is moſt attended to, and the catching of the wild ones the moſt profitable; inſomuch, that it brings a pretty conſiderable revenue to that country *. It is conducted on an extenſive plan

" which they hold the ends; at evening theſe nets being raiſed as " high as the Ducks flight, theſe birds ſhoot acroſs in multitudes, " and with ſuch force, that they ſometimes break through the bar- " rier, but are oftener caught.

" Theſe Ducks ſerve as a barometer and a weather-cock to the " Kamtſchadales, for they pretend that theſe birds turn and fly al- " ways againſt the wind which is to blow." *Hiſt. Gen. des Voy. tom.* xix. *p.* 274.

" Ducks are remarkably numerous in Poland, eſpecially on the " river Styr in Volhinia, for there one hundred and twenty or an " hundred and eighty, enticed by Buck wheat, are often taken at " once by a net." *Rzaczynſki.*

* " A good part of the wild Ducks, and other birds of the ſame " kind, which ſupply the markets of Paris, is brought from Picardy. " The quantity caught each winter in the two paſſages, is aſtoniſh- " ing. This ſport begins in the Laonois, a few leagues from Laon: " from thence to the ſea there is a continued chain of marſhes or of " meadow, that are overflowed in the winter, the extent ſcarcely leſs " than thirty leagues: when the rivers Oiſe and Serre ſwell over " their banks, their waters unite, and cover all the interjacent coun " try. The river Somme likewiſe ſpreads over an immenſe country " in its inundations. The fowling for Ducks conſtitutes therefore " a branch of trade in Picardy: I have been aſſured that it was " farmed at thirty thouſand livres (about £. 1,250.) on the ſingle " pool of St. Lambert, near La Fere; it is true that this pool is " ſeven or eight leagues in compaſs, and perhaps the right of fiſh- " ing was included. When I reſided in that province, there were " barks freighted from ten to fifty crowns, according to the advan- " tageouſneſs of their ſituation; and I am beſides aſſured that there " were ſome of theſe duck-boats furniſhed with nets to the value " of three thouſand livres (£. 125.)

" Viewing

plan in the inlets or little creeks, diſpoſed by nature or cut artificially along the margins of lakes, and into the thick cluſters of reeds. But no where is this ſpecies of ducking carried on with greater preparation, or more agreeable ſucceſs, than on the beautiful pool of Arminvilliers in Brie. I ſhall here give the deſcription which was ſent to us by M. Rey, ſecretary to his grace the duke of Penthievre.

" On one ſide of this pool, ſhaded with reeds " and ſkirted by a ſmall wood, the water forms " a deep creek in the grove, a ſort of little ſhel- " tered haven, where perpetual calm prevails. " From this haven canals are cut into the heart " of the wood, not in ſtraight lines, but in twiſt- " ed arches: theſe, called *horns*, are pretty broad " and deep at their mouth, but gradually con- " tract both in breadth and depth as they extend

" Viewing theſe vaſt marſhes from the neighbouring heights, I " perceived that great glades were formed, by cutting the ruſhes " between two waters with a bill or hook; theſe glades are nearly " of a triangular ſhape, and it is in the corners that the nets are " ſet; they ſeemed to be a ſort of large weel-nets, that would ſink " on letting go the counterpoiſe which keeps them on the ſurface " of the water; I am at leaſt certain, that the Ducks are drowned " in them: often have I ſeen thirties of them ſpread on the moſs, " to dry in the ſun, in order to prevent, I was told, the fleſh from " contracting a muſty ſmell from the wet feathers; I then learnt " that they drowned the Ducks in the nets; they added, that they " employed little tawny dogs, much like foxes, to collect them and " drive them into the nets: the Ducks collect round a fox, from a " ſort of antipathy, as they do about an owl or any other call-bird; " theſe little dogs are trained to lead them whither they have been " taught." *Extract of the Memoir communicated by M. Hebert.*

" and

" and wind among the trees, and at laſt draw to " a dry point.

" The canal, from its origin to near its mid- " dle, is covered with a cradle-net, at firſt pretty " wide and high, which narrows and deſcends " as the canal contracts, and terminates at its " point in a weel, which ſhuts like a purſe.

" Such is the great ſnare fitted and prepared " for the numerous flocks of Ducks, mixed with " pochards, golden-eyes, and teals, which come " to alight on this pool in the middle of Octo- " ber. But to draw them to the creek and the " fatal *horns*, required ſome ſubtle contrivance; " and this contrivance has been long concerted " and practiſed.

" In the midſt of the grove and of the canals, " dwells the Ducker, who thrice every day goes " from his little houſe to ſcatter the grain, on " which he feeds the whole year above an hun- " dred Ducks, half tame, half wild, that, ſwim- " ming conſtantly in the pool, never fail, at the " accuſtomed hour, and at the ſound of a whiſtle, " to riſe and fly vigorouſly to the inlet, and wind " up the canals where their food waits them.

" Theſe are the *traitors*, as the Ducker calls " them, which mingling on the pool with the " wild flocks, lead them to the inlet, and thence " decoy them into the *horns*; while, concealed " behind a row of reed-hurdles, the Ducker pro- " ceeds throwing grain before them, and entices " them under the mouth of the cradle-nets;

S " then

" then ſhowing himſelf through the intervals of " the hurdles, diſpoſed obliquely, and to conceal " him from the Ducks that advance, but diſcloſe " him to the ſight of ſuch as have got before, " which in their trepidation ruſh headlong into " the labyrinth, and drive pellmel into the " weel. The half tame ones ſeldom enter; they " are uſed to the diverſion, and return again to " repeat the decoy *."

In the autumnal paſſage, the wild ducks roam at large on the lakes, and remote from the ſhores; and there they ſpend a great part of the day reſting themſelves, or ſleeping. " I have " obſerved them," ſays M. Baillon, " with a " perſpective glaſs on our largeſt pools, which " ſometimes appear entirely covered with them. " Their heads lay motionleſs under their wings, " till they all took flight half an hour after ſun- " ſet."

In fact, the Ducks ſhow more activity in the night than in the day: they feed, they journey, they arrive and depart chiefly in the evening and

* Willughby deſcribes exactly the ſame mode of ducking as practiſed in the counties of Lincoln and Norfolk in England, and where they take, he ſays, four thouſand Ducks probably in the courſe of the winter. He ſays alſo, that to collect them, a tawny dog is uſed: moreover, a great number of Ducks muſt breed in thoſe fenny countries, ſince, according to his account, the greateſt capture is made when the Ducks are in moult, at which time the boats have only to puſh them forward into the nets ſtretched on the pools. [An ample deſcription of the method of catching Ducks in the Lincolnſhire fens, may be ſeen in the Britiſh Zoology.—*T.*]

and in the night; and moſt of thoſe which are ſeen in broad day have been forced to fly by ſportſmen or birds of prey. In the night, the ruſtling of their wings marks their courſe. The clapping of their wings is the moſt noiſy at their riſing; and hence Varro gives the Duck the epithet *Quaſſagipenna*.

As long as the ſeaſon continues mild, the aquatic inſects and ſmall fiſh, the frogs which have not yet crept under the mud, the ſeeds of the bull-ruſh, the water lentil, and ſome other bog plants, afford abundant ſubſiſtence to the Ducks. But towards the end of December or the beginning of January, if the great pieces of ſtanding water are frozen, they remove to running rivers, and afterwards reſort to the edge of woods to gather acorns, and ſometimes even they alight among the fields ſown with corn; and if the froſt laſt eight or ten days, they diſappear, and return not till the thaws in the month of February: at that time, they are ſeen to arrive in the evening with the ſouth winds, but in ſmaller numbers *, their flocks being probably thinned by the loſſes ſuſtained during the winter †. Their ſocial inſtinct ſeems to

* "The difference is great between thoſe which arrive and thoſe "which retire; I have been able to make the compariſon in Brie "for ſix or ſeven years; perhaps not the half re-paſs, and yet their "number keeps up, and every year as many return." *M. Hebert.*

† "It has often come into my head to compare the population "of

to be impaired by the diminution of their numbers; and they no longer keep company with each other. They paſs diſperſed, fly in the night-time, lurk among the ruſhes during the day. They halt no longer in a place than a contrary wind conſtrains them. They ſeem already to join in pairs, and they haſten to the northern countries, where they breed and ſpend the ſummer.

In that ſeaſon, they may be ſaid to cover all the lakes and all the rivers of Siberia * and

" of the wild Ducks with that of the rooks, the crows, &c. Of " theſe one would be tempted to think that more retire than arrive, " and that, becauſe they retire in flocks. They are never killed, " they have very few enemies, and they take the ſureſt precautions " for their ſafety. The rigours of our winters cannot affect their " temperament, which is adapted to cold; in the end, the earth " muſt be covered with them. Yet their multitude, though it " might ſeem to be innumerable, is fixed; which proves, I think, " that they are not, as uſually believed, favoured with a longer life " than other birds, and if they make only one annual hatch, as I " am well aſſured, their population cannot be immenſe.

" I ſuppoſe that the wild Duck lays fifteen or ſixteen eggs, and " hatches them: allowing one half for accidents, addle eggs, &c. I " would reckon the multiplication at eight young to each pair. " Suppoſing the deſtruction during winter to reduce this again to " an half, the ſpecies might ſtill, we ſee, maintain its numbers. " More than the half are killed in Picardy, but very few in Brie " and in Breſſe, where there are many pools. When I limit each " hatch to eight young, I make but a moderate allowance; the " marſh buzzard deſtroys many, as I am certain; and the fox, it is " ſaid, concerts his meaſures ſo well as always to catch a few." *M. Hebert.*

* In the plain of Mangaſea, on the Jeniſea, there are innumerable flocks of geeſe and Ducks of different kinds. *Gmelin.*—The Barabin Tartars live on milk, fiſh, . . . game, and eſpecially the Ducks and the divers, which abound in this diſtrict. *Idem.*

Lapland:

Lapland *: they advance as far north even as Spitzbergen † and Greenland ‡. "In Lap-"land," ſays M. Hœgſtroem, "theſe birds ſeem "diſpoſed, if not to drive away the men, at leaſt "to fill up their place: for as ſoon as the Lap-"landers go in the ſpring to the mountains, the "flocks of wild Ducks fly to the weſtern ſea; "and when the Laplanders deſcend again in au-"tumn to inhabit the plain, theſe birds have "already retired." Many other travellers give the ſame account §: "I do not believe," ſays Regnard, "that there is a country in the world "more abounding with Ducks, teals, and other "water fowls, than Lapland. The rivers are "all covered with them; . . . and in the month "of May their neſts are in ſuch plenty, that the "deſert ſeems filled with them." Yet ſome pairs of theſe birds, which circumſtances have

* I believe that there is no country in the world which abounds more with Ducks, ſwans, divers, teals, &c. than Lapland. *Regnard.*

† In the ſouth haven of Spitzbergen, there are many little iſlands, which have no other names than *the birds iſlands*, becauſe the eggs of Ducks and kirmews are gathered on them. *Hiſt. Gen. des Voy. tom.* i. *p.* 270.

‡ When the winter, ſetting in earlier than uſual, ſurprizes them in theſe inhoſpitable ſhores, great numbers periſh. In the winter of 1751, the iſlands round the Daniſh miſſion at Greenland were ſo covered with wild Ducks, that they were taken by the hand, having been driven to the coaſt. *Crantz.*

§ In the northern lakes, the Ducks are ſo numerous as to ſeem to cover almoſt the whole water; they are ſeldom diſturbed by the fowler, as the ſport is much more abundant in the wood than on the water. *Olaus Magnus.*

 prevented

prevented from joining the bulk of the ſpecies, remain in our temperate countries, and breed in our marſhes. It is only on theſe ſtragglers that obſervations could be made with regard to the peculiarities in the amours of theſe birds, and the attention they beſtow on rearing their young in the wild ſtate.

After the firſt gentle airs, towards the end of February, the males begin to court the females, and ſometimes fight with each other through rivalſhip. The pairing laſts about three weeks. The male ſeems diligent in ſeeking out a proper place for the depoſiting the fruits of their loves: he points it out to the female, who conſents, and takes poſſeſſion. The ſpot is generally a thick tuft of ruſhes, raiſed and inſulated in the middle of the marſh. The female pierces this tuft, deepens it, and moulds it into the ſhape of a neſt, by preſſing down the ruſhes which incumber it. But though the wild Ducks, like other water-fowls, prefer the vicinity of water for breeding *, yet ſome neſts are found pretty remote, among heaths, or in the cultivated fields on the cocks of ſtraw gathered by the labourer, or even in the foreſts on mutilated oaks, and in old forſaken neſts †. Each neſt contains

* Ariſtotle, *lib.* vi. 7.

† The wild Duck is very cunning; ſhe does not always make her neſt by the edge of water, nor even on the ground; they are often found in the middle of heaths, at the diſtance of a quarter of a league from the water: they have been known to lay in the neſts of magpies and crows, on very lofty trees. *Salerne.*

uſually

usually from ten to fifteen eggs, and sometimes eighteen: their albumen is greenish, and their yolk red *. It is remarked that the old Ducks lay more, and begin earlier, than the young ones.

Every time the female rises from her eggs, even for a short interval, she covers them with the down that she pulls from her body to clothe her nest. She never descends upon them from the wing, but alights an hundred paces beyond the spot, and walks to it warily, observing whether any foes be nigh; but when once she is seated on the eggs, the approach even of a man will not flush her.

The male seems to take no share in covering the eggs; only he keeps at a short distance, and accompanies the female when she goes in search of food, and protects her from the importunities of other males. The incubation lasts thirty days: all the young are hatched in one day; and on the succeeding the mother descends from the nest, and calls them to the water. Timorous or chilly, they hesitate to enter, and some even retire; but the boldest plunges after its mother, and the rest soon follow. When they have once quitted their nest, they return no more. If it is situated far from the water, or too elevated, the father † and the mother ‡ take them in the bill,

* Belon.

† According to M. Hebert.

‡ According to M. Lottinger.

 and

and tranſport them one after another *. In the evening, the mother gathers them together, and withdraws them among the reeds, where ſhe cheriſhes them under her wings during the night. All day they watch, on the ſurface of the water and on the graſſy mead, for gnats, which are their firſt food. They are ſeen to dive, to ſwim, and to make a thouſand evolutions on the water, with equal quickneſs and facility.

Nature, while ſhe early invigorates the muſcles neceſſary for ſwimming, ſeems to neglect for ſome time the formation, or at leaſt the growth of their wings: theſe continue near ſix weeks ſhort and miſhapen. The duckling has acquired half its ſize, is feathered under the belly and along the back, before the quills of the wings begin to appear; and it can hardly attempt to fly till three months. In this ſtate it is called *hallebran* in French, a name derived apparently from the German *halber-ente*, or half duck †: and as theſe *hallebrans* are unable to fly, they afford an eaſy and ſucceſsful ſport on the pools and marſhes that are ſtocked with them ‡. Probably

* This was known to Belon.

† This appellation was given as early as the time of Aldrovandus.

‡ "I ſhall here deſcribe what a gentleman of my acquaintance "practiſed on a marſh, between Laon and Notre Dame de Lieſſe. "The bottom of this marſh is vitrifiable ſand, which is never miry. "In the months of June and July, the water does not reach above "the waiſt in the deepeſt parts, where grows a ſort of low reeds, "not

bly these grown ducklings are the same which the Laplanders fell with sticks on their lakes *.

The same species of wild Ducks which visit us in winter, and inhabit the northern regions of our continent in summer, occurs in the corresponding regions of the new world †; their migrations,

" not close, yet affording a proper retreat to the young *hallebrans*. " This gentleman, clothed with a simple linen vest, went into this " marsh, accompanied with his game-keeper and a domestic servant: " he had caused the reeds to be cut into very long strips, seven or " eight feet wide, like alleys in a forest, or trenches in a marsh. He " kept along these openings, while his people were beating the " marsh; and when they lighted on some troops of *hallebrans*, " they gave him notice. The *hallebrans* are not able to fly until the " 15th of August; they fled swimming, and the people pursued, kill- " ing some in their progress; the rest were forced to cross the alleys " made in the reeds; it was in this passage that the expert fowler " killed them at his ease: those which escaped were made to re-pass, " and another discharge was made, always profitable; the more so, " as these *hallebrans* or young ducks are excellent eating." *Extract of the Memoir communicated by M. Hebert.*

* " The use of sticks for hunting with is unknown in our temperate " climates; here (in Lapland) in the extraordinary abundance of game, " they use indifferently sticks or whips. The birds which we took " in greatest numbers were Ducks and divers, and we admired the " dexterity of our Laplanders in killing them: they followed them " with their sticks, without seeming to notice them; they approached " gradually, and when, being sufficiently near, they saw them swim- " ming in the communication between two pools, they threw a stick " at them, which crushed their head against the bottom or the stones, " with a promptness that our sight could scarce follow: if the " Ducks took flight before they were approached, they brought " down several by the stroke of a whip." *Regnard.*

† At Louisiana the wild Ducks are larger, more delicate, and better tasted, than those of France, but in other respects entirely similar; they are so numerous, that we may reckon a thousand for one of ours. *Dupratz.*—I received this year from Louisiana many birds

grations, and their autumnal and vernal paſſages ſeem to obſerve the ſame order, and to be performed in the ſame time *: nor need we wonder that birds which prefer the arctic tracts, and which poſſeſs vigour of wing, ſhould tranſport themſelves from the boreal parts of the one continent into the other. But we ſuſpect that the Ducks ſeen by navigators, and found in many of the iſlands in the South Sea †, are not of the

birds ſimilar to ſpecies of the ſame genus, which occur in France and in the various parts of Europe, and particularly a Duck exactly like our wild Duck; it had no difference in the plumage, and only ſeemed to be rather larger. The inhabitants have themſelves perceived ſuch a reſemblance between this Duck and that of Europe, as to have named it the French Duck. *Dr. Mauduit*.—*Metzanauhtli*, or Moon Duck, is a ſort of Duck like the domeſtic one, and variegated with the ſame colours; it lives on the Mexican lake. *Fernandez*.—The Canadian Ducks are like thoſe which we have in France. *Leclerc*.

* About the end of April, the Ducks arrive in abundance at Hudſon's Bay. *Hiſt. Gen. des Voy. tom.* xiv. *p.* 657.—In the very ſhort and piercing days of December, at Hudſon's Bay, one kills as many partridges as one chooſes: towards the end of April, geeſe, buſtards, Ducks, and many other birds, arrive, and ſtay about two months. *Lade*.

† Ducks on the coaſt of Diemen's land, in the 43d degree of latitude. *Cook*.—Wild Ducks at Cape Forward, in Magellan's Strait. *Wallis*.—In the bay of Cape Holland, in the ſame Strait. *Idem*.—In great plenty at Port Egmont. *Byron*.—At Tanna, a pool contained multitudes of rails and wild Ducks. *Cook*.—In croſſing a rivulet on our way (at Otaheite), we ſaw ſome Ducks; as ſoon as they got to the other ſide, Mr. Banks fired upon them, and killed three at one ſhot: this incident ſpread terror among the Indians. *Idem*.—We killed (at Famine Bay, in Magellan's Strait) a great number of birds of different kinds, and particularly geeſe, Ducks, teals, &c. *Wallis*.—Two great freſh-water lakes (at Tinian) preſented a multitude of Ducks and teals, and many whiſtlers. *Anſon's Voyage*.

common

common kind; and we apprehend that they belong rather to ſome of the ſpecies hereafter to be deſcribed, and which are indeed peculiar to thoſe climates: at leaſt, we ſhould preſume that ſuch is the caſe, till we know more particularly the ſpecies of theſe Ducks which occur in the ſouthern Archipelago. We are certain that thoſe which, at St. Domingo, have the name of wild Ducks, are different from ours *; and from ſome hints, with regard to the birds of the torrid zone †, we are perſuaded that the ſpecies of our wild Duck has not penetrated there, unleſs the tame ſort has been introduced ‡. But whatever

* What are called *wild Ducks* in St. Domingo differ widely from the true wild Duck of Europe in bulk in plumage, and in taſte; nor is the teal the ſame with that of Europe. *Chevalier Deſhayes.*—The wild Ducks of Cayenne are the ſame with thoſe known in Europe by the name of Barbary Ducks or Muſcovy Ducks. *M. Bajou.*

† "There are in this country (on the coaſt of Guinea) two ſorts "of wild Ducks; during the time that I was there I ſaw only two "of the firſt ſpecies ... They differed not in ſize or in figure from "other Ducks, but their colour was of a very beautiful green, with "the bill and legs of a fine red; their colour was ſo rich and fine, "that, if they had been offered to ſale alive, I would not have ſcru-"pled to have given an hundred franks or more ... It is about four "months ſince I ſaw one of the ſecond kind, which had alſo been "killed by ſome of our people, and which had the ſame figure "with the preceding; its legs and its bill yellow, and its body half "green half gray, ſo that it was far from being ſo handſome." *Boſman.*

‡ Tame Ducks were not known on the coaſt of Guinea till within theſe few years." *Boſman.*—The Dutch were conducted to the apartment of the Ducks (in the palace of the king of Tubaon, at Java); they found theſe to be like thoſe in Holland, except that they were ſomewhat bigger, and moſtly white; their eggs are twice as large as thoſe of our fineſt hens. *Hiſt. Gen. des Voy. tom.* viii. *p.* 137.

ever be the ſpecies which inhabit theſe ſouthern regions, they ſeem not ſubject to thoſe migrations, which, in our climates, reſult from the viciſſitude of the ſeaſons *.

In all countries, men have been ſolicitous to domeſticate, to appropriate a ſpecies ſo uſeful as that of our Duck †; and not only has it become common, but foreign kinds, originally equally wild, have been multiplied, and have produced new tame breeds. For example, that of the Muſcovy Duck, from the double profit of its plumage and its fleſh, and from the facility of raiſing it, has grown one of the moſt uſeful fowls, and one the moſt diffuſed in the new world.

To rear Ducks with profit, and form numerous and proſperous flocks, they require, like the geeſe, a place near water, and where ſpacious open banks and turfy ſtrands afford them room to feed, reſt, and play. Not but Ducks are often ſeen confined and kept dry within the incloſure of a court-yard; but this mode of life is not congenial to their nature; they generally pine and degenerate in that ſtate of captivity;

* At Tonquin, ſmall houſes are built for the Ducks, where they lay their eggs; they are ſhut up every evening, and let out every morning ... The nnmber of wild Ducks, of water hens, and of teals, is immenſe: theſe birds come to ſeek their food here in the months of May, of June, and of July, and then they fly only in pairs; but from October to March you will ſee great flocks together that cover the country, which is low and marſhy. *Dampier.*

† Belon.

vity; their feathers rumple and rot; their feet are hurt on the gravel; their bill ſhivers with frequent rubbing, all is ſpoiled and injured, becauſe all is conſtrained; and Ducks thus raiſed can neither yield ſo good a down, nor propagate ſo ſtrong a race as thoſe which enjoy a part of their native liberty, and live in their proper element. If the place does not naturally afford any current or ſheet of water, a pond ought to be dug, in which the ducks may dabble, ſwim, waſh, and dive, exerciſes abſolutely neceſſary to their vigour, and even their health. The ancients, who beſtowed more attention than we on the intereſting objects of rural œconomy, and of a country life, thoſe Romans, who with the ſame hands held the plough * and bore the laurels of victory, have on this head, as on many others, left us uſeful inſtructions.

Columella † and Varro dwell with complacency on the ſubject, and deſcribe at full length the diſpoſition of a yard proper for Ducks. It contains a pond with a ſmall iſland; the water branches in rills over the turf; buſhes intermix their ſhade: and the whole is laid out in ſo artful and pictureſque a manner, that it might form an ornament to the fineſt country-houſe ‡.

The

* *Gaudebat terra vomere laureato & triumphali Aratore.* Plin.

† Rei Ruſtic. *lib.* viii. 15.

‡ "In the middle a pool is dug . . . whoſe brink ſlopes gently "into the water . . . in the centre riſes an iſlet planted with "various aquatic ſhrubbery, which may afford ſhady retreats for "the

The water muſt not be infeſted with leeches, for theſe would fix on the feet of the ducklings, and occaſion their death. To rid the pool of ſuch pernicious inhabitants, tench or other fiſh are thrown in to feed on them *. In all ſituations, whether on the banks of a ſtream or on the margin of ſtagnant water, baſkets muſt be placed at intervals with covered tops, and containing a commodious apartment that may invite theſe birds to neſtle. The female lays every two days, and has ten, twelve, or fifteen eggs; ſhe will even produce thirty or forty, if ſhe be abundantly fed, and the eggs repeatedly removed. She is of an ardent nature, and the male is jealous. He uſually appropriates two or three females, which he leads, protects, and fecundates. When the Drake is unprovided with theſe miſtreſſes, his

" the birds . . . Around, the water ſpreads without interruption, " that the Ducks may freely play in the warm ſun, and ſportively contend in ſwimming . . . The banks are clothed with " herbage . . . In the ſurrounding walls are cut holes for the " birds neſtling in, and theſe are ſcreened with buſhes of box and " myrtle . . . Adjacent, a continued pipe is ſunk along the ground, " by which their food, mixed with water, is every day conveyed to them; for this kind of birds require their aliments to " be diluted . . . In the month of March, ſtraws and ſprigs ſhould " be ſtrewed in the aviary, with which they may build their neſts " . . . and he who wiſhes to form a *neſſotrophium* of birds, may gather the eggs about marſhes, and ſet them under coop-hens; for " the young being thus hatched and educated, will loſe their wild " nature . . . but having laid on a lattice-work, let the aviary be covered with nets, to prevent the tame birds from eſcaping, or the " eagles and hawks from annoying them."

* Tiburtius, in the Memoirs of Stockholm.

luſt

luſt often takes a wrong direction *; nor is the Duck more reſerved in admitting the careſſes of ſtrangers †.

The time of incubation is above four weeks ‡; and that time is the ſame if a hen ſit on the eggs. The hen is no leſs tender to the ducklings than their proper mother: when ſhe firſt leads them to the brink of water, they fondly recognize their element, and obey the impulſe of nature, regardleſs of the earneſt and reiterated calls of their nurſe, who remains diſconſolate and tormented on the bank ||.

Ducklings are firſt fed with the ſeeds of millet or panic, and a little barley may ſoon be added §. Their natural voracity diſplays itſelf

* A Drake of my court having loſt his Ducks, took a liking to the hens; he t od ſeveral, of which I was witneſs; thoſe which he had trod could not lay, and it was neceſſary to perform a ſort of Cæſarean operation to extract the eggs, which were ſet to hatch: but whether from want of care, or from want of fecundation, they produced nothing. *M. de Querhoënt.*

† I ſaw, two years in ſucceſſion, a Duck pair with a ſheldrake, and produce hybrids. *M. Baillon.*

‡ It appears that the Chineſe hatch Duck eggs, like thoſe of hens, by means of artificial heat, according to the following notice of Francis Camel: *Anas Domeſtica* ytic *Luzonienſibus, cujus ova Sinæ calore fovent & excludunt.* Philoſ. Tranſ. N° 285.

|| *Super omnia eſt admiratio anatum ovis ſubditis gallinæ, atque excluſis; primò non planè agnoſcentis fœtum, mox incertos incubitus ſollicitè convocantis; poſtremo lamenta circa ſtagnum, mergentibus ſe pullis, naturâ duce.* Plin. *lib.* x. 55.

§ *Gratiſſima eſca terreſtris leguminis, panicum & milium, nec non & hordeum: ſed ubi copia eſt, etiam glans ac vinacea præbeantur. Aquatilibus etiam cibis, ſi ſit facultas, datur cammarus, & rivalis alecula, vel ſi quæ ſunt incrementi parvi fluviorum animalia.* Columella, *De Re Ruſtica*, lib. viii. 15.

almoſt

almoſt at their birth; young or old they are never ſated; they ſwallow whatever they meet with *, whatever is offered; they crop graſs, gather ſeeds, gobble inſects, and catch ſmall fiſh, their body plunged perpendicularly, and only their tail out of the water; they ſupport themſelves in this forced attitude more than half a minute, by continually ſtriking with their feet.

They acquire in ſix months their full ſize, and all their colours. The Drake is diſtinguiſhed by a ſmall curl of feathers that riſes on the rump †: his head, too, is gloſſed with a rich emerald green, and his wing decorated with a brilliant ſpangle. On the middle of the neck there is a white half collar; the fine purple brown of the breaſt, and the colours on other parts of the body, are diſpoſed in pleaſing gradations, and upon the whole form a beautiful plumage.

Yet we muſt obſerve, that theſe choice colours never ſhew all their vivacity but in the males of the wild kind: they are always duller and more indiſtinct in the tame Ducks, as the ſhape is alſo heavier and leſs elegant; ſo that an eye a little accuſtomed may diſtinguiſh between them. In that kind of fowling where tame Ducks ſearch the wild ones, and bring them within aim of the fowler, it is cuſtomary to pay the Ducker a price agreed on for each tame Duck killed by

* Aldrovandus.

† Aldrovandus and Belon.

miſtake.

miſtake. But the experienced fowler ſeldom errs, though the tame Ducks are choſen of the ſame colour with the wild ones: for not only are the tints more vivid in theſe, but their feathers are ſmoother and cloſer, their neck ſlenderer, their head finer, the lineaments more delicately traced, and all their motions diſplay the eaſe, ſtrength, and dignity, which freedom inſpires. "When "I viewed this picture from my ſentry-box," ſays M. Hebert, ingeniouſly, "I fancied a ſkil-"ful painter had delineated the wild Ducks, "while the tame Ducks ſeemed the production "of his ſcholars." The young ones hatched in the houſe from wild Ducks' eggs, before they diſcover their fine colours, are already diſtinguiſhed by their ſtature and their elegance of form. Nay, the difference is much more perceptible when the wild Duck is brought to our table: its ſtomach is always rounded, whilſt it forms a ſenſible angle in the tame Duck, which laſt is ſurcharged with fat, while the fleſh of the former is delicate and juicy. Purveyors know them eaſily by the legs, of which the ſcales are finer, equal, and gloſſy; by the membranes, which are thinner; by the nails, which are ſharper and more ſhining; and by the thighs, which are more ſlender than in the tame Duck.

The male, in all the water-fowl with a broad bill and palmated feet, is always larger than the female *; contrary to what obtains among the

* Belon had before made this obſervation.

birds of prey. In the Ducks and teals alſo, the males are robed with the richeſt colours, while the females are only of an uniform brown or gray *; and this difference, which is very conſtant in the wild kinds, remains impreſſed on the tame breeds, as far at leaſt as the variations and alterations of colour, occaſioned by croſſing the wild and the tame, have permitted †.

In

* Edwards makes this obſervation.

† It has been remarked, that in flocks of wild Ducks, there are ſome different from the reſt, and which reſemble the tame ones in the ſhape of their body and the colours of their plumage: this baſtard breed proceed from thoſe which the inhabitants near marſhes raiſe every year in great numbers, and of which they always leave a certain proportion on the marſhes; their method of rearing them is equally ſimple and curious.

" The females," ſays M. Baillon, " are ſet to hatch in the houſes; " every place agrees with them, for they are much attached to their " eggs; they are allowed twenty-five a-piece: ſome eggs are alſo " hatched by turkies and hens, and the young immediately diſtri- " buted to the Ducks.

" On the morning after the birth, each inhabitant marks his own; " one cuts the firſt nail of the right foot, another, the ſecond, ano- " ther bores a hole in ſuch a part of the ſkin of the foot, &c. Every " perſon retains his mark; it is perpetuated in his family, and known " by the whole village.

" As ſoon as the ducklings are marked, they are carried with " their mothers to the marſh; there they rear themſelves, and with- " out trouble; it is only neceſſary to drive away the ravenous birds, " particularly the buzzards, which deſtroy many. There are per- " ſons who thus put ſeven or eight hundred in the water every " year.

" At the end of May and later, the inhabitants aſſemble to take " them again with nets; each knows his own; poulterers come from " a diſtance to buy them: a certain number are always preſerved " in the marſh, both to ſerve in winter as a call to the wild ones, " and to multiply the ſpecies in the ſpring following; each perſon

" habituates

In fact, like all the other tame birds, the Ducks have undergone the effects of domestication. The colours of their plumage have been diluted, and sometimes even entirely effaced or changed. Some are more or less white, brown, black or mixed; others have assumed ornaments foreign to the species; such as the crested breed; another, still more deformed by domestication, has its bill twisted and bent *. In some, the

" habituates them to return to his house; they are attracted by " throwing barley to them, of which they are very fond.

" Many of these desert during the ains of October and November, " and mix with the wild ones which arrive at this season; they " pair, and this union produces the bastards, which are distinguish-" able both by their form and by their plumage . . .

" These bastards have usually their bill longer, their head and " neck thicker than the wild ones, but slenderer than the tame; " they are usually stouter, as it happens when breeds are crossed . . .

" I have frequently seen Ducks perfectly white pass with flocks " of wild ones; these are probably the deserters . . . It is not how-" ever impossible but this bird may assume the white colour in the " north; yet I doubt this, because it is migratory: it might turn " white during the winter, if it remain always, or for a great length " of time ... but it departs every year at the beginning of autumn, " and advancing into the temperate regions, in proportion as the " cold is felt, it flies from the cause which whitens other birds: the " more severe the winter, the more numerous are their migrations. " We saw white Ducks in 1765 and 1775, but they were only as " one among a thousand.

" It is possible that this colour may be the effect of degeneration, " as in other animals; for I have seen several white Ducks that " were impotent: the white females, more common than the males, " are commonly smaller, weaker, and sometimes less prolific than " the rest. I have had two barren Ducks in my court-yard, which " were extremely white, and their eyes red."

* *Anas Adunca.* Linn. and Gmel.
Anas Rostro Incurvo. Briss.
The Hook billed Duck. Ray, Will. and Alb.

consti-

conſtitution is altered, and betrays all the marks of degeneracy; they are feeble, indolent, inclined to exceſſive fat, and the young delicate and difficult to raiſe. Friſch, who makes this obſervation, ſays alſo, that the white Ducks are conſtantly ſmaller and weaker than the other ſorts. He adds, that when the breed is croſſed between individuals of different colours, the young generally reſemble the father in the tints of the head, back, and tail; which happens alſo in the mixture of a foreign Drake with the common Duck. With reſpect to Belon's opinion, that the wild kind contains a greater and a ſmaller breed, I can find no proof of it; and moſt probably he was led into that notion by the compariſon of individuals of different ages.

Not but the wild kind exhibits ſome varieties, merely accidental, or derived perhaps from their intercourſe on the pools with the tame ſort. In fact, Friſch obſerves, that both intermingle and pair; M. Hebert remarks, that he often found in the ſame flock of Ducks reared near great pools ſome young which reſemble the wild, have a ſavage, independent inſtinct, and fly away in the autumn *. But what the wild Drake here

* "In the laſt place, I remarked two of this ſort in my court-"yard, fed with others of the ſame age: I told the ſervants, and "gave orders that they ſhould clip the wings; they neglected to "do this, and on a fine day they diſappeared, after reſiding two "months in this little court, where they wanted nothing, and where "they could ſee neither the fields nor the horizon." *Sequel of the notes communicated by M. Baillon.*

operates

operates with the tame Duck, the tame Drake may operate with the wild Duck, ſuppoſing that ſometimes ſhe yields to his ſolicitation: and hence might reſult thoſe differences in bulk * and in colours †, which has been noticed between ſome of the wild kind ‡.

All of them, wild as well as tame, are ſubject to an almoſt ſudden moulting, in which their great feathers drop in a few days, and often in a ſingle night §: indeed all birds with flat bills and palmated feet ſeem ſubject to a quick ſhedding of their plumage ‖. This happens to the males

* Salerne and Ray.

† *The wild black Duck* in Friſch.—We ourſelves ſaw, on the pool of Armainvilliers, of which all the Ducks have the livery of the wild ones, two varieties, the one called *red*, whoſe flanks are of a fine brown bay; the other was a male, which had not the collar, but inſtead of it all the lower part of the neck, and the creſcent on the breaſt, of a fine gray.

‡ M. Salerne ſpeaks of a wild Duck entirely white, killed in Sologne; but the bulk which he attributes to it, makes it doubtful whether it really was a Duck. "It was white," he ſays, "and as "white as ſnow, but what was moſt ſtriking, it was as large as a "middle-ſized gooſe."

§ According to M. Baillon.

‖ "I have often obſerved with aſtoniſhment, ſheldrakes, brents, "and whiſtlers, rid themſelves in two or three days, or even in a "ſingle night, of all the feathers of their wings." *Sequel of the notes communicated by M. Baillon.*—In the ſummer ſeaſon, the Indian or Muſcovy Ducks loſe entirely all their feathers; they are obliged to remain in the water and among the mangroves, where they run a riſk of being devoured by ſerpents, alligators, quachis, and other ravenous animals. The Indians go to hunt them at this time in the places where they know that they are numerous; they return

 with

males after pairing, and to the females after hatching; it appears to be occasioned by the waste of strength in the amours, and in the laying and incubation. "I have often observed at "the time of moulting," says M. Baillon, "that they were restless for some days previous, "and seemed to be tormented with great itch-"ings. They concealed themselves to cast their "feathers. Next day and the following ones "these birds were dispirited and bashful: they "seemed conscious of their feebleness, dared "not to spread their wings, and when pursued "they seemed to have forgotten the use of them. "This time of dejection lasted thirty days for "the Ducks, and forty for the barnacles and "geese. Their cheerfulness was restored with "their feathers, and then they bathed much, "and began to flutter. More than once I lost "them for not having noticed the time when "they essayed to fly: they disappeared during "the night: I heard them attempting the mo-"ment before; but I avoided appearing, because "they would all have taken flight."

The interior organization of the Ducks and geese exhibits some peculiarities. The *trachea arteria*, before it divides to enter the lungs, dilates

with their canoes loaded with these Ducks: I found five or six in a creek which had no feathers in their wings; I killed one, the rest escaped among the mangroves. *Memoir sent from Cayenne, by M. de la Borde, king's physician in that colony.*

lates into a ſort of bony and cartilaginous veſſel, which is properly a ſecond *larynx*, placed below the *trachea* *, and which ſerves perhaps as an air-magazine while the bird dives †, and gives undoubtedly to its voice that loud and raucous reſonance which characterizes its cry. The ancients had a particular word to denote the voice of Ducks ‡; and the ſilent, reſerved Pythagoras adviſed that they ſhould be kept remote from the habitation of his ſage, who was to be abſorbed in meditation §. But every man, philoſopher or not, who is fond of the country, muſt be pleaſed with what conſtitutes its greateſt charm, that is, the motion, life, and noiſe of nature, the ſinging of birds, the cries of fowls, varied by the frequent and loud *kankan* of Ducks; it chears and animates the rural abode; it is the clarion and trumpet among the flutes and hautbois; it is the muſic of the ruſtic regiment.

And it is the females, as in a well-known ſpecies, that are the moſt noiſy and the moſt loquacious: their voice is higher, ſtronger, more ſuſceptible of inflexions than that of the male, which is monotonous and always hoarſe. It

* Hiſt. de l'Acad. *tom.* ii. *p.* 48.—Mem. 1700, *p.* 496.

† Willughby and Aldrovandus.

‡ *Anates tetrinire.* Ant. Philomel.

§ Geſner.

has been remarked, that the female does not ſcrape the ground like the hen, yet ſcrapes in ſhallow water to lay bare the roots, or diſentangle inſects or ſhell-fiſh.

Both ſexes have two long *cæca*. The male organ of generation is twiſted into a ſpiral form *.

The bill of the Duck, like that of the ſwan, and of the ſeveral kinds of geeſe, is broad, thick, indented at the edges, clothed within with a ſort of fleſhy palate, filled with a thick tongue, and terminated at its point by a horny nail, of a harder ſubſtance than the reſt of the bill. The tail in all theſe birds is very ſhort, the legs placed much back, and almoſt concealed in the abdomen. From this poſition of the legs, proceeds the difficulty of walking and of keeping their equilibrium on land, which occaſions aukward motions, a tottering ſtep, a heavy air which paſſes for ſtupidity, whereas the facility of their evolutions in the water evinces the force, the delicacy, and even the ſubtlety of their inſtinct †.

The

* In certain moments it is pretty long and pendulous, which has led country people to think that the bird, having ſwallowed an adder, this hangs out at the anus.—See *Friſch*.

† "We had a very tame ferret, which, for its gentleneſs, was "careſſed by all our ladies; it was moſt of its time on their knees. "One day when we were in the ſaloon, a ſervant entered, holding "in his hand a tame Duck, which he let looſe on the floor; the "ferret immediately darted after the Duck, which no ſooner per- "ceived

The fleſh of the Duck is ſaid to be heating and of difficult digeſtion *; yet it is much uſed, and the fleſh of the wild Duck is finer and better taſted than that of the tame. The ancients knew this as well as we do, for Apicius gives no leſs than four different ways of ſeaſoning it. Our modern Apiciuſes have not degenerated, and a pie of Amiens Ducks is a diſh familiar to all the gluttons of the kingdom.

The fat of the Duck is uſed in topical remedies; and its blood is ſaid to counteract poi-

"ceived it, than he ſquatted his whole length; the ferret fell upon "him, and ſought to bite his neck and head; in an inſtant the "Duck ſtretched out his body, and feigned death; the ferret then "ſmelled the bird from the head to the feet, and perceiving no ſigns "of life, it left the body, and returned to us: the Duck now ſeeing his enemy retire, roſe gently on his toes, ſeeking to get upon "his feet; but the ferret, ſurprized at this reſurrection, ran and "threw him down, and did the ſame a third time. Several days in "ſucceſſion we amuſed ourſelves by repeating this little ſpectacle; "I cannot ſufficiently expreſs the ſort of intelligence perceived in "the conduct of the Duck; ſcarcely had he extended his head and "his neck on the floor, and had got rid of the ferret, than he began "to trail his head in ſuch manner as to be able to examine the proceedings of his enemy; then he raiſed his head gently and repeatedly, took to his feet and fled ſwiftly; the ferret returned to the "charge, and the Duck played again the ſame trick." *Extract of a letter written from Coulomiers, by M. Huvier to M. Hebert.*

* *Comedi de ipſâ & calefecit me: dedi calefacto, & incaluit amplius; & rurſus refrigerato, & calefecit denuo.* Serapio, *apud Aldrov.* —*Caro multi alimenti; auget ſperma & libidinem excitat.* Willughby.—Salerne, after ſaying "its fleſh is little eſteemed at our "tables," ſays, two lines after, "its fleſh is accounted better than "that of the gooſe."

 ſon,

ſon, even that of the viper *: this blood was the baſis of the famous antidote of Mithridates †. It was indeed believed that the Ducks in Pontus feeding on all the poiſonous plants which that country produces, their blood muſt have the virtue of countervailing the diſmal effects of venom. We ſhall obſerve by the way, that the denomination *Anas Ponticus* of the ancients refers to no particular ſpecies, as ſome naturaliſts have ſuppoſed, but the common ſpecies of wild Duck which frequented the borders of the Pontus Euxinus, as well as other ſhores.

Naturaliſts have endeavoured to introduce order, and eſtabliſh ſome general and particular diviſions in the great family of Ducks. Willughby diſtributes their numerous ſpecies into the *marinæ*, or thoſe which inhabit the ſea, and the *fluviatiles* ‡, or thoſe which frequent the rivers and freſh waters. But as moſt of theſe ſpecies live by turns both on ſalt and freſh water, and paſs indifferently from the one to the other, the diviſion of this author is inexact, and becomes defective in the application; nor are

* Galen.

† Belon.

‡ "Ducks are either *marine* or *fluviatile* . . . the *marine* have "their bills broader, eſpecially the upper mandible, and more "turned up; the tail ſomewhat long, not ſharp, the hind-toe broad, "or enlarged with a membrane: in the *fluviatile*, the bill is "ſharper and narrower; the tail ſharp; the hind-toe ſmall." *Willughby.*

the

the characters which he gives sufficiently constant. We shall therefore arrange them according to the order of their bulk, dividing them first into the *Ducks* and *Teals*; the former comprehending all the species of Ducks which equal or surpass the common sort, the latter including all the small species, whose bulk exceeds not that of the ordinary teal.

[A] Specific chatacter of the Mallard or Wild Duck, *Anas Boschas*: "It is cinereous; the middle feathers of the tail (in the " male) curled back; its bill straight; its collar white." Specific character of the Tame Duck, *Anas Domestica*: "It is variegated; " the middle feathers of the tail (in the male) curled back; its " bill straight."—The quantities of Ducks of various kinds that are caught in the fens of Lincolnshire are prodigious: above thirty thousand have been caught in one season in only ten decoys. The time for taking them is restricted by act of parliament to the space between the end of October and the beginning of February.

The MUSK DUCK.

LE CANARD MUSQUE'. *Buff.*

Anas Moſchata. Linn. Gmel. Ray, Briſſ. &c.
Groſſe Cane de Guinée. Belon.
Anas Indica. Geſner and Aldrovandus.
Anas Lybica. Johnſt. Charl. Will. &c.
Anas Cairina. Aldrov. Johnſt. Charl.
Anas Muſcovitica. Charleton.
The Muſcovy Duck, the Cairo Duck, the Guinea Duck, the Indian Duck. Will. Alb. and Lath. *

THIS Duck is ſo called, becauſe it exhales a pretty ſtrong odour of muſk †. It is much larger than our common Duck, and is even the biggeſt of all the Ducks known ‡: it is two feet long from the point of the bill to the end of the tail. All its plumage is of a brown black, gloſſed with green on the back, and interſected by a broad white ſpot on the coverts of the wing. But in the females, according to Aldrovandus,

* In German *Indianiſcher-entrach, Turkiſch-ente:* In Italian *Anatre d'India, Anatre di Libya.*

† Ray.—"The Indian Duck is peculiar to this country (Louiſiana); it has on both ſides of its head caruncles of a brighter red than thoſe of the turkey; the fleſh of the young ones is very delicate and well-taſted, but that of the old ones ſmells of muſk; they are as tame as thoſe of Europe." *Dupratz.*

‡ Ray.

 the

Nº 240

THE MUSCOVY DUCK.

the fore ſide of the neck is mixed with ſome white feathers. Willughby ſays, that he ſaw one entirely white; yet as Belon has remarked, the fact is, that ſometimes the male, as well as the female, is entirely white, or more or leſs variegated with white: and this change of the colours into white is pretty frequent in the domeſticated breeds. The character, however, that diſtinguiſhes the Muſk Duck is a broad piece of naked ſkin, red, and ſprinkled with *papillæ*, which covers the cheeks, extends behind the eyes, and ſwells on the root of the bill into a red caruncle, which Belon compares to a cherry. On the back of the head of the male hangs a bunch of feathers ſhaped like a creſt; this is wanting in the female *, which is alſo rather ſmaller, and has not the tubercle on the bill. Both have ſhort thighs and thick legs, the nails large, and that of the inner toe hooked: the upper mandible is marked on the edges with a deep indenting, and terminates in a ſharp curved nail.

This large Duck has a hollow voice, ſo low that it can ſcarce be heard, except when angry. Scaliger was miſtaken in aſſerting that it is mute. It walks ſlowly and heavily; yet in the wild ſtate it perches on trees †. Its fleſh is good, and even much eſteemed in America, where great numbers are raiſed; which has

* Aldrovandus. † Marcgrave.

given

given occasion to its appellation in France, the *Inaian Duck*. Yet we are uncertain from what country this bird was introduced among us, since it is not a native of the north *, and the name of *Muscovy Duck* is erroneous. We know only that they first appeared in France in the time of Belon, who termed them *Guinea Ducks*; and at that period, Aldrovandus says, they were brought from Cairo into Europe: and we may learn from Marcgrave, that the species occurs in its wild state in Brazil; for this large Duck is evidently the same with his *wild Duck of the bulk of a goose* †, and also the same with the *ypeca-guacu* of Piso. With respect to the *ipecati-apoa* of these two authors, we cannot doubt, from the bare inspection of the figures, that it is a different species, which Brisson ought not to have confounded with this.

According to Piso, this large Duck fattens equally well, whether confined to our farm-yards, or permitted to enjoy freedom on the rivers. It is also recommended by its great fertility; the female lays many eggs, and can hatch at almost every time of the year ‡; the male is very ardent

* Linnæus.

† "It is entirely black, except the beginning of the wings, "which is white; the black has however a green cast; on the head "is a crest consisting of black feathers, and above the origin of the "upper mandible is a wrinkled flesh bump. There is a red skin "also round the eyes." *Marcgrave.*

‡ Belon.

in

in his amours, and surpasses the rest of his kind by the size of his genital organ *. All females suit his appetite, nor does he despise those of inferior species. He pairs with the common Duck, and the progeny of this union are said to be unprolific, perhaps from prejudice †. We have also been told of the copulation of the Musk Drake with the goose ‡: but that intercourse is probably very rare, while the former is common in the French colonies of Cayenne and St. Domingo §; where these large Ducks live and propagate like the others in the state of domestication. Their eggs are quite round; those of the young females are greenish, but in the succeeding hatches they assume a paler colour ||. The odour of musk which these birds diffuse pro-

* Belon.

† *Idem.*

‡ " M. de Tilly, an inhabitant of the district of Nippes, a very " good observer, and of unimpeached credit, assures me, that he saw " at M. Girault's, who lives at *Acul-des-savanes*, birds which pro" ceeded from this copulation, and which partook of both species; " but he could not tell me whether these hybrids propagated upon " one another, or upon the geese or ducks." *Note sent from St. Domingo, by M. Lefebvre Deshayes.*

§ " At St. Domingo, there are Ducks whose plumage is en" tirely white, except the head, which is of a very fine red. The " Spaniards have carried thither Musk Ducks, which is the only " kind they rear, both on account of their bulk, and of the beauty " of their plumage: they have several layings in the year; and it " is remarked, that the ducklings bred between them and the fe" male Ducks of the island never propagate." *Oviedo and Charlevoix.*

|| Willughby.

ceeds,

ceeds, according to Barrere, from a yellowiſh liquor ſecreted by the glands of the rump.

In the wild ſtate, as they are found in the overflowed ſavannas of Guiana, they neſtle on the trunks of rotten trees; and after the young are hatched, the mother takes them one after another by the bill, and throws them into the water *. It appears that the alligators deſtroy many of them; for ſeldom do the families of ducklings contain five or ſix, though the eggs are much more numerous. They feed in the ſavannas upon the ſeeds of a ſort of graſs called *wild rice*; they fly in the morning to theſe immenſe overflowed meadows, and return in the evening to the ſea. They paſs the hotteſt hours of the day perched on branching trees. They are ſhy and miſtruſtful; can ſcarcely be approached, and are as difficult to ſhoot as moſt of the other water-fowl †.

* This fact has been confirmed to me by the ſavages, who have it in their power to verify ſuch obſervations. *M. de la Borde.*

† Extract from the journal of an expedition performed by M. de la Borde, into the interior parts of Guiana. *Journal de Phyſique, du mois de Juin* 1773.

[A] Specific character of the Muſk Duck, *Anas Moſchata:* "Its "face is naked and pimpled."

The WIGEON.

LE CANARD SIFFLEUR, ET LE VINGEON, ou GINGEON. *Buff.* *

Anas Penelope. Linn. Gmel. Gefn. and Aldrov.
Anas Fiftularis. Gefn. Aldrov. Johnft. Klein, and Briff.
Anas Clangofa. Barrere.
The Wigeon, Whewer, or *Whim.* Will. Alb. Penn. and Lath.

A CLEAR, whiftling voice, which may be compared to the fhrill notes of a fife, diftinguifhes this duck from all the reft, whofe voice is hoarfe and almoft croaking †. As it whiftles on wing, and very frequently, it is often heard and difcovered at a great diftance. It flies ufually in the evening, or even the night. It has a fprightlier air than the other ducks; it is very nimble, and perpetually in motion. It is fmaller than the common duck, and nearly equal to the fhoveler. The bill is very fhort, not larger than that of the golden-eye; it is blue, and its tip is black: the plumage on the top of the

* i. e. *The whiftling Duck, Vingeon* and *Gingeon* (both corrupted from the Englifh *Wigeon).* In German *Pfeiff-Ente,* or Fifing-Duck. —The *Penelops* of the Greeks feems to have been a kind of Duck; but we cannot decide whether it was a Wigeon or a pochard.

† Salerne and Dampier miftook this voice for the ruftling of their wings.

the head and neck is of a fine rufous; the crown of the head is whitish; the back is fringed and wreathed delicately with little blackish lines in zig-zags on a white ground; the first coverts form on the wing a large white spot, and the following a little spangle of bronze-green; the under surface of the body is white, but both sides of the breast and the shoulders are of a fine purple rufous: according to M. Baillon, the females are somewhat smaller than the males, and continue always gray *, and do not, like the females of the shovelers, assume, as they grow old, the colours of the males. This observer, equally accurate and attentive, and at the same time very judicious, has communicated to us more facts relating to the water-fowls than are to be found in all the professed naturalists: he has discovered, from a series of observations, that the Wigeon, the pintail, the gadwall, and the shoveler, are hatched gray, and retain that colour till the month of February; so that, at first, the males cannot be distinguished from the females, but in the beginning of March their feathers colour, and nature bestows on them the powers and ornaments suited to the season of love; she afterwards disrobes them of their apparel about the end of July: the males retain little or nothing of their handsome colours;

* "The female is clouded with cinereous, except the breast "and the belly, which are white; it has no spot on the wings." *Fauna Suecica.*

gray

No. 241

THE COMMON WIGEON, THE MALE.

Nº 242

THE COMMON WIGEON, THE FEMALE.

gray and dark feathers ſucceed to thoſe with which they were decorated; their voice dies away and is loſt like that of the females, and half the year all ſeem condemned to ſilence and inſenſibility.

It is in this diſmal ſtate that theſe birds commence, in the month of November, their diſtant voyage, and many are caught in this firſt paſſage. It is then ſcarce poſſible to diſtinguiſh the old from the young, eſpecially thoſe of the pintails; the gray garb being more complete in that ſpecies than in others.

When all theſe birds return into the north, about the end of February or the beginning of March, they are decorated with their fineſt colours, and are inceſſantly heard to whiſtle or ſcream. The adults now pair, and none remain in our marſhes but a few ſhovelers, which can be obſerved to lay and hatch.

The Wigeons fly and ſwim always in bodies * Every winter a few companies paſs in moſt of our provinces, even thoſe the moſt diſtant from the ſea, ſuch as Lorraine † and Brie ‡; but they

* Schwenckfeld and Klein.

† Obſervations of M. Lottinger.

‡ "Though I never killed, nor even knew this ſort of duck in "Brie, I am aſſured that it appears there at two paſſages: having "ſeen it very near on the pool in the orangery of the Palais-Royal "at Paris, I recollected to have ſeen on our lakes, though at a diſ-"tance, ducks with red heads and white faces, which were un-"doubtedly the ſame." *Obſervation of M. Hebert.*

 paſs

paſs in much greater numbers on our coaſts, particularly thoſe of Picardy.

“ The north and north-eaſt winds,” ſays M. Baillon, “ bring to us Wigeons in great flocks. “ They ſpread on our marſhes, where one part “ of them ſpend the winter, another advances “ farther ſouth.

“ Theſe birds fly very well during the night, “ unleſs it is quite dark. They ſeek the ſame “ paſture as the wild ducks, and like theſe feed “ on the ſeeds of ruſhes and other herbs, inſects, “ ſnails, frogs, and worms. The more violent “ the wind, the greater the number of theſe “ ducks that are ſeen roving. They keep at a “ good diſtance from the ſea and the mouths of “ rivers, notwithſtanding the rigour of the wea- “ ther, and they are very patient of cold.

“ They retire regularly about the end of “ March with the ſouth winds: none remain “ here: I think they advance to the north, hav- “ ing never ſeen their eggs or neſts. I may “ obſerve, however, that theſe birds are hatched “ gray, and that prior to the moulting there is “ no difference, with reſpect to plumage, between “ the males and the females: for often on their “ firſt arrival I found young ones almoſt gray, “ and only half covered with the feathers cha- “ racteriſtic of their ſex.

“ The Wigeon,” adds M. Baillon, “ is eaſily “ reconciled to domeſtication; it eats readily “ bread, and barley, and fattens when ſo fed; “ it "it requires much water, in which it inceſſantly "frolics by night as by day. I have had them "ſeveral times in my yard, and was always de- "lighted with their ſprightlineſs."

The ſpecies of the Wigeon or whiſtling duck occurs in America as well as in Europe. We have received ſeveral ſpecimens from Louiſiana under the name of *jenſen duck* and *gray duck* *. They ſeem to be the ſame with the *vingeons* or *gingeons* in the French ſettlements at St. Domingo and Cayenne. They are found in all the intermediate latitudes †: and they have the ſame natural habits ‡, unleſs in ſo far as they are affected by

* I have received from Louiſiana a duck which the French ſettled in that country call the *gray duck*; it correſponds to the European duck which M. Briſſon denominates *the whiſtler duck.* Between the gray duck of Louiſiana and the whiſtler duck of Europe, there are ſome ſlight differences; yet not ſufficient to diſcriminate their ſpecies: the gray duck is rather larger; it has along the neck on each ſide a greeniſh ſtripe wanting in the whiſtling duck of Europe: the plumage is the ſame in both, except a few ſtrokes or ſhades which may vary in different individuals: but the form of the bill, its colour, the colour of the legs, the ſhape of the tail, which is pointed, the whole habit of body, and much the greateſt part of the plumage, are ſimilar in the gray duck of Louiſiana, and in the whiſtling duck of Europe. I believe, therefore, that I may very ſafely refer them to the ſame ſpecies. *Extract of the notes communicated by Dr. Mauduit.*

† The whiſtler ducks are not quite ſo large as our common ducks; but they differ not from theſe in their colour or their figure: when they fly they make a ſort of whiſtling with their wings, which is tolerably pleaſant; they perch on trees. *Dampier.*

‡ We muſt except that which Father Dutertre aſcribes to the Wigeon of the Antilles, viz. that they leave the rivers and pools at night, and come to dig up the yams in the gardens.

 climate;

climate; yet we dare not pronounce whether the whiſtling duck and the *vingeon* be the ſame ſpecies. Our doubts with reſpect to this and other ſubjects would have been cleared up, had not the war, among other loſſes which it has occaſioned to natural hiſtory, deprived us of a ſeries of coloured drawings of St. Domingo birds, made on that iſland with the utmoſt care by the Chevalier Deſhayes, correſpondent of the king's cabinet. Fortunately a duplicate of the papers of that obſerver, as ingenious as he is laborious, have come into my hands: and we cannot do better than give an extract, but without venturing to decide whether this bird is preciſely the whiſtling duck.

" The *gingeon*, which at Martinico is termed " the *vingeon*," ſays the Chevalier Deſhayes, " is " a particular kind of duck, which is not diſ- " poſed to make diſtant voyages like the wild " duck, but uſually limits its excurſions to the " paſſing from one pool or marſh to another, or " to make depredations of ſome field of rice " near their haunts. It ſometimes perches on " trees; but, as far as I could obſerve, this hap- " pened only in the rainy ſeaſon, when its ordi- " nary retreat during the day was ſo deluged, " that no aquatic plant appeared to conceal or " ſhelter it, or when the extreme heat obliged " it to ſeek the cool ſhade amidſt the thick fo- " liage.

" One

" One might be tempted to take the *vingeon* " for a nocturnal bird, for it is ſeldom ſeen in " the day; but as ſoon as the ſun is ſet, it riſes " from among the flags and reeds, and makes " for the open ſides of the pools, where it " dabbles and paſtures like other ducks. It " would be difficult to ſay how it is employed " through the day: we can hardly obſerve it " without being perceived. But we may pre- " ſume that, though it lurks among the reeds, " it does not paſs its time in ſlumber. We " may draw this inference from tame *vingeons*, " which, like other fowls, ſeek not to ſleep in " the day-time, till after they are ſated.

" The *gingeons* fly in flocks like the ducks, " even in the love ſeaſon. This inſtinct, which " prompts them to aſſociate, ſeems to be pro- " duced by fear; and it is ſaid, that like the " geeſe, they always plant a ſentinel, when en- " gaged in ſearch of food. If the guard per- " ceives any motion, he gives notice by a parti- " cular cry, reſembling a cadence or rather a " hoarſe bleating; inſtantly the *gingeons* deſiſt " from their gobbling, raiſe their heads, and " look with a ſteady, earneſt aſpect: if the noiſe " ceaſes, they reſume their feeding; but if the " ſignal is redoubled, and announces real danger, " the alarm is communicated by a ſhrill, pierc- " ing cry, they all mount and follow the ſenti- " nel, who firſt takes flight.

"The *gingeon* is a noiſy bird: when a flock "is feeding, a continual murmuring is heard, "like a low ſmothered laugh. This gabbling "betrays them, and directs the fowler. Even "when they fly, there is always ſome one of the "body which whiſtles; and as ſoon as they have "alighted on the water, their chuckling is re-"newed.

"They lay in January; and in March the "young are ſeen. Their neſts are nothing re-"markable, except that they contain many "eggs. The negroes are very expert at finding "theſe neſts, and the eggs hatch well if placed "under ſitting hens. In this way tame *gin-"geons* are obtained; but it would be a world "of difficulty to domeſticate ſuch as are taken a "few days after their birth: for already they "have contracted the wild, ſhy temper of their "parents; while thoſe hatched under hens re-"ceive a part of the ſocial familiar diſpoſition. "The young *gingeons* have more agility and "vivacity than ducklings: at firſt they are "covered with a brown down; they grow very "faſt, and in ſix weeks they attain their full "ſize, and the feathers of the wings begin to "ſprout *

"Thus,

* "One could not believe to what lengths the wild Wigeons "carry the paternal affection. M. le Gardeur, lately member "of the Chamber of Agriculture at St. Domingo, and who joins "to a very accompliſhed mind much knowledge in Natural Hiſ-"tory,

" Thus, with very little pains, we may procure tame *gingeons*; but, if we may judge from almoſt all that have, we can ſcarcely expect that they will multiply in the domeſtic ſtate; yet I know ſome tame *gingeons* which have laid, covered, and hatched.

" It would be an extremely valuable acquiſition to obtain a domeſtic breed of theſe birds; becauſe their fleſh is excellent, and eſpecially that of ſuch as have been tamed, not having the marſh taſte of the wild ones. And another reaſon for reducing this ſpecies to domeſtication, would be the advantage in extinguiſhing, or at leaſt of weakening, thoſe in the wild ſtate; for they often deſolate our crops, and ſeldom do the fields of rice, near pools, eſcape their ravages. In ſuch ſituations, the ſportſman waits for them in the evening by moon-light: they are alſo caught with nooſes, and hooks baited with earth-worms.

" The *gingeons* feed not only upon rice, but on all other grain uſually given to fowls; ſuch as maize and different kinds of millet. They alſo crop graſs, and catch ſmall fiſh and crabs.

" Their cry is a real whiſtle, which may be

" tory, aſſured me, that he ſaw them dart with the utmoſt rancour, pecking a negro who ſought to plunder their brood; they annoyed him ſo much as to retard the taking of the young, which in the mean time eſcaped and concealed themſelves as much as was poſſible." *Sequel of the Memoir of the Chevalier Deſhayes.*

 " imitated

"imitated ſo exactly with the mouth, as to de-"coy the flocks when they paſs. The ſportſ-"men fail not to counterfeit this whiſtle, which "runs rapidly over all the notes of the octave, "from the baſe to the treble, reſting on the laſt "note, which is prolonged.

"The *gingeon* carries its tail low, and bent "to the ground, like the pintado; but on enter-"ing the water, it raiſes its tail. Its back is "higher and more arched than that of the "duck: its legs are much longer in proportion: "its eye is livelier, and its tread firmer: it has a "better carriage, and holds its head high like "the gooſe. Theſe characters, together with "its habit of perching upon trees *, ſufficiently "diſtinguiſh it. This bird with us has not near "ſo thick a plumage as the ducks in cold coun-"tries.

"The *gingeons*," M. Deſhayes continues, "far from copulating with the muſk or com-"mon ducks, as theſe have done with each "other, ſeem, on the contrary, to be the de-"clared enemies of all poultry, and league toge-"ther to attack the ducks and geeſe. They "always ſucceed in routing theſe, and in obtain-"ing the object of the quarrel, that is, the

* To this ſpecies we ought probably to refer the *branch duck*, which occurs in many narratives. "There are no leſs than twenty-"two kinds of ducks in Canada, of which the moſt beautiful and "the beſt are called *branch ducks*, becauſe they perch on branches of "trees; their plumage is variegated with much brilliancy." *Hiſt. Gen. des Voy. tom.* xv. *p.* 227.

"grain

" grain which is thrown to them, or the pool " in which they dabble. It muſt be owned, that " the diſpoſition of the *gingeon* is miſchievous and " quarrelſome; but as its force equals not its " ſtrength, we cannot but wiſh, though it " ſhould diſturb the peace of the court-yard, to " propagate in the domeſtic ſtate this ſpecies of " duck, ſo ſuperior in quality to all the reſt."

[A] Specific character of the Wigeon, *Anas Penelope*: " Its tail " is ſomewhat ſharp; its vent black; its head brown; its front white; " its back waved with cinereous."

The CRESTED WHISTLER.

Anas Rufina. Gmel.
Anas Fiſtularis Criſtata. Briſſ.
Anas Capite Rufo Major. Ray.
Anas Criſtata Flaveſcens. Marſigli and Klein.
Anas Erythrocephalos. Rzacynſki.
The Great Red-headed Duck. Will.
The Red Creſted Duck. Lath. *

THIS whiſtling duck has a creſt, and is as large as the wild duck; all its head is clothed with fine rufous feathers, delicate and ſilky, raiſed on the front and the crown of the head in a hairy tuft, reſembling the

* In Italian *Capo Roſſo Maggiore*, or, *Greater rufous-headed*: in German *Brandt-ende (fire duck)*, *Rott-köpf (red-head)*, *Rott-hals* (red-threat).

frizzled

frizzled tete lately worn by our ladies: the cheeks, the throat, and the compaſs of the neck, are rufous like the head; the reſt of the neck, the breaſt, and the under ſide of the body, are black or blackiſh, which on the belly is lightly waved or clouded with gray; ſome white appears on the flanks and the ſhoulders, and the back is brown gray; the bill and the iris are of a vermilion colour.

This ſpecies, though leſs common than the preceding, has been ſeen in our climates by ſeveral obſervers.

[A] Specific character of the *Anas Rufina*: "It is black; its "head, and the upper part of its neck, brick-coloured; its top ruſty "and creſted (in the male); its wings white below, and at the mar-"gin; its tail duſky."

The WHISTLER WITH RED BILL AND YELLOW NOSTRILS.

Anas Autumnalis. Linn. and Gmel.
Anas Fiſtularis Americana. Briſſ.
Anas Fera mento cinnabarino. Marſigli and Klein.
The Red-billed Whiſtling Duck. Edw. and Lath.

IT is probable that this ſpecies, as well as the preceding ones, has received the name of *Whiſtler* from the whiſtling of its voice or of its

 wings.

wings. To the appellation given by Edwards of *red-billed,* we add the circumſtance that it has *yellow noſtrils,* to diſtinguiſh it from the foregoing ſpecies, whoſe bill is alſo red. This Whiſtler is tall, but not larger than a coot. Though it has not vivid or brilliant colours, it is a very beautiful bird of its kind: a cheſnut brown ſpread on the back is clouded with flame-colour or deep orange; the lower part of the neck has the ſame tint, which melts into gray on the breaſt; the coverts of the wings are waſhed with ruſty on the ſhoulders, next aſſume an aſh hue, then a pure white; its quills are blackiſh brown, and the primaries are marked on the middle of their outer ſurface with white; the belly and tail are black; the head is covered with a ruſty cap, which ſtretches with a long blackiſh track to the top of the neck; all the circumference of the face and neck is clothed with gray feathers.

This ſpecies is found in North America, according to Briſſon; yet we received it from Cayenne.

[A] Specific character of the *Anas Autumnalis*: "It is gray; its "wing-quills, its tail, and its belly, are black; there is a fulvous and "white ſpangle on the wings."

The BLACK-BILLED WHISTLER.

Anas Arborea. Linn. and Gmel.
Anas Fistularis Jamaicensis. Briss.
Anas Fistularis Arboribus insidens. Sloane.

WE adopt the name given by Edwards, as more precise than any indication drawn from climate. The legs and neck appear proportionally longer than in the other ducks: its bill is black or blackish; its plumage is brown, clouded with rusty waves; its neck is speckled with little white streaks; the front, and the sides of the head behind the eyes, are tinged with rufous; and the black feathers on the top of the head recline like a crest.

According to Sir Hans Sloane, this duck, which is seen frequently in Jamaica, perches and makes a sort of whistling. Barrere says, that it is a bird of passage in Guiana; that it feeds in the savannas, and is excellent meat.

[B] Specific character of the *Anas Arborea:* "It is brown; its "head somewhat crested; its belly spotted with white and black."

No. 243

THE GAD-WALL DUCK, THE FEMALE.

The GADWALL.

LE CHIPEAU, *ou* LE RIDENNE. *Buff.*

Anas Strepera. Linn. Gmel. Gesn. and Klein.
Anas Platyrinchos rostro nigro & plano. Aldrov. Johnst. and Ray.
The Gadwall, or *Gray.* Will. Penn. and Lath. *

THIS is not so large as the wild duck; its head is finely speckled or dotted with dark brown and white, and the blackish tint predominates on the top of the head and the upper side of the neck; the breast is richly festooned or scaled, and the back and the flanks are all vermiculated with these two colours; on the wing there are three spots or bars, the one white, the other black, and the third of a fine reddish chesnut. M. Baillon has observed, that of all the ducks, the Gadwall preserves the longest the fine colours of its plumage, but at last, like the others, it assumes a gray garb after the love season. The cry of this duck resembles much that of the wild duck; nor is it more raucous or louder, though Gesner seems to have meant to characterize it by applying the epithet *strepera*; which has been adopted by ornithologists.

* In German *Schnarr-endte, Schnatter-endte,* or, *snarling* or *chattering duck*; sometimes *Leiner.*

The

The Gadwall is as alert in diving as in ſwimming, and it eſcapes a ſhot by plunging under water; it ſeems timorous, and flies little during the day; it lurks ſquatted among the ruſhes, and ſeeks not its food except early in the morning or in the evening, and even a good while after night has come on. They are then heard flying in company with the whiſtlers, and, like theſe, are caught by the decoy of tame ducks. "The "Gadwalls," ſays M. Baillon, "arrive on our coaſts "of Picardy in the month of November, with "the north-eaſt winds; and when theſe winds "blow ſome days, they paſs on without halting. "About the end of February, with the firſt "ſouth winds, they are ſeen repaſſing on their "return to the north.

"The male is always larger and more beau-"tiful than the female; like the male pochards "and whiſtlers, it has the under ſide of the tail "black, which part of the plumage is in the "females conſtantly gray.

"The females bear great reſemblance in all "theſe ſpecies; yet ſome practice will enable "us to diſtinguiſh them. The female Gadwalls "become of an intenſe rufous as they grow "old.

"The bill of this bird is black; its legs are "of a pale clay-yellow, with black membranes, "and the under ſide of each joint of the toes is "alſo black. The male meaſures twenty inches "from the bill to the tail, and nineteen inches

"to

" to the extremity of the nails; its alar extent " is thirty inches. The female differs only fif- " teen lines in all the dimenſions.

" I fed in my court ſeveral months," continues Baillon, " two Gadwalls, male and female: " they would eat no grain, but ſubſiſted on " bran and ſoaked bread. I had alſo wild ducks " which refuſed grain, and others which lived " on barley from the firſt days of their confine- " ment. This difference, I imagine, is owing " to the nature of the places where theſe birds " were bred: thoſe which come from the de- " ſert marſhes of the north muſt be unac- " quainted with barley and wheat, and therefore " it is not ſurpriſing that they ſhould reject ſuch " food: thoſe, on the contrary, which were " hatched in cultivated countries, are led in the " night into the corn-fields by their parents; " they are thus accuſtomed to live on grain, " and readily recognize it in the farm-yard; " while the others will often die of want, though " the reſt of the poultry, picking up the ſeeds " before them, might inſtruct them in the uſe " of this food."

[A] Specific character of the Gadwall, *Anas Strepera*: " It has " on its wing a ſpeckle of rufous, of black, and of white."

The SHOVELER.

Le Souchet, *ou* Le Rouge. *Buff.*

Anas Clypeata. Linn. Gmel. and Briſſ.
Anas Latiroſtra Major. Geſner and Aldrov.
Anas Latiroſtra. Schwenckf. and Klein.
Anas Platyrynchos Altera (male). Ray and Will.
Anas Platyrynchos (female). Ray and Will.
Anas Vireſcens. Marſig.
Phaſianus Marinus. Charleton.
The Blue-winged Shoveler (fem.) Cateſby.

This duck is remarkable for its ſhort bill *, round and ſpread at the end, like a ſpoon; whence are derived its various names. It is rather ſmaller than the wild duck; its plumage is rich in colours, and ſeems to merit the epithet *very beautiful*, which Ray beſtows on it. The head and the upper half of the neck are of a fine green; the coverts of the wing, near the ſhoulder, are of a pale blue, the following are white, and the laſt form on the wing a bronze green ſpangle: the ſame colours mark, though more faintly, the wing of the female, which has beſides only the dull colours of a

* In German *Breit-ſchnabel (broad-bill), Schall-endtle (ſhell-duck)*: in Daniſh *Krop-and*: in Norwegian *Stock-and*: in Greenland it is called *Kertlutock*; which ſignifies *broad-bill.*

white

white gray, and rufty, mailed and feftooned with blackifh; the breaft and the under fide of the neck of the male are white, and all the under furface of the body is of a fine rufous; yet fometimes the belly is white. M. Baillon affures us, that the old Shovelers retain fometimes their beautiful colours, and that tinged feathers grow at the fame time with the gray, which cover them every year after the love feafon: and he obferves juftly, that this fingularity of the Shovelers and gadwalls may miflead nomenclators with refpect to the number of the fpecies of thefe birds. He fays alfo, that aged females, which he faw, had, like the males, colours on their wings, but that, during their firft year, they were entirely gray. Their head retains always its colour. We fhall here alfo give the excellent remarks which he has obligingly communicated on the Shoveler in particular.

"The form of the bill of this beautiful bird," fays M. Baillon, "denotes its manner of living; "its two broad mandibles have edges fur- "nifhed with a fort of indenting or fringe, that "allows only the dirt to efcape, but holds the "worms, the flender infects, and the crufta- "ceous animals, which it fearches among the "mud by the margin of water: it has no other "food *. I have feveral times opened them at

* We muft add flies, which it catches alertly as it flutters on the water; whence the name *Anas Mufcaria*, which Gefner has given to it.

" the end of winter and during froſt; I found " no herbage in their ſtomach, though the want " of inſects muſt have obliged them to recur to " that ſpecies of food. They are found then " near ſprings only: they grow very lean: " they recruit again in the ſpring, by eating " frogs.

" The Shoveler dabbles inceſſantly, chiefly " in the morning and evening, and even very " late at night: I think that it ſees in the " duſk. It is ſavage and gloomy: it can ſcarce " be reconciled to domeſtication: it conſtantly " rejects bread and grain. I had a great num-" ber, which died after having been long fed by " cramming into the bill, without ever learning " to eat by themſelves. I have at preſent two " in my garden, which I have fed in that way " more than a fortnight. They are now living " on bread and ſhrimps: they ſleep almoſt the " whole day, and lie ſquat by the box-borders: " in the evening, they run about a great deal, " and they bathe repeatedly in the night. It is a " pity that ſo beautiful a bird has not the cheer-" fulneſs of the garganey or ſheldrake, and " cannot become an inhabitant of our court-" yards.

" The Shovelers arrive in our diſtricts about " the month of February. They diſperſe in the " marſhes, and a part of them hatch there every " year. I preſume that the reſt advance towards " the ſouth, becauſe theſe birds become rare here

" after

" after the firſt northerly winds that blow in " March. Thoſe which are bred in the coun- " try, depart about the month of September: " it is very uncommon that any are ſeen in the " winter, and I thence conjecture that they " avoid the approach of cold *.

" They neſtle here in the ſame places with " the ſummer teals; they chooſe, like theſe, " large tufts of ruſhes in ſpots almoſt inacceſſi- " ble, and they arrange their neſt after the ſame " faſhion. The female lays ten or twelve eggs, " of a ſomewhat pale rufous: ſhe covers them " twenty-eight or thirty days, as ſportſmen " have told me; but I am myſelf inclined to " think, that the incubation laſts only twenty- " four or twenty-five days, ſince theſe birds " hold a middle rank between the ducks and " the garganeys, with reſpect to ſize.

" The young are hatched with a gray ſpotted " down, like the ducklings, and are extremely " ugly. Their bill is then almoſt as broad as " their body, whoſe weight ſeems to oppreſs " them: they almoſt conſtantly reſt on their " breaſt. They run and ſwim as ſoon as they " burſt from the ſhell. Their parents lead them, " and appear attached to them; they inceſſantly " guard againſt the ravenous birds: on the leaſt " apprehenſion of danger, the family ſquat " among the graſs, and the parents throw them-

* However, they are ſeen in Scania and Gothland, according to Linnæus.

" selves into the water, and plunge over head.

" The young Shovelers become first gray like " the females: the first moulting gives them their " fine feathers, but they turn bright not until " the second."

With respect to the colour of the bill, observers are not agreed. Ray says, that it is quite black; Gesner, as cited by Aldrovandus, asserts, that the upper mandible is yellow; Aldrovandus makes it to be brown: all that we can infer is, that the colour of the bill varies from age or other circumstances.

Schwenckfeld compares the clapping of the Shoveler's wings to the clattering of castanets; and M. Hebert told us, that he could not better compare its cry, than to the creaking of a hand-rattle, turned round with little shakes. It is likely that Schwenckfeld mistook its voice for the noise of its flight. The Shoveler is the best and most delicate of the ducks; it grows very fat in winter; its flesh is tender and juicy; this is said to be always red, though well dressed, and that the bird has hence received the name of *rouge*, particularly in Picardy, where many are killed in the long chain of marshes that extend from the vicinity of Soissons to the sea.

Brisson, following the other ornithologists, gives a variety of the Shoveler; but the only difference is, that its belly, instead of being chestnut rufous, is white.

The

The *yacapatlahoac* * of Fernandez, a duck which that naturalist characterises by its remarkable broad bill, and by the three contrasted colours of its wings, appears to be a Shoveler: and we shall class with it the *tempatlahoac* of the same author, which Brisson makes his *Mexican wild duck* †: for Nieremberg terms it *avis latirostra*, or *broad-bill*; and Fernandez takes care to remark, that many persons call the *yacapatlahoac* by the same name, *tempatlahoac*. Our opinion is corroborated by the observations of Dr. Mauduit, which leave no doubt that the Shoveler is found in America. "The individuals of this "species," says he, "are liable in Europe to "variations of plumage, and some have a mix-"ture of gray feathers, which occur not in the "others. I remarked, in seven or eight Sho-"velers sent from Louisiana, the same diversity "that might be found in an equal number of "birds killed at random in Europe; which

* *Anas Mexicana.* Gmel.
Anas Clypeata Mexicana. Briss.
The Mexican Shoveler. Lath.

—— "It is a kind of wild duck, having its bill long and broad, "especially at the extremity ... its wings partly white, partly glossy, "and brown green ... The Spaniards call it the royal duck; and "some also give it the name of *tempatlahoac*." *Fernandez.*

† *Anas Clypeata*, 3 var. Linn. and Gmel.
Boschas Mexicana. Briss.
The Broad-billed Bird. Will.

—— "The broad-billed bird ... a kind of wild duck ... its "wings first sky-blue, then bright white, and afterwards shining with "a green lustre, and their tips on either side fulvous." *Fernandez.*

" proves that the Shoveler of Europe and that " of America are abſolutely the ſame ſpecies *."

* *Note communicated by M. Mauduit.*

[A] Specific character of the Shoveler, *Anas Clypeata:* " The " end of its bill is dilated and rounded; its nail curved inwards."

The PINTAIL.

LE PILET, *ou* CANARD A LONGUE QUEUE.

*Buff**.

Anas Acuta. Linn. and Gmel.
Anas Longicauda. Briſſ.
Anas Caudacuta. Geſn. Aldr. Johnſt. Will. Klein, &c.
Tzitzihoa. Fernandez.
The Sea-Pheaſant, or *Cracker.* Will. and Alb. †

THIS is excellent game, and a very beautiful bird. Though it has not the reſplendent colours of the ſhoveler, its plumage is very handſome, of a light gray, waved with little black ſtreaks, which might be ſaid to be traced with a pencil: the great coverts of the wings are marked with broad ſtripes of jet-black and ſnowy-

* i. e. *The Long-tailed Duck,*

† At Rome this duck is called *Coda Lancea,* or *lance-tailed:* in German it has the names of *Faiſan-ente, Meer-ente, See-vogel (pheaſant duck, ſea-duck, ſea-bird),* and in ſome places *Spitz-ſchwantz (pointed-tail)*: in Swediſh *Ala, Aler, Ahl-fogel.*

No. 244

THE PINTAIL WIGEON.

ſnowy-white; on the ſides of the neck are two white bars like ribbands, which readily diſtinguiſh it, though at a diſtance. The proportions of its body are longer and more taper than in any other ſpecies of duck; the neck is remarkably long, and very ſlender; the head is ſmall and cheſnut-colour; the tail is black and white, and terminates in two narrow filaments, which might be compared to thoſe of the ſwallow; it is not carried horizontally but half-cocked. Its fleſh is in every reſpect preferable to that of the wild duck; it is not ſo black, and the thigh, which in the wild duck is commonly hard and tendinous, is as tender as the wing in the Pintail.

" The Pintail," M. Hebert tells us, " is ſeen " in Brie during both paſſages: it lives on the " large pools: its cry is heard pretty far off, *hi* " *zouë zouë*; the firſt ſyllable is a ſharp whiſtle, " the ſecond a murmur, deeper, and leſs ſono- " rous.

" The Pintail," adds this excellent obſerver, " ſeems to form the ſhade between the ducks " and the garganeys, and, in many reſpects, it " approaches the latter: the diſtribution of its " colours reſembles more that in the garganey, " and it has alſo the bill of that bird."

The female differs from the male as much as the wild duck differs from the drake. Like the male, it has its tail long and pointed, and might otherwiſe be confounded with the wild duck; but the length of its tail is ſufficient alone to distinguish it from all the other ducks. The two filaments which project from the tail, have given occasion to the German name, *pheasant-duck*, and the English, *sea-pheasant*, which are very improperly applied. The appellation of *winter-duck*, which it receives in the north, seems to prove that it bears the most intense cold; and, in fact, Linnæus assures us that it is seen in Sweden in the depth of winter *. The species seems to be common to both continents: for it is evidently the Mexican *tzitziboa* of Fernandez; and Dr. Mauduit received one from Louisiana, under the name of *pintailed duck (canard paille-en-queue)*. Thus, though a native of the north, it advances into the hot climates.

* *Fauna Suecica.*

* [A] Specific character of the Pintail, *Anas Acuta*: "Its tail is "sharpened and elongated, black below; there is a white line on "either side behind the head; its back is cinereous and waved." Great flocks of these ducks visit the Orknies in winter; also the west of Ireland in the month of February, and are there reckoned delicate food.

The LONG-TAILED DUCK from NEWFOUNDLAND.

Anas Glacialis. Linn. and Gmel.
Anas Longicauda ex insula N. Terræ. Briss.
The Swallow-tailed Sheldrake. Ray and Will.

THIS Duck is very different from the preceding in its plumage, and has no resemblance to it, except in the long shafts that project from its tail.

The coloured figure of Edwards represents those parts brown, which in the duck called *Miclon* are black in our *Planches Enluminées*; yet we may perceive that both these birds are the same, by the two long shafts which project from the tail, and by the fine disposition of the colours: white covers the head and the neck as far as the top of the breast and back; there is only a band of orange fulvous, which descends from the eyes on both sides of the neck; the belly, and also two bunches of long, narrow feathers, lying between the back and the wing, are of the same white with the head and the neck; the rest of the plumage is black as well as the bill; the legs are of a blackish red, and a small edging of membrane may be observed running along

along the margin of the inner toe, and below the little hind-toe: the length of the two ſhafts of the tail increaſes the total bulk of this duck; yet it is ſcarce equal to a common duck.

Mr. Edwards ſuſpects, with every probability, that his *long-tailed duck from Hudſon's Bay* is the female of this. The ſize, the figure, and even the plumage, are nearly the ſame; only the back of the latter is leſs variegated with white and black, and the plumage is on the whole browner.

This ſubject, which appears to be a female, was caught at Hudſon's Bay, and the other was killed in Newfoundland; and as the ſame ſpecies is recognized in the *havelda* of the Icelanders and of Wormius, we may conclude that, like many others of the genus, it is an inhabitant of the remoteſt countries of the north. It occurs alſo in the north-eaſt of Aſia; for it is the *ſawki* of the Kamtſchadales, which they alſo name *kiangitch* or *aangitch*, that is, *deacon* *, becauſe they find that this duck ſings like a Ruſſian deacon.—So it ſeems that a Ruſſian deacon ſings like a duck!

* Hiſt. Gen. des Voy. *tom.* xix. *pp.* 273 & 355.

[A] Specific character of the Long-tailed Duck, *Anas Glacialis*: "Its tail is ſharpened and elongated; its body black, and below "white." It breeds in the remoteſt parts of the north, and viſits our ſhores only in the ſevereſt winters.

Nº 245

THE SHELDRAKE.

The SHELDRAKE.

Le Tadorne. *Buff.*

Anas Tadorna. Linn. and Gmel.
Vulpanſer. Geſner, Aldrov. and Klein.
Anas Maritima. Geſner, and Aldrov.
Tadorna. Johnſt. Sibbald, Ray, and Briſſ.
The Sheldrake, or *Burrough-duck.* Will. Alb. &c. *

WE are convinced that the *fox-gooſe* of the ancients (*χηναλωπηξ*, or *vulpanſer*) is the ſame with the Sheldrake. Belon has heſitated and even varied about the application of theſe names: in *his obſervations*, he refers them to the gooſander, and in his book *of the nature of birds*, he appropriates them to the barnacle. But we may eaſily aſcertain, from one of thoſe natural properties which are more deciſive than all the conjectures of erudition, that theſe names apply ſolely to the Sheldrake; for it is the only bird which reſembles the fox in a ſingular circumſtance, that of lodging in a hole: it uſually invades and poſſeſſes itſelf of the rabbits' burrows, and there it lays and breeds.

* In Greek Χηναλωπηξ, from *χην*, *a gooſe,* and *αλωπηξ*, *a fox:* in Latin *Vulpanſer,* which is only a tranſlation of the preceding; and alſo *Anas Strepera:* in German *Berg-enten (mountain-duck), Fuchs-gans (fox-gooſe):* in Swediſh *Ju-goas.*

Ælian

Ælian ascribes also to the *vulpanser* the instinct of presenting itself, like the partridge, before the feet of the sportsman, to avert the danger from its young. This was the general opinion of the ancients; since the Egyptians, who ranked this bird among the sacred animals, figured it, in their hieroglyphics, as the emblem of the generous tenderness of a mother *. In fact, it will be seen from our observations, that the Sheldrake exhibits precisely the same marks of maternal affection.

The appellations bestowed on this bird in the north, that of *fox-goose*, or rather *fox-duck* in Germany, that of *mountain-duck* in Saxony, and that of *burrow-duck* in England, mark, equally with the ancient names, its singular habit of living in burrows the whole time of its incubation. These appellations are even more accurate; since the Sheldrake belongs to the genus of ducks, not to that of geese. It is rather larger than the common duck, and its legs are somewhat taller; but in other respects, in its figure, its port, and its structure, it preserves the resemblance. It differs from the duck, only because its bill is more raised, and the colours of its plumage more vivid and beautiful, and appear more brilliant at a distance. Its fine plumage is broken into large spaces of three colours, white, black, and cinnamon-yellow:

* *Vid.* Pieri, in Orum, *lib.* xx.

the

the head, and as far as the middle of the neck, are black, gloſſed with green; the lower part of the neck is encircled by a white collar, and below is a broad zone of cinnamon-yellow, which covers the breaſt, and forms a little band on the back; this ſame colour tinges the lower belly; below the wing, on each ſide of the back, a black bar extends on a white ground; the great and middle quills of the wing are black, the ſmall ones have the ſame ground colour, but are gloſſed with ſhining green; the three quills next the body have their outer edge of cinnamon-yellow, and their inner of white; the great coverts are black, and the ſmall ones white. The female is ſenſibly ſmaller than the male, which it reſembles even in the colours; only the greeniſh reflections of the head and wings are leſs apparent than in the male.

The down of theſe birds is very fine and ſoft *: the feet and their membranes are fleſh-coloured; the bill is red, but its tip, and the noſtrils, are black; the upper mandible is much arched near the head, depreſſed into a concavity on the noſtrils, and raiſed horizontally at the end into a round ſpoon, edged with a pretty deep and ſemi-circular groove. The *trachea* has a double ſwelling at its partition †.

* "The feathers are very ſoft, as in the eider." *Linnæus.*

† Willughby.

Pliny

Pliny commends the flesh of the Sheldrake, and says, that the ancient Britons knew no better game*. Athenæus ranks its eggs next to those of the peacock, as being the second in point of goodness. It is highly probable that the Greeks raised Sheldrakes, for Aristotle remarks that some of their eggs are addle †. We had never an opportunity of tasting either their flesh or their eggs.

It appears that the Sheldrakes inhabit the cold as well as the temperate climates, and that they have penetrated into the regions of the Pacific Ocean ‡: yet the species is not equally dispersed through all the coasts of our northern countries §.

Though the Sheldrakes have been called sea ducks ||, and in fact do prefer the sea shores, some are found on the rivers ¶ or lakes considerably inland; but the bulk of the species never leaves our coasts **. Every spring, some flocks arrive on those of Picardy, where one of

* *Suaviores epulas, olim, vulpansere non noverat Britannia.* Plin. *lib.* x. 22.

† *Lib.* iii. 1.

‡ On the coast of Van Diemen's land, in the forty-third degree of latitude, I reckoned among the sea-fowl, ducks, teals, and Sheldrakes. *Cook.*

§ They are found only in Gothland. *Fauna Suecica.*

|| *Anas Maritima.* Gesner.

¶ Schwenckfeld.

** Salerne speaks of a couple of Sheldrakes that were seen on the pool of Sologne.

our

S

our beſt correſpondents, M. Baillon, has ſtudied the natural habits of theſe birds, and made the following obſervations; which we are happy to publiſh.

"The ſpring," ſays M. Baillon, "brings to "us the Sheldrakes, but always in ſmall num-"ber. As ſoon as they arrive, they ſpread "among the ſand plains near the ſea: each "pair wanders among the warrens, which are "there interſperſed, and ſeek a burrow among "thoſe of the rabbits. They ſeem very nice in "chooſing this ſort of lodgement, for they enter "an hundred before they find one to ſuit them. "It is remarked, that they never fix on a bur-"row but ſuch as ſinks more than a fathom "and a half deep, and runs with an aſcent into "ridges or hillocks, its mouth opening to the "ſouth, and viſible from the top of ſome diſ-"tant ſand-bank.

"The rabbits give place to theſe new gueſts, "and enter no more.

"The Sheldrakes make no neſt in theſe holes. "The female lays her firſt eggs on the naked "ſand, and after ſhe has extruded her comple-"ment, which is ten or twelve for young birds, "and twelve or fourteen for old ones, ſhe wraps "them in a very thin down, which ſhe plucks "from her own body.

"During the whole time of incubation, "which is thirty days, the male remains con-"ſtantly on the ſand-bank, and only leaves it

"twice

" twice or thrice a-day, to procure subsistence " on the sea. In the morning and evening, the " female quits her eggs, to provide also for her " wants: then the male enters the burrow, " especially in the morning; and on the female's " arrival, he returns to his sand bank.

" If in the spring we see a Sheldrake thus " on watch, we may be sure to find the nest; " we have only to wait till the hour when it " goes into the burrow. But if it perceive itself " to be discovered, it flies away in the opposite " direction, and expects its female at sea. In " their return they hover long over the warren, " till the danger is removed.

" The day after the young are hatched, the " parents conduct them to the sea, and usually " adjust matters so that they arrive when the " tide is full. By this management, their pro-" geny sooner reach the water; and from that " moment they appear no more on land. It is " difficult to conceive how these birds can, the " first days after their birth, preserve themselves " in an element, whose furious waves so often " destroy the adults of all kinds.

" If a fowler meet the little family on their " journey, the parents fly away: the mother, " however, affects to reel and fall an hundred " paces off; she trails on her belly and strikes " the earth with her wings, and by this trick " she draws the fowler after her. The brood " remain motionless till the return of the pa-" rents;

" rents, and if a perſon lights on them, he may " take them all; nor will any try to eſcape.

" I have witneſſed all theſe facts: I have fre- " quently taken, and ſeen taken, the eggs from " the Sheldrake's neſt. We dug in the ſand, " following the burrow to its end: there we " found the mother ſitting on her eggs; we car- " ried them, with their downy coat, in a thick " woollen cloth, and ſet them under a duck. " The adopted mother rears the foreign brood " with much care, provided none of her own eggs " are left with her. The young Sheldrakes have " at firſt their back white and black, and their " belly very white. But they ſoon loſe this " livery, and become gray: then the bill and " the legs are blue; about the month of Sep- " tember they begin to aſſume their beautiful " feathers; but it is not before the ſecond year " that their colours gain all their luſtre."

" I have reaſon to think, that the male is " not completely grown and fit for propagating " before this ſecond year *; for it is not till " then that the blood-coloured tubercle ap-

* " The life of the Sheldrake, which is pretty long, ſeems to " confirm the conjecture concerning its ſlow growth: laſt winter I " had one that died eleven years old; it would have lived longer, " but it became very miſchievous, and domineered over all the inha- " bitants of the court-yard, except a muſk duck, ſtronger than itſelf, " with which it fought inceſſantly: we thought to preſerve the " weaker by ſhutting it up; but it died a ſhort time after, ra- " ther from the languor of its confinement than from old age." *Note of M. Baillon.*

" pears, which decorates their bill in the sea-
" son of love, and at other times is obliterated:
" this new sort of production seems to have
" some sympathy with the parts of generation.

" The wild Sheldrake lives on sea-worms,
" on sand-hoppers innumerable, and, no doubt,
" on fish-fry, and on little shell-fish, which are
" thrown up by the waves, and float on the froth.
" The raised form of its bill gives it great ad-
" vantage in gathering these different substances,
" by skimming, so to say, the surface of the
" water, much more lightly than the duck.

" The young Sheldrakes reared under a duck
" are soon reconciled to the domestic state, and
" live in court-yards like the ducks. They are
" fed with crumbs of bread and with grain. The
" wild Sheldrakes are never seen assembled in
" flocks, like the ducks, the teals, and the wi-
" geons. The male and female never part;
" they are observed constantly together, either
" on the sea or the sands: they rest satisfied
" with each other's company; and in pairing
" they seem to tie an indissoluble knot *. The

* " Domestication, which softens the natural disposition, at the " same time corrupts it: I saw in my court-yard a male Sheldrake " pair two years successively with a light-coloured duck, and yet be- " stow always the same caresses on his own female; he was then six " year old. This intercourse produced hybrids, which had nothing " of the Sheldrake but the cry, the bill, and the legs; their colours " were those of the duck; the only difference was, that a yellow tint " appeared under the tail. I have kept three years a female of these " hybrids; it has never listened to the addresses either of the drakes " or of the Sheldrakes." *Note of M. Baillon.*

" male

" male appears prone to jealouſy; and yet, not-
" withſtanding the ardour of theſe birds in love,
" I have never been able to obtain one hatch
" from any female: one alone laid a few eggs by
" chance, and they were addle. They are com-
" monly of a very light flaxen colour, without
" any ſpots; they are as large as ducks eggs,
" but rounder.

" The Sheldrake is ſubject to a ſingular diſor-
" der: the luſtre of its feathers tarniſhes, they
" become dirty and oily, and the bird dies, after
" languiſhing near a month. Being curious to
" learn the cauſe of this malady, I opened ſeve-
" ral, and found the blood melted down, and
" the principal bowels choaked with a reddiſh
" lymph, viſcous and fœtid. I attribute the diſ-
" eaſe to the want of ſea-ſalt, which I believe
" to be neceſſary to theſe birds, at leaſt from
" time to time, to divide by its points the red
" particles of the blood, and to preſerve the
" union with the *ſerum*, by diſſolving the viſ-
" cous humours, whch the ſeeds that ſupport
" them in the court-yards accumulate in the
" inteſtines."

Theſe obſervations, detailed by M. Baillon, leave very little to be added to the hiſtory of the Sheldrakes. We reared a pair of them under our eyes; they ſeemed not to have a wild diſpoſition; they readily allowed themſelves to be caught; they were kept in a garden, where they had liberty during the day; and when they were taken and held in the hand, they made ſcarce any efforts to eſcape: they ate bread, bran, corn, and even the leaves of plants and ſhrubs; their ordinary cry was much like that of a duck, but was leſs extended and much leſs frequent, for they were very ſeldom heard: they had alſo a ſecond cry, *uute, uute*, which they utter when caught ſuddenly, and which ſeemed to be only the expreſſion of fear: they bathed very often, eſpecially in mild weather, and before rain; they ſwam rocking on the water, and when they reached the land, they ſtood on their feet, clapt their wings, and ſhook themſelves like ducks; they alſo frequently preened their plumage with the bill. Thus the Sheldrakes, which reſemble much the ducks in the ſhape of their body, reſemble them alſo by their natural habits, only they are nimbler in their motions, and diſcover more cheerfulneſs and vivacity: they have beſides over all the ducks, even the moſt beautiful, a privilege of nature, which belongs to them alone; that is, they retain conſtantly, and at all ſeaſons, the charming colours of their plumage. As they are not difficult to tame, and as their rich garb is conſpicuous at a diſtance, and has a very fine effect on pieces of water, it is to be wiſhed that we could obtain a domeſtic breed of theſe birds: but their temper and conſtitution ſeem to fix them on the ſea, and to repel them from freſh pools; the experiment could therefore

fore be made only on lands ſituated very near ſalt water.

[A] Specific character of the Sheldrake, *Anas Tadorna*: "Its "bill is flat; its front compreſſed; its head greeniſh-black; its body "variegated with white." Theſe birds remain in England the whole year: they lay fifteen or ſixteen eggs, which are white and roundiſh: their fleſh is very rank.—Mr. Pennant writes the name *Shieldrake*; and perhaps the form of its bill might ſuggeſt a *ſhield*.

The POCHARD.

LE MILLOUIN. *Buff.*

Anas Ferina. Linn. and Gmel.
Penelope. Johnſt. Charleton, and Briſſon.
Anas Fera fuſca, vel media. Geſn. Aldrov. Will. &c.
Anas Fuſca. Johnſt. Marſ. and Schwenckf.
The Poker, Pochard, Red-head Wigeon. Will. and Ray *.

THE Pochard is ſtiled by Belon *the rufous-headed duck*. In fact, its head, and part of its neck, is of a rufous brown, or cheſnut; that colour cut round at the bottom of the neck, is ſucceeded by black or blackiſh brown, which is likewiſe cut round on the breaſt and the top of

* In Brie it is called *Moreton*: in Burgundy *Rougeot*: in Catalonia *Buixot*: in the Bologneſe *Collo Roſſo (red-neck)*: in Germany *Rot-hals (red-throat)*; *Rot-ente (red duck)*; *Mittel-ente (middle-duck)*; *Braun koëpfichte-ente (brown-headed duck)*: in Sileſia *Braun-ente*: in Denmark *Brun Nakke (brown-neck)*: in Norway *Rod Nakke (red-neck)*.

the back: the wing is gray, tinged with blackiſh, and without any ſpangle; but the back and the ſides are prettily worked with a very fine fringe, which runs tranſverſely in little black zigzags on a ground of pearl-gray. According to Schwenckfeld, the head of the female is not rufous like that of the male, and has only ſome ruſty ſpots.

The Pochard is as large as the ſheldrake, but is more unwieldy; its round ſhape gives it a heavy air; it walks with difficulty and ungracefully, and is obliged from time to time to flap its wings, in order to preſerve its equilibrium on land.

Its cry reſembles more the hollow hiſs of a large ſerpent than the voice of a bird. Its bill, broad and ſcooped, is very proper for dabbling in the mud, like the ſhovelers and the morillons, to ſearch for worms, ſmall fiſh, and cruſtaceous animals. Two male Pochards, which M. Baillon kept a winter in his court-yard, remained almoſt conſtantly in the water; they were very ſtrong and courageous on that element, and would ſuffer none of the other ducks to approach them, but drove them away with their bill. Theſe, however, in their turn, beat them when they came on land, and the Pochards could then make no defence, but eſcaped to the water. Though they were tame, and even grown familiar, they could not be long preſerved, becauſe they could not walk without hurting their feet; the gravel of the garden-walks were as pernicious as the pavement

ment of the court; and, notwithſtanding the care which M. Baillon took of theſe two Pochards, they lived not more than ſix weeks in their captivity.

"I believe," ſays this good obſerver, "that "theſe birds belong to the north. Mine conti- "nued in the water during the night, even "when the froſt was intenſe; they alſo agitated "it, to prevent its freezing round them."

The Pochards, he adds, as well as the ſhovelers and the golden-eyes, eat much, and digeſt as quickly as the duck. They lived at firſt only on ſoaked bread, afterwards they ate dry, but ſwallowed it in that ſtate with difficulty. I could never habituate them to grain. The ſhovelers alone ſeem fond of the ſeeds of the bulruſh.

M. Hebert, who, as an attentive and even ingenious ſportſman, has found other pleaſures in fowling than that of killing, has made on theſe birds, as on many others, intereſting obſervations. "It is the ſpecies of the Pochard," ſays he, "which, next to that of the wild duck, "appears to me the moſt numerous in the coun- "tries where I have gone a-fowling. They ar- "rive with us in Brie about the end of Octo- "ber, in flocks from twenty to forty. Their "flight is more rapid than that of the duck, "and the noiſe made by their wings is quite "different. The troop forms a cloſe body in "the air, but not diſpoſed like the wild ducks "in triangles. On their arrival, they are reſt-

 "leſs;

" leſs; they alight on the large pools, and, the " inſtant after, they riſe, make ſeveral wheels in " the air; a ſecond time they alight, but their " ſtay is equally ſhort; they diſappear, and re- " turn in an hour, and yet do not ſettle. When " I killed one, it was always by chance, and " with very coarſe ſhot, and when they whirled " in the air. They were all remarkable for a " large rufous head, whence they are called " *rougeot* in Burgundy.

" It is not eaſy to get near them on the large " pools; they alight not on the brooks in froſty " weather, nor on the little pools in autumn; " and many of them cannot be killed, except " on the *duckeries* of Picardy. However, they " are pretty common in Burgundy, and at Dijon " they are ſeen in the cooks ſhops almoſt the " whole year. I killed one in Brie in the " month of July, when the weather was ex- " tremely hot: it fluſhed at the ſide of a pool, " in the middle of a wood, and in a very ſoli- " tary ſpot: it was attended by another, which " made me think that they were paired, and " that ſome couples of this ſpecies breed in " France on the large marſhes."

We ſhall add, that this ſpecies has penetrated into diſtant countries, for we received from Louiſiana a Pochard exactly like what is found in France; and beſides the ſame bird may be recognized in the *quapacheanauhtli* of Fernandez, which Briſſon has, for that reaſon, called the

Mexican

Mexican Pochard *. With regard to the variety of the French Pochard described by that ornithologist, we must content ourselves with what he has said: for we are unacquainted with this variety.

* *Anas Fulva.* Gmel.
Penelope Mexicana. Briss.
The Mexican Pochard. Lath.

Specific character: "It is fulvous; its back, its shoulders, its "wings, and its rump, are striped transversely with fulvous and "brown; its tail is variegated with black and white."

[A] Specific character of the Pochard, *Anas Ferina:* "It is "waved with cinereous; its head is brown; the bar on its breast, "its vent, and its rump, are black." The Pochards are reckoned delicate eating, and are sold in the London markets under the name of *dun-birds.* They are found also through the whole extent of North America.

The MILLOUINAN.

THIS beautiful bird, for our knowledge of which we are indebted to M. Baillon, is as large as the pochard *(millouin)*, and its colours, though different, are disposed in the same manner: we have therefore called it the *Millouinan.* Its head and neck are covered with a large black *domino* with copper-green reflections, cut round on the breast and the top of the back: the mantle is finely worked with a small black hatching,

hatching, running lightly in the ground of pearl-gray: two pieces of the ſame work, but cloſer, cover the ſhoulders; the rump is worked in the ſame way: the belly and ſtomach are of the fineſt white. On the middle of the neck may be obſerved the obſcure trace of a rufous collar: the bill of the *Millouinan* is neither ſo long nor ſo broad as that of the pochard.

The individual which we deſcribe was killed on the coaſt of Picardy; and I have ſince received from Louiſiana another, preciſely ſimilar if not ſomewhat ſmaller. It is not, we have ſeen, the only ſpecies of duck which is common to both continents; yet this *Millouinan* has not hitherto been remarked or deſcribed, and, no doubt, ſeldom appears on our coaſts.

The GOLDEN EYE.

LE GARROT. *Buff.*

Anas Clangula. Linn. and Gmel.
Clangula. Geſner, Johnſt. and Klein.
Anas Platyrinchos. Aldrov.*

THE Golden Eye is a little duck whoſe plumage is black and white, and its head remarkable for two white ſpots placed at the corners

* In Lorraine it is called the *Hungarian Duck:* in Alſace the *Magpie Duck:* in Italy *Quattr' Occhi:* in Germany *Kobel-ente, Straus-ente,*

ners of the bill, which at a diſtance appear like two eyes, ſituated near the other two, in the black hood gloſſed with green, which covers the head and the top of the neck. Hence the Italian name *Quattr'Occhi*, or *four eyes*. The Engliſh have termed it *Golden Eye*, becauſe its iris is of a golden-yellow. Its tail and back are black, as well as the great quills of the wing, of which moſt of the coverts are white: the lower part of the neck, with all the fore ſide of the body, is of a fine white: the legs are very ſhort, and the membranes which connect the toes extend to the tips of the nails, and are there faſtened.

The female is rather ſmaller than the male, and differs entirely in its colours, which, as generally obſerved in all the ducks, are duller and paler in the females: thoſe of the female Golden Eye are gray or browniſh, which in the male are black; and thoſe white gray, which in the other are of a fine white: nor has ſhe the green reflection on the head, or the white ſpot at the corner of the bill *.

The flight of the Golden Eye, though pretty low, is very ſtiff, and makes the air to whiſtle †.

ente, *Quaker-ente*, *Eiſs-ente*; and in the neighbourhood of Straſburg *Weiſſer-dritt-vogel*: in Sweden *Knippa*; and in the province of Skonen *Dopping*: in Norway *Ring-eye*, *Hviin-and*, *Lund-and*.

* Aldrovandus.

† *Idem*.

It

It does not ſcream in taking wing, and ſeems not ſo ſhy as the other ducks. Small flocks of Golden Eyes are ſeen on our pools during the whole winter; but they diſappear in the ſpring, and no doubt go to neſtle in the north; at leaſt, Linnæus ſays, in the *Fauna Suecica*, that this duck is ſeen in ſummer in Sweden, and in that ſeaſon, which is alſo that of breeding, it lives in the hollows of trees.

M. Baillon, who tried to keep ſome Golden Eyes in the domeſtic ſtate, has juſt communicated the following obſervations.

" Theſe birds," ſays he, " loſt much fleſh in " a ſhort time, and hurt themſelves under the " feet when I allowed them to walk at liberty. " They lay for the moſt part on their belly; but " if other birds attacked them, they made a ſtout " defence: I can even ſay, that I have ſeen few " birds ſo rancorous. Two males which I had " laſt winter, tore my hand with their bill, as " often as I laid hold of them. I kept them in " a large ozier cage, that they might be habitu- " ated to captivity, and might ſee the other " fowls rambling about the court. But they " betrayed in their priſon only the marks of " impatience and rage, and darted againſt the " bars at the other birds which approached. I " ſucceeded, with much difficulty, in teaching " them to eat bread, but they conſtantly refuſed " every ſort of grain.

" The

" The Golden Eye," adds this attentive obſerver, " like the pochard and the morillon, " walks under conſtraint and difficulty, with " effort, and ſeeming pain. Yet theſe birds come " from time to time on ſhore, but only to remain " there in tranquillity and repoſe, ſtanding or " lying on the ſtrand, and to enjoy a pleaſure " which is peculiar to themſelves. Land-birds " feel the neceſſity of bathing at inrervals, whe-" ther to clean their plumage of the duſt which " inſinuates into it, or to give dilatation to their " body, which facilitates their motions; and they " announce, by their chearfulneſs on quitting the " water, the agreeable ſenſation which they feel. " In the aquatic birds, on the contrary, in thoſe " which remain long in the water, their feathers " become through time penetrated and moiſ-" tened, and permit the humidity to ſteal inſenſi-" bly to the ſkin: then they have occaſion for an " air bath to dry and contract their relaxed limbs; " they come, for this purpoſe, on ſhore, and the " ſprightlineſs of their eyes, and the ſlow balancing " of their head, expreſs their agreeable ſenſation. " But the Golden Eyes, and likewiſe the pochards " and the morillons, are ſatisfied with that gratifi-" cation; they never willingly come to land, and " eſpecially avoid walking on it, which ſeems to " cauſe extreme fatigue: in fact, accuſtomed as " they are to move in the water by ſhort darts, " produced by the briſk and ſudden motion of

" their

" their feet, they bring this habit with them on " land, and walk by ſprings, ſtriking the ground " ſo forcibly with their broad feet, that their " pace is attended with the ſame noiſe as the " clapping of hands; they uſe their wings to " preſerve the equilibrium, which they loſe every " minute; and if they be haſtened, they make a " bound, throwing their legs back, and fall on " their breaſt: their feet alſo are torn and cut in " a ſhort time by rubbing on the gravel. It " appears, therefore, that theſe birds, deſtined " ſolely for the water, can never augment the " colonies planted in our court-yards."

[A] Specific character of the Golden Eye, *Anas Clangula*: "It " is variegated with black and white; its head violet and ſwelled; " a black ſpot on the corner of the mouth." Linnæus ſays, that it dives excellently for fiſh and ſhell-fiſh, that it eats frogs voraciouſly, that it often builds its neſt in trees with graſs, and lays from ſeven to ten white eggs, and that its fleſh is agreeable.—This bird viſits the meres of Shropſhire in winter.

No. 246

THE MORILLON

The MORILLON.

Anas Glaucion. Linn. and Gmel.*

THE Morillon is a handſome little duck, which, when ſeen at reſt, exhibits theſe colours; a broad blue bill, a large black *domino*, a mantle of the ſame, and white on the ſtomach, the belly, and the top of the ſhoulders: this white is free and unadulterated, and all the black is ſhining, and heightened with fine purple and greeniſh-red reflections; the feathers on the back of the head riſe into a bunch: often the lower part of the black domino on the breaſt is waved with white: and, in this ſpecies, as in others of the genus of ducks, the colours are liable to certain variations, but which are only individual †.

When the Morillon flies, its wing appears ſtriped with white: this effect is produced by ſeven feathers, which are partly of that colour ‡. The inſide of the legs and thighs are reddiſh, and the outſide black; the tongue is fleſhy, and ſwelled at the root, which ſeems parted in two:

* In Brie it is called the *Jacobine*: in Germany *Scheel-ente* (*ſquinting-duck*), *Schilt-ente* (*ſhield-duck*), *Lepel-ganz*.

† Ray.

‡ Belon.

there

there is no gall-bladder. Belon regards the Morillon as the *glaucium* of the Greeks, *not having found*, he ſays, *a bird with eyes of ſuch a glaucous colour*. Indeed, the *glaucium* of Athenæus was ſo called becauſe its eyes were ſea-green.

The Morillon frequents the pools and rivers *, and yet occurs alſo on the ſea †; it dives pretty deep ‡, and feeds on little fiſh, cruſtaceous animals, and ſhell-fiſh, or on the ſeeds of aquatic plants ||, eſpecially thoſe of the common ruſh. It is leſs ſhy, and not ſo apt to fluſh, as the wild duck: it may be approached within gun-ſhot on pools, or, ſtill better, on rivers, when the froſt prevails. When it riſes, it does not fly to great diſtances §.

M. Baillon has communicated his obſervations on this ſpecies in the ſtate of domeſtication. "The colour of the Morillon," he ſays, "its "manner of balancing its head as it walks, and "of holding its head almoſt erect, give it an air "the more ſingular, as the beautiful light-blue "of its bill, applied always on its breaſt, and its "large brilliant eyes, are ſtrongly contraſted with "the black of its plumage.

"It is pretty chearful, and dabbles like the "duck whole hours. I eaſily tamed ſeveral in

* Belon.

† *Fauna Suecica.*

‡ Belon.

|| *Idem.*

§ Obſervation of M. Hebert.

"my

"my court: they became ſo tame in a ſhort "time, that they entered the kitchen and the "rooms: they were heard before they could be "ſeen, becauſe of the noiſe made at each ſtep, "clapping their broad feet on the ground and "the floor. They were never ſeen to take any "unneceſſary perambulations; which proves "what I have before ſaid, that this ſpecies walks "only when urged by its wants. Their feet "were indeed ſoon peeled on pavement; yet they "grew lean very ſlowly, and might have lived a "long time, had not the other fowls tormented "them.

"I procured," adds M. Baillon, "more than "thirty Morillons, to ſee whether the tuft, "which is very apparent in ſome individuals, "conſtitutes a particular ſpecies: I found that "it was an ornament of all the males *.

"The young ones are at firſt of a ſmoky "gray: this livery remains till after moulting; "and they acquire not all their fine brilliant "black till the ſecond year; at that time, alſo, "their bill becomes blue. The females are al-"ways leſs black, and have no tuft."

* I have killed ſome which had on the crown of the head a few feathers that were longer and broader than the reſt, which formed a ſort of inconſpicuous tuft: I have killed others that had not a veſtige of it. *Note communicated by M. Hebert.*

[A] Specific character of the Morillon, *Anas Glaucion:* "Its "body is blackiſh; its breaſt cloudy; a white linear ſpangle on the "wings."

The LITTLE MORILLON.

Anas Fuligula. Linn. Gmel. Klein. Johnst. &c.
Glaucium Minus. Briss.
Fuligula. Gesner, Aldrov. &c.
Anas Cristata. Ray.
The Tufted Duck. Will. Penn. and Lath. *

After what we have said of the diversity that prevails in the plumage of the Morillons, we are much inclined to refer to the same accidental causes the difference of bulk which has made the little Morillon be reckoned a distinct and separate species. That difference is indeed so small, that we might strictly disregard it †, or at least attribute it to the varieties which necessarily obtain among individuals, occasioned by the diversity of age and of the seasons of growth. Yet most ornithologists have described this little Morillon as a different species from the other; and, as we cannot contradict them by positive facts, we shall here mention our

* In Swedish *Wigge*: in German *Woll-enten*; and by some *Rusgen*: at Venice it is called *Capo Negro*, or *black-head.*

† "*The Morillon*—from the end of the bill to that of the tail it "fourteen inches nine lines; to the end of the nails fifteen inches."

"*The Little Morillon*—from the end of the bill to that of the "tail is twelve inches six lines; to the end of the nails fourteen "inches ten lines." *Brisson.*

our doubts, which we believe to be not ill-founded. Belon even, whom the reſt have followed, and who was the firſt author of this diſtinction, ſeems to furniſh a proof againſt his own opinion: for after having ſaid of his *little diver*, which is our little Morillon, that " it is an handſome tight bird, round and ſhort, its eyes ſo " yellow and ſhining, that they are brighter " than poliſhed iron," and that with a plumage ſimilar to that of the Morillon, it has likewiſe a white line acroſs the wing; he adds, " that it is " far from being a true Morillon, for it has a " tuft on the back of the head like the gooſander " and the pelican, while the Morillon has none." But Belon is here miſtaken, and this character of the tuft is another reaſon that this bird ſhould be claſſed with the true Morillon.

Briſſon gives ſtill another variety of this ſpecies, under the name of the *little ſtriped Morillon* *; but it is certainly a variety from age.

* *Anas Marila.* Linn. and Gmel.
Glaucium Minus Striatum. Briſſ.
Fuligula Geſneri. Ray, and Will.
The Scaup Duck. Penn. and Lath.

Specific character: " It is black, its ſhoulders waved with cine" reous; its belly, and the ſpangle on its wings, are white." It owes its name to its feeding on *ſcaup*, or broken ſhell-fiſh.

[A] Specific character of the Tufted Duck, *Anas Fuligula:* " Its creſt is hanging, its body black; its belly, and the ſpangle on " its wings, are white."

The SCOTER.

LA MACREUSE. *Buff.*

Anas Nigra. Linn. Gmel. Will. and Briff.
The Whilk. Philof. Tranfact.
The Black Diver, or *Scoter.* Will.

IT has been pretended that the Scoters are engendered, like the barnacles, in fhells or in rotten wood *. We have fufficiently refuted thefe fables, with which natural hiftory is here, as in other parts, too much tinctured. The Scoters lay, neftle, and hatch, like other birds. They prefer for their habitation the moft northern countries, whence they defcend in great numbers along the coafts of Scotland and England, and arrive on the coafts of France in winter, to afford a very indifferent fort of game, but which is eagerly expected by our monks and nuns, who, being entirely denied the ufe of flefh, and reftricted to fifh, are indulged with thefe birds, from a notion that their blood is as cold as that of fifh: but in fact their blood is juft as warm as that of other aquatic birds; though

* Hence the name *Scoter*; Scotland being the principal fcene of this fabulous tranfmutation of the barnacles.—*T.*

indeed

Nº. 247

THE SCOTER DUCK.

indeed the black, dry, and hard fleſh of the Scoter may be deemed a diet of mortification.

The plumage of the Scoter is black: it is nearly as large as the common duck, but it is ſhorter and more compact. Ray obſerves, that the tip of the upper mandible is not terminated by a horny nail, as in all the other ſpecies of this genus: in the male, the baſe of that part, near the head, is conſiderably ſwelled, and exhibits two tubercles of a yellow colour; the eye-lids are of the ſame colour; the toes are very long, and the tongue is very large; the *trachea* has no labyrinth, and the *cæca* are very ſhort in compariſon of thoſe of the other ducks.

M. Baillon, that intelligent and laborious obſerver, whom I have ſo often had occaſion to cite on the ſubject of water-fowl, has ſent me the following obſervations:

" The north and north-weſt winds bring " along our coaſts of Picardy, from the month " of November to that of March, prodigious " flocks of Scoters: the ſea, ſo to ſpeak, is co- " vered with them. They are ſeen flying in- " ceſſantly from place to place, and by thou- " ſands: they appear and diſappear on the wa- " ter every minute: as ſoon as a Scoter dives, " the whole troop imitates it, and emerges again " a few moments after. When the ſouth and " ſouth-eaſt winds blow, about the month of " March, they are driven from our coaſts, and " entirely diſappear.

"The favourite food of the Scoters is a kind " of bivalve ſhell, ſmooth and whitiſh, four lines " broad and about ten long, which are found " cluſtered in many deep ſhoals: there are " pretty extenſive banks of them, which are " left bare by the ebb tide. When the fiſher- " men remark that the Scoters dive for the " *vaimeaus* (the term which they apply to theſe " ſhells) they ſpread their nets horizontally but " very looſe, above theſe ſhell-fiſh, and two or " three feet at moſt from the ſand; a few hours " after, the tide flowing in covers the nets, and " the Scoters following the reflux two or three " hundred paces from the beach, the firſt that " perceives a *vaimeau* dives, and all the reſt, co- " pying the example, entangle themſelves among " the floating meſhes, or if ſome, more ſhy, go " a little aſide and paſs under the nets, they riſe " after having fed, and ſoon inwrap themſelves " like the reſt: they are all drowned, and when " the ſea has retreated, the fiſhermen go to diſ- " engage them from the nets, on which they " are ſuſpended by the head, the wings, or the " feet.

"I have ſeveral times ſeen this ſort of fiſh- " ing: a net of an hundred yards long and three " yards broad caught ſometimes twenty or " thirty dozen in a ſingle tide; but to balance " this good fortune, the nets are often ſtretched " twenty times without catching one; and at

" times

" times they are carried away or rent by por-
" poiſes or ſturgeons.

" I never ſaw any Scoters fly any where but " above the ſea, and I have always remarked " that their flight was low and gentle, and of " ſmall compaſs: they ſcarce ever riſe, and " while on wing their feet often drag in the " water. It is probable that the Scoters are as " prolific as the ducks, for the number which " arrives every year is prodigious; and notwith- " ſtanding the multitudes that are caught, they " ſeem not to diminiſh."

Having enquired of M. Baillon his opinion with regard to the diſtinction between the male and female of this ſpecies, and to the gray Scoters or *griſettes*, which ſome have ſaid to be females; he gave me this anſwer:

" The *griſette* is certainly a Scoter, and has " exactly the figure. Theſe *griſettes* are always " ſeen in company with the other Scoters; they " feed on the ſame ſhell-fiſh, ſwallow them en- " tire, and digeſt them in the ſame manner. " They are caught in the ſame nets, and they " fly as badly and in the ſame way, and this is " peculiar to theſe birds, which have the bones " of the wings more turned backwards than the " ducks, and the cavities, in which the two " thigh-bones are ſunk, very near each other; a " ſtructure which gives them great facility in " ſwimming, and makes them at the ſame time " very aukward in walking: and ſurely no ſpe-

"cies of ducks has the thighs placed in this "manner. Laſtly, the taſte of their fleſh is the "ſame.

"I opened three of theſe *griſettes* in winter, "and they were found to be females.

"On the other hand, the number of gray Sco-"ters is much inferior to that of the black; "often not ten occur among an hundred of the "others, caught in the ſame net. How could "the females be ſo few in this ſpecies?

"I freely own that I have not ſought ſuffi-"ciently to diſtinguiſh the males from the fe-"males. I ſtuffed a great number; I choſe the "blackeſt and the largeſt, and they were all "found to be males, except the *griſettes*. I be-"lieve, however, that the females are ſomewhat "ſmaller, and not ſo black, or at leaſt they have "not that velvet ſurface which makes the black "of the male plumage ſo deep."

It appears to us from this detail, that as the female Scoters are not quite ſo black as the males, and more inclined to gray, the *griſettes*, or the Scoters which verge on gray, are too few to repreſent all the females of the ſpecies, and are in fact the younger females, which require time to aſſume all the black of their plumage.

After this firſt anſwer, M. Baillon ſent us alſo the following notes, which are all intereſting. "I have had," ſays he, "this year, 1781, for "ſeveral months in my court, a black Scoter.

"I fed

"I fed it with ſoaked bread and ſhell-fiſh. It "was become very familiar.

"I believed till then that the Scoters could "not walk, and that their conformation deprived "them of that power. I was the more per- "ſuaded, as I had ſeveral times, in ſtorms, ga- "thered, on the ſea-ſhore, Scoters, penguins, "and puffins alive, which could drag themſelves "along only by help of their wings. But theſe "birds had, no doubt, been much beaten by the "waves; and that circumſtance, which I had "overlooked, confirmed me in my error. I "was ſurprized to remark, that the Scoter "walks well, and faſter even than the pochard; "it balances itſelf in the ſame manner at each "ſtep, holding its body almoſt erect, and ſtrik- "ing the ground with each foot alternately and "with force; its pace is ſlow; if preſſed, it "tumbles, becauſe the efforts which it makes "deſtroys its equilibrium: it is indefatigable in "the water; it runs on the waves like a petrel, "and as nimbly; but, on land, the celerity of "its motions are of no avail; mine ſeemed "quite out of its natural element.

"Indeed, it had a very aukward air; each "movement gave its body fatiguing jogs; it "walked only from neceſſity: it uſually lay "down or ſtood ſtraight like a ſtake, its bill "leaning on its ſtomach: it always ſeemed to "be melancholy: I never once ſaw it bathe "joyous, like the other water-fowls, with which

"my

" my court is filled; it never entered the ſhal-
" low trough which is placed level with the
" ſurface of the ground, but to eat the bread
" which I threw to it: when it had eaten and
" drank it remained motionleſs; ſometimes it
" dived to the bottom, to gather the crumbs
" which fell down; if any bird came into the wa-
" ter and approached it, the Scoter endeavoured
" to drive away the intruder; if this made any
" oppoſition or reſiſtance, it dived, and after
" making two or three turns at the bottom of
" the trough, it flew out of the water, making
" a ſort of whiſtling, very ſoft and clear, like the
" firſt tone of a German flute: this is the only
" cry I ever knew it make, which it repeated as
" often as a perſon approached it.

" Being deſirous to know if the bird could
" continue long under water, I held it down by
" force; it made conſiderable efforts after two
" or three minutes, and ſeemed to ſuffer much;
" it bounded up as quick as a cork. I believe
" it could remain longer, becauſe it deſcends
" often to the depth of thirty feet in the ſea to
" gather the oblong bivalves on which it feeds.

" Theſe ſhell-fiſh are whitiſh, four or five
" lines broad, and near an inch long. It does
" not amuſe itſelf like the ſea-pie in opening
" them, the ſhape of its bill not being, as in that
" bird, adapted for the purpoſe: it ſwallows
" them whole, and digeſts them in a few hours.
" I gave ſometimes more than twenty to a ſingle
" Scoter;

" Scoter; and it received them till its *œſophagus* " was filled up to the bill: then its excrements " were white. They aſſumed a green tinge " when the bird was fed with bread, but were " always liquid. I never ſaw it eat herbs or the " ſeeds of plants, like the wild ducks, the teals, " the wigeons, and others of this genus: the ſea " is its only element. It flies as ill as it walks: " I have often amuſed myſelf in viewing thro " a ſpy-glaſs the numerous flocks on the ſea; I " never ſaw them riſe and fly to any diſtance; " they fluttered inceſſantly above the ſurface of " the water.

" The feathers of this bird are ſo ſmooth and " cloſe, that the bird, on coming out of the wa- " ter, can ſhake itſelf dry.

" The ſame cauſe which proved fatal to ſo " many other birds in my court, occaſioned alſo " the death of my Scoter. The ſoft and tender " ſkin of its feet were perpetually bruiſed by the " gravel; a *callus* formed on each joint of the " toes; in time they were worn to ſuch a de- " gree, that the nerves were diſcloſed; it durſt " no longer walk, or go to the water, each ſtep " increaſing its wounds: I put it in my garden " on the graſs under a cage, but it would not " eat; and it died in my court a few days " after."

[A] Specific character of the Scoter or Black Diver, *Anas Nigra:* " Its body is entirely black." It is frequent in the lakes and rivers of Siberia: it has a fiſhy taſte.

The DOUBLE SCOTER.

La Double Macreuse. *Buff.*

Anas Fusca. Linn. and Gmel.
Anas Nigra Major. Briss.
Anas Nigra. Aldrov. Gesn. Klein, &c.
The Great Black Duck. Will.
The Velvet Duck. Penn. and Lath.

AMONG the great number of Scoters which come in winter on the coasts of Picardy, some are remarked much larger than the rest, and therefore called the *Double Scoters.* Besides this difference of size, they have a white spot on the side of the eye, and a white bar on the wing; while the plumage of the others is entirely black. These characters are sufficient to constitute a second species, which appears to be much less numerous than the first, but resembles it in structure and habits. Ray observed in the stomach and the intestines of these large Scoters, fragments of shells; the same probably that, Baillon says, is the principal food of the Scoter.

[A] Specific character of the Velvet Duck, *Anas Fusca:* "It "is blackish; its lower eye-lid, and the spangle on its wings, are "white." It is frequent in Siberia, and even in Kamtschatka: it lays eight or ten eggs.

The BROAD-BILLED SCOTER.

LA MACREUSE A LARGE BEC. *Buff.*

Anas Perſpicillata. Linn. and Gmel.
Anas Nigra Major Freti Hudſonis. Briſſ.
The Great Black Duck from Hudſon's Bay. Edw.
The Black Duck. Penn. and Lath.

THIS is undoubtedly a Scoter, and perhaps belongs to the ſame ſpecies with the preceding. It is well characterized by the breadth of its flat, ſhort bill, edged with an orange ſtreak, which incircling the eye, ſeems to delineate ſpectacles *. This large Scoter viſits England in the winter; it alights in the meadows, where it feeds on graſs. Edwards thinks that he can diſcover it in one of the figures of a ſmall collection publiſhed at Amſterdam in 1679, by *Nicolas Viſcher*, in which it is denominated *turma anſer (troop-gooſe)*; a term which probably alludes to its bulk, which exceeds that of the common duck, and indicates at the ſame time that theſe birds are ſeen in flocks: and as they occur in Hudſon's Bay, the Dutch might have

* Hence the Linnæan epithet for the ſpecies, *Perſpicillata*, from *Perſpicillum*, a pair of ſpectacles.—*T.*

obſerved

obſerved them in Davis's Straits, where they carry on the whale-fiſhery.

[A] Specific character of the Black Duck, *Anas Perſpicillata:* "It is black; its top and nape white; a black ſpot on the bill behind the noſtrils." It breeds in July along the ſhores of Hudſon's Bay: it builds its neſt with graſs, and lines it with feathers: it lays from four to ſix eggs, which are white. It paſtures on graſs.

The BEAUTIFUL CRESTED DUCK.

Anas Sponſa. Linn. and Gmel.
Anas Æſtiva. Briſſ.
Yſtactzonyayauhqui. Fernandez.
The American Wood Duck. Brown.
The Summer Duck. Cateſ. Penn. and Lath.

THE rich plumage of this beautiful Duck ſeems to be a ſtudied attire, a gala ſuit, to which its elegant head-dreſs adds grace and luſtre *: a piece of beautiful rufous, ſpeckled with little white daſhes, covers the back of the neck and the breaſt, and is neatly interſected on the ſhoulders by a ſtreak of white, accompanied by a ſtreak of black; the wing is covered with feathers of a brown that melts into black with rich reflections of burniſhed ſteel; and thoſe of the flanks are very delicately fringed and vermi-

* Hence Linnæus calls it *Sponſa*, or the bride.—*T.*

culated

culated with little blackiſh lines on a gray ground, and are prettily ſtriped at the tips with black and white, of which the ſtreaks are diſplayed alternately, and ſeem to vary according to the motion of the bird: the under ſide of the body is pearly white-gray; a ſmall white collar riſes into a chin-piece below the bill, and ſends off a ſcallop below the eye, on which another long ſtreak of the ſame colour paſſes like a long eye-lid; the upper ſide of the head is decorated with a ſuperb tuft of long feathers, white, green, and violet, which fall back like hair, in bunches parted by ſmaller white bunches. The front and the cheeks dazzle with the luſtre of bronze: the iris is red; the bill the ſame, with a black ſpot above, and the horny tip is of the ſame colour; the baſe is hemmed with a fleſhy brim of yellow.

This beautiful Duck is ſmaller than the common duck, and the female is as ſimply clothed as the male is pompouſly attired. She is almoſt all brown; "having, however," ſays Edwards, "ſomething of the creſt of the male." This obſerver adds, that he received ſeveral of theſe charming ducks alive from Carolina; but he does not inform us whether they propagated. They like to perch on the talleſt trees; whence ſeveral travellers ſtyle them *branch ducks*. Cateſby calls them *ſummer ducks* *; from which we may infer,

* "The moſt beautiful birds that I have ſeen in this country "(at Port-Royal in Acadia, or Nova Scotia) are the *branch ducks*, "ſo

infer, that they reside during the summer in Virginia and Carolina *: in fact, they breed there, and place their nests in the holes made by the woodpeckers in large trees near water, particularly on the cypress; the parents carry their young into the water on their back, and these on the least symptom of danger cling by the bill.

" so called because they perch; nothing is finer or better mingled " than the endless variety of colours that compose their plumage: " but I was still more surprized to see them perched on a pine, a " beech, or an oak, and to see them hatch their young in a hole of " some of these trees, which they rear till they are able to leave " the nest, and, according to instinct, follow their parents to the " water. They are very different from the common sort, called " *black*, and which in fact are almost entirely of that colour, with- " out being variegated like ours: the branch ducks have a more " slender body, and are likewise more delicate eating." *Voyage au Port-Royal de l'Acadie, par M. Dierville; Rouen*, 1708, *p.* 112.— " There is a kind which we call *branch ducks*, which roost on trees, " and whose plumage is very beautiful on account of the agreeable " diversity of colours which form it." *Nouvelle Relation de la Gaspesie, par le P. Le Clerc; Paris*, 1698, *p.* 485.

* According to Du Pratz, they are seen the whole-year in Louisiana. " The branch ducks are somewhat larger than our " teals; their plumage is exceedingly beautiful, and so changing " that painting cannot imitate it; they have on the head a beau- " tiful crest of the brightest colours, and their red eyes appear like " flames. The natives deck their calumets or pipes with the skin " of the neck: their flesh is very good, but when too fat it has " an oily taste. This species of duck is not migratory, it is found " in all seasons, and it perches, which the rest do not; hence it is " called *the branch duck*."

[A] Specific character of the Summer Duck, *Anas Sponsa*: " Its " crest is hanging and double; it is variegated with green, with " blue, and with white." It nestles in the holes bored by the woodpeckers in trees near water: and when the young are hatched, it carries them to the stream. This bird seems to retire to Mexico in winter. It is esteemed delicate food.

The LITTLE THICK-HEADED DUCK.

Anas Bucephala. Linn. and Gmel.
Anas Hyberna. Briff.
The Buffel-headed Duck. Catefby, Penn. and Lath.

THIS little Duck is of a middle fize between the common duck and the garganey. All its head is clothed with a tuft of unwebbed feathers, agreeably tinged with purple, and heightened by reflections of green and blue. This thick tuft increafes confiderably the bulk of its head; and hence Catefby ftiles it the *Buffel-headed Duck*. It frequents the frefh waters in Carolina. Behind the eye is a broad white fpot; the wings and the back are marked with longitudinal fpots, black and white alternately; the tail is gray; the bill is lead-colour, and the legs are red.

The female is entirely brown; its head uniform, and without a tuft.

This Duck appears in Carolina only in winter; but that is no reafon why Briffon fhould give it the appellation of *winter Duck*; for it muft live elfewhere in the fummer, and in fuch countries it might with equal propriety be named the *fummer Duck*.

[A] Specific character of the *Anas Bucephala*: "It is whitifh; "its back and wing-quills black; above and below its head there "is a fwelling of a filky glofs."

The COLLARED DUCK of NEWFOUNDLAND.

Anas Histrionica. Linn. and Gmel.
Anas Torquata ex insula Novæ Terræ. Briss.
Anas Brimond. Olaff.
Stone Duck. Hist. Kamtschatka.
The Dusky and Spotted Duck. Edw.
The Harlequin Duck. Penn. and Lath.

THIS Duck, though small, short, and round, and of a dusky plumage, is yet one of the handsomest birds of the genus: besides the white streaks which intersect the brown of its garb, the face looks like a mask, with a long black nose and white cheeks; and this black of the nose extends as far as the top of the head, and there joins to two large rufous eye-lids of a very bright bay-colour. The black *domino*, which covers the neck, is edged and intersected below by a little white ribband, which probably induced the fishers at Hudson's Bay to style it *lord* *. Two other little white bands, fringed with black, are placed on each side of the breast, which is iron gray; the belly is dun-gray; the flanks are bright rufous, and the wing exhibits a spangle of purple-blue or burnished steel: there is also a white speckle behind the ear, and a little white serpentine line on the side of the neck.

* Edwards.

The

The female has none of these decorations; her garb is a blackish brown-gray on the head and the mantle; a white gray on the fore side of the neck and of the breast; and a pure white on the stomach and the belly. The bulk is nearly that of the morillon, and the bill is very short and small in proportion.

This species is the same with Steller's *anas picta capite pulchre fasciato* *, or the *mountain Duck* of Kamtschatka, and the *anas histrionica* of Linnæus, which appears in Iceland, according to Brunnich, and occurs not only in the north-east of Asia, but even on lake Baikal, according to Georgi's account, though Krachenninikoff considers this species as peculiar to Kamtschatka †.

* i. e. The painted Duck with a beautifully striped head.

† He says, that in autumn the females are found on the rivers, but not the males. He adds, that these birds are very stupid, and are easily caught in clear water; for when they see a man, they dive, and may be killed at the bottom with strokes of a pole. *History of Kamtschatka.*

[A] Specific character of the Harlequin Duck, *Anas Histrionica*: "It is brown, variegated with white and blue; its ears, a double "line on its temples, its collar, and a bar on its breast, are white." It breeds on the banks of swift streams among the low shrubs: and in winter it repairs to the open sea. It is clamorous, and its flight is lofty and rapid.

The BROWN DUCK.

Anas Minuta. Gmel.

WERE it not for the too great difference in bulk, the reſemblance, almoſt complete, of the plumage would have induced us to refer this ſpecies to *the little brown and white Duck from Hudſon's Bay* of Edwards. But this is only as large as the *ſarcelle*, and the Brown Duck is intermediate between the wild Duck and the golden-eye. It is probable that the individual delineated is only the female of this ſpecies; for it wears the duſky livery appropriated to all the female Ducks. A blackiſh brown ground on the back, and ruſty-brown, clouded with white gray, on the neck and the breaſt; the belly white, with a white ſpot on the wing, and a broad ſpot of the ſame colour between the eye and the bill, are all the daſhes in its plumage. It is probably the ſame with what Rzaczynſki mentions in theſe words, *Lithuania Poleſia alit innumeras anates, inter quas ſunt nigricantes**. He adds, that theſe blackiſh Ducks are known to the Ruſſians by the name of *uhle*.

* i. e. Poliſh Lithuania breeds innumerable Ducks, among which is a blackiſh ſort.

[A] Specific character of the *Anas Minuta:* " It is brown; its " ears white; its primary wing-quills blackiſh."

The GRAY-HEADED DUCK.

Anas Spectabilis. Linn. and Gmel.
Anas Freti Hudſonis. Briſſ.
The King Duck. Penn. and Lath.

WE prefer the appellation of *Gray-headed Duck*, given by Edwards, to that of *Hudſon's Bay Duck*, employed by Briſſon: in the firſt place, becauſe there are many other Ducks in Hudſon's Bay; and in the ſecond place, becauſe an epithet founded on a ſpecific character is always preferable to one drawn from the country. This gray-headed Duck is hooded remarkably with a blueiſh cinereous cowl, falling in a ſquare piece on the top of the neck, and parted by a double line of black points, like inverted commas, and by two plates of pale green which cover the cheeks: the whole is interſected by five black muſtachoes, three of which project to a point on the top of the bill, and two others extend behind under the corners: the throat, the breaſt, and the neck, are white; the back is blackiſh brown, with a purple reflection; the great quills of the wing are brown; the coverts are purple or deep violet, ſhining, and each feather terminated by a white point, of which the ſeries forms a tranſverſe line: there is alſo a

 large

large white ſpot on the ſmall coverts of the wing, and another of a round form on each ſide of the tail; the belly is black; the bill is red, and its upper mandible is parted into two brims, which ſwell, and, to uſe the words of Edwards, nearly reſemble beans. It is the moſt ſingular part, he adds, of the conformation of this bird, which exceeds the ſize of a domeſtic duck. Yet we muſt obſerve, that *the female collared duck from Newfoundland* of our *Planches Enluminées* is much analogous to the gray-headed Duck of Edwards; the chief difference conſiſts in this, that the tints of the back are blacker in the plate of that naturaliſt, and that the cheek is painted greeniſh.

[A] Specific character of the King Duck, *Anas Spectabilis*: "Its bill is compreſſed at the baſe; a black feathery keel; its head "ſomewhat hoary." This bird is very common in Greenland, and affords the natives much down.

The WHITE-FACED DUCK.

THE firſt peculiarity that ſtrikes us in this bird is, that its face is entirely white, contraſted by a black veil that covers the head, and, including the fore ſide and the top of the neck, falls behind: the wing and tail are blackiſh; the

the reſt of the plumage is finely interwoven with waves and feſtoons of blackiſh, ruſty and rufous, of which the tint, deeper on the back, runs into a brick red colour on the breaſt and the lower part of the neck. This Duck, which is found at Maragnon, is larger and more corpulent than our wild duck.

The MAREC and MARECA, BRAZILIAN DUCKS.

1. *Anas Bahamenſis.* Linn. Gmel. Briſſ. and Klein.
 The Ilathera Duck. Cateſby, Penn. and Lath.
2. *Anas Braſilienſis.* Gmel. Ray, Will. and Briſſ.
 The Mareca Duck. Lath.

MARECA is, according to Piſo, the generic name of the Ducks in Brazil; and Marcgrave applies it to two ſpecies, which ſeem not far removed from each other; and for this reaſon we place them together, diſtinguiſhing them however by the names of *Marec* and *Mareca.* The firſt, ſays this naturaliſt, is a duck of ſmall ſize, with a brown bill, and a red or orange ſpot on each corner; the throat and the cheeks are white, the tail gray, the wing decorated with a green ſpangle and a black border. Cateſby, who has deſcribed the ſame bird at

 Bahama,

Bahama, ſays, that this ſpangle on the wing is edged with yellow; but the name of *Bahama duck*, uſed by Briſſon, is the leſs founded, as Cateſby expreſsly remarks, that it appears there very ſeldom, having never ſeen any except the ſubject which he deſcribes.

The Mareca, Marcgrave's ſecond ſpecies, is of the ſame ſize with the other, and its bill and tail are black; a ſpangle ſhines with green and blue on the wing, on a brown ground; a ſpot of yellowiſh white is placed, as in the other, between the corner of the bill and the eye; the legs are vermilion, which, even after cooking, tinges with a fine red. The fleſh of this laſt is, he adds, ſomewhat bitter; that of the former is excellent, yet the ſavages ſeldom eat it, fearing, they ſay, that feeding on an animal that appears unwieldy, they ſhould become themſelves leſs fit for running *.

* Coreal, *Voyage aux Indes Orientales, Paris*, 1722.

[A] Specific character of the Mareca Duck, *Anas Braſilienſis*: "It is brown, below gloſſy cinereous; an ochry-white ſpot between "the bill and the eyes; its chin white; its tail wedge-ſhaped and "black."—Specific character of the Ilathera Duck, *Anas Bahamenſis*: "It is gray; its bill lead-coloured; a fulvous ſpot on its "ſide, a green and yellow ſpot on its wings." It perches on trees, and does not migrate into the north.

The SARCELLES.

THE form which nature has the moſt ſhaded, varied, and multiplied in the water-fowls, is that of the duck. After the great number of ſpecies in that *genus* which we have enumerated, comes a ſubordinate *genus*, almoſt as extenſive as the primary one, and which ſeems to preſent the ſame ſubjects on a ſmaller ſcale. This ſecondary kind is that of the *Sarcelles*, which we cannot better paint in general terms than by ſaying, that they are ducks much ſmaller than the others. But the analogy obtains not only in their natural habits, their ſtructure, and the proportions of their form *; but alſo in the diſpoſition of their plumage, and even in the great difference of colour that takes place between the males and the females.

The Sarcelles were often ſerved up at the Roman tables †: they were ſo much eſteemed, that pains was taken to rear them, like ducks, in the domeſtic ſtate ‡. We ſhould, no doubt,

* Belon.

† *Idem.*

‡ *Nam clauſæ paſcuntur, Anates, Querquedulæ, Boſchides, Phalorides, ſimileſque volucres quæ ſtagna & paludes rimantur.* Columella, *de Re Ruſtica.*

ſucceed

ſucceed alſo; but the ancients apparently employed more care on their poultry-yards, and in general beſtowed much greater attention than we to rural œconomy and agriculture.

We proceed to deſcribe the different ſpecies of Sarcelles, ſome of which, like certain ducks, have ſpread to the extremities of both continents *.

The COMMON SARCELLE.

FIRST SPECIES.

Anas Querquedula. Linn. and Gmel.
Querquedula. Geſner, Klein. Briſſ.
Boſcas. Geſner.
The Garganey. Will. Penn. and Lath. †

Its figure is that of a little duck, its ſize that of a partridge; the plumage of the male, though inferior in the brilliancy of its colours to

* In the plains of Chili, according to Frezier.—On the coaſt of Diemen's land. *Cook.*—In the bay of Cape Holland, at Magellan's Straits. *Wallis.*—In great plenty at Port Egmont. *Byron.*

† In Greek Βοσκας, which Charleton derives from Βοσκω, *to paſture*; M. de Buffon objects, that this appellation is not characteriſtic, for all ducks may be ſaid to paſture. The modern Greeks apply the name *pappi* to all the different ſpecies of ducks. In Italian this bird is called *Sartella, Cercedula, Cercevolo, Garganello:* in Spaniſh *Cerceta:* in German *Murentlein (mumbler), Mittle-entle*

No. 248

THE MALE GARGANEY.

No 249

THE FEMALE GARGANEY.

to that of the drake, is no leſs rich in agreeable reflections, which it would be impoſſible to deſcribe. The fore ſide of the body preſents a beautiful breaſt-plate woven with black or gray, and, as it were, mailed with little truncated ſquares, incloſed in larger, and all diſpoſed with ſo much neatneſs and elegance, that the moſt charming effect is produced. The ſides of the neck and the cheeks, as far as under the eyes, are worked with ſmall ſtreaks of white, vermiculated on a rufous ground: the upper ſide of the head is black, and alſo the throat; but a long white ſtreak, extending from over the eye, falls below the nape: long feathers, drawn to a point, cover the ſhoulders, and recline on the wing in white and black ſtripes; the coverts which reſt on the wings are decorated with a little green ſpangle: the flanks and the rump exhibit hatches of blackiſh-gray on white-gray, and are ſpeckled as agreeably as the reſt of the body.

The attire of the female is much ſimpler: clothed entirely with gray and dun-gray, it hardly ſhows ſome traces of waves or feſtoons on

entle (middle duck), Scheckicht-endtlin (thieviſh-duck): in Low Dutch *Crak-kaſona*; and in ſome parts, as in the neighbourhood of Straſburg, *Kernell*, according to Geſner: in Norwegian *Krak-and*: in Ruſſian *Tchirka*. At Madagaſcar, it is called *Sirire*. In ſome provinces of France *Garſotte*, according to Belon; in others *Halbran*; in the Orleanois, Champagne, and Lorraine, *Arcanette*; in the Milaneſe, and in Picardy, *Garganey*.

on its garb. It has no black on the throat *, like the male; and in general there is ſo much difference between the ſexes in the Sarcelles, as in the ducks, that inexperienced ſportſmen miſtake them, and apply the improper names *tiers*, *racannettes*, *mercannetes*. In ſhort, naturaliſts ought on this, as on other occaſions, to beware of falſe appellations, and not to multiply ſpecies from the mere difference of the colours which are found in theſe birds; it would even be very uſeful, to prevent error, that both the male and female be figured in their true colours.

In the pairing ſeaſon, the male utters a cry like that of the rail; yet the female ſeldom makes her neſt in our provinces †, and almoſt all theſe birds leave us before the 15th or 20th of April ‡. They fly in bands in the time of their migrations, without preſerving, like the ducks, any regular order: they take their flight from above the water, and proceed with great rapidity.

* *Fauna Suecica.*—" There is as much difference between the " male and the female of the Sarcelle as between the ducks and the " drakes. . . . Generally the females are gray round the neck, and " yellowiſh below the belly; brown on the back, the wings, and " the rump." *Belon.*

† Salerne ſays, that he never ſaw its neſt in that part of the Orleanois where he obſerved.

‡ As the Sarcelle ſeldom appears but in winter, Schwenckfeld thence derives its name: *Querquedula, quoniam querquero, id eſt frigido & hyemali tempore, maxime apparet.* [Varro ſays, that it is a ſort of diminutive from the Greek Κερκις, which ſignifies *a weaver's ſhuttle*; on account either of its rapid flight or its whiſtling voice. —*T.*]

They

They do not often bathe, but find their proper food on the ſurface of the lakes, or near the margin: flies, and the ſeeds of aquatic plants, are what they prefer.

Geſner found in their ſtomach little ſtones mixed with theſe aliments; and Friſch, who kept two months a couple of theſe birds taken young, has given us the following detail of their mode of living in this ſort of incipient domeſtication: " I preſented firſt to theſe Sarcelles," he ſays, " different ſeeds, and they would touch " none; but ſcarce had I ſet beſide their water-" trough a baſon, filled with millet, than they " both ran to it. At every bill-full which they " took, each went to the water, and they car-" ried as much of it in a ſhort time as com-" pletely to ſoak the millet. Yet the grain was " not moiſtened ſufficiently to their mind, and " I ſaw my Sarcelles buſy themſelves in carry-" ing millet and water to the ground of their " pen, which was of clay, and when the bottom " was ſoftened and tempered enough, they began " to dabble, and made a pretty deep cavity, in " which they ate their millet mixed with earth. " I put them in a room, and they carried, in the " ſame way, though to little purpoſe, the millet " and water to the deal-floor. I led them on the " graſs, and they ſeemed to do nothing but dig " for ſeeds, without eating the blades, or " even earth-worms. They purſued flies and " ſnapped them like ducks. When I delayed

" to

" to give them their accuſtomed food, they " called for it with a feeble hoarſe cry, *quoak*, " repeated every minute. In the evening they " lay in the corners, and even during the day, " when a perſon went near them, they hid them- " ſelves in the narroweſt holes. They lived thus " till the approach of winter; but when the ſe- " vere cold ſet in, they died ſuddenly."

[A] Specific character of the Garganey, *Anas Querquedula*: " It has a green ſpot on the wings, and a white line above the " eyes.

The LITTLE SARCELLE.

SECOND SPECIES.

Anas Crecca. Linn. and Gmel.
Phaſcas. Geſner.
Querquedula. Id.
Querquedula Major. Johnſt.
Querquedula Minor. Briſſ.
Pepatzca. Fernandez.
The Common Teal. Ray, Penn. and Lath. *

THIS Sarcelle is ſmaller than the firſt, and differs beſides by the colours of its head, which is rufous, and ſtriped with a broad ſtreak

* In German *Tröſſel*, *Kriech-enten (crawl-duck) Kruk-entle (crutch-duck) Graw-entlin (gray-duck)*; and the female *Brunn-kæpficht-entlin (brown-headed duck)*: in Swiſs *Mour-entle, Sor-entle, Söke*: in Poliſh *Cyranka*: in Swediſh *Arta, Kræcka*: in Daniſh *Krik-ard*: in Norwegian *Heſtelort-and*: in Dutch *Täling*: in Mexican *Pepatzca*.

of

of green edged with white, that extends from the eyes to the occiput: the reſt of the plumage is pretty much like that of the common ſarcelle or garganey, except that its breaſt is not richly mailed, but only ſpeckled.

This little Sarcelle breeds on our pools, and continues in the country the whole year. It conceals its neſt among the large bulruſhes, and builds it with their ſtalks, their pith, and with a heap of feathers: this neſt, conſtructed with much care, is pretty wide, and reſts on the ſurface of the water, ſo as to riſe and fall with it. The eggs amount to ten or twelve, and are about the ſize of a pigeon's; they are dirty white, with hazel-ſpots. The females take the whole management of the incubation; the males ſeem to leave them and aſſociate together during that time, but in autumn they return to their families. The teals are ſeen on the pools in cluſters of ten or twelve; and in winter they reſort to the rivers and unfrozen ſprings; there they live on creſſes and wild chervil. On pools they eat the ruſh-ſeeds, and catch ſmall fiſh.

They fly very ſwiftly; their cry is a ſort of whiſtle, *vouire, vouire*, which is heard on the pools as early as the month of March. Hebert aſſures us, that this little Sarcelle is as common in Brie as the other is rare, and that great numbers are killed in that province. According to Rzaczynſki, they are caught in Poland by means of of nets ſtretched from one tree to another; the Teals throw themſelves into theſe nets as they riſe from the pools about the duſk of the evening.

Ray, from the name, *the Common Teal*, which he beſtows on our little Sarcelle, ſeems not to have known the common Sarcelle: Belon, on the contrary, was acquainted with no other; and though he applies to it indiſcriminately the two Greek names *boſcas* and *phaſcas*, the latter ſeems to have referred peculiarly to the little Sarcelle; for Athenæus ſays, that the *phaſcas* is larger than the little *colymbis*, which is the little grebe. This ſpecies has obtained a communication with the new world by way of the north; ſince it is evidently the *pepatzca* of Fernandez, and ſeveral that we have received from Louiſiana differ not from thoſe of Europe.

[A] Specific character of the Common Teal, *Anas Crecca:* "It "has a green ſpangle on the wings, and a white line above and "below the eyes." It is found as far north as Greenland, where it lays from thirteen to nineteen eggs. The teals of America are not ſo prolific.

The SUMMER SARCELLE.

THIRD SPECIES.

Anas Circia. Linn. and Gmel.
Querquedula Æstiva. Briss.
The Summer Teal. Will. Alb. and Lath.

WE should have classed this species with the preceding, if Ray, who appears to have examined both, had not separated them; and we can only copy his account of the bird. "It "is," says he, "somewhat smaller than the com-"mon teal, and is, without exception, the least "of the whole genus; its bill is black; all its "mantle brown cinereous, with the tips of the "feathers white on the back; on the wing is "a bar about the breadth of a finger, black, "with reflections of emerald-green, and edged "with white; all the fore side of the body is "white washed with yellowish, spotted with "black on the breast and the lower belly; the "tail is pointed; the legs are blueish, and their "webs black."

M. Baillon has sent me some notes on a *Summer Sarcelle*, by which he means the little

* In German *Birckilgen*.

Sarcelle of the preceding article, and not the Summer Teal described by Ray; but we cannot forbear inserting his observations, which are important.

"We here (at *Montreuil-sur-mer*) call the "Summer Sarcelle *criquard* or *criquet*; this "bird is well made, and has much grace; its "form is rounder than that of the com-"mon sarcelle or garganey; it is also more "decorated, its colours are more varied and "better contrasted; it has sometimes little blue "feathers, which are not seen but when the "wings are opened. Few water-birds are so "chearful and sprightly as this Sarcelle; it is "almost continually in motion, and bathes in-"cessantly: it is very easily tamed; I kept "some several years in my court, and I still "have two which are very familiar.

"These handsome Sarcelles join to all their "qualities an extreme gentleness. I never saw "them fight either among themselves or with "other birds: they make no defence even when "attacked. As delicate as they are gentle, the "least accident hurts them; the agitation into "which they are thrown if chased by a dog, is "sufficient to occasion their death: when they "cannot escape by the aid of their wings, they "remain extended on the spot, exhausted and "expiring. Their food is bread, barley, wheat, "and bran: they also catch flies, earth-worms, "slugs, and insects.

"They

" They arrive on our marſhes that lie near " the coaſt, about the firſt days of March: I " believe that the ſouth wind brings them. They " do not keep in flocks, like the other Sarcelles " and the whiſtlers. They are ſeen roving on " all ſides, and they pair ſoon after their arrival. " In April they ſeek, in ſlimy ſpots ſcarce ac- " ceſſible, large tufts of ruſhes, or herbs, very " cloſe, and ſomewhat raiſed above the level of " the marſh. They obtain a lodgment by re- " moving the ſtalks that encumber them, and " by continual treading they form a little cavity, " four or five inches in diameter, of which they " line the bottom with dry herbs: the top is " well covered by the thickneſs of ruſhes, and " the entrance is hid by the ſtalks which were " laid there; this entrance, for the moſt part, " faces the ſouth. The female depoſits from " ten to fourteen eggs of a white ſomewhat tar- " niſhed, and almoſt as large as pullets' eggs. I " diſcovered that the time of incubation is, as in " hens, from twenty-one to twenty-three days.

" The young are hatched covered with down, " like the ducklings: they are very alert, and, a " few days after birth, they are conducted by " their parents to the water. They ſeek worms " under the graſs and in the mud. If any rave- " nous bird chance to paſs, the mother makes a " faint cry, and the whole family ſquats, and re- " mains motionleſs till another cry recalls them " to their activity.

 " Their

"Their firſt feathers are gray, like thoſe of "the females. It is then very difficult to diſ-"tinguiſh the ſexes, nor is the difficulty remov-"ed till the love-ſeaſon; for it is a fact peculiar "to this bird, which I have frequently had an "opportunity of verifying, and which I ſhall "here relate:—I commonly procured theſe Sar-"celles about the beginning of March; at that "time the males were arrayed in their moſt "beautiful feathers; the ſeaſon of moulting ar-"rived, they became as gray as the females, and "continued in that ſtate till the month of Ja-"nuary; in the ſpace of a month their feathers "aſſumed another tinge. The preſent year I "have again admired this change; the male "which I have now is as beautiful as it can be, "and I ſaw it as gray as the female. It would "ſeem that nature has attired it for the ſeaſon "of love.

"This bird is not a native of the northern "countries; it is ſenſible to cold. Thoſe which "I had retired regularly to ſleep in the hen-"houſe, and kept themſelves in the ſun or near "the kitchen-fire. They all died of accidents, "moſt of them from the pecks which they re-"ceived from ſtronger birds. However, I have "reaſon to believe that they do not live long, "ſince their full growth is completed in two "months, or thereabout."

[A] Specific character of the Summer Teal, *Anas Circia*: "The ſpangle on its wing is of a various colour; there is a white "line

"line above the eye-brows; its bill and legs are cinereous." This Teal inhabits the lakes and rivers of Europe, and the Caſpian ſea. It is not migratory. Linnæus ſays, that it hatches in from thirty to thirty-three days.

The EGYPTIAN SARCELLE.

FOURTH SPECIES.

Anas Africana. Gmel.
The African Teal. Lath.

THIS Sarcelle is nearly as large as the garganey; but its bill is rather larger and broader: its head, neck, and breaſt, are of a rufous brown, glowing and intenſe; all its mantle is black; there is a ſtreak of white on the wing; the ſtomach is white, and the belly is of the ſame rufous brown with the breaſt.

The female in this ſpecies has nearly the ſame colours as the male, only they are not ſo deep, or ſo finely contraſted; the white of the ſtomach is interſperſed with brown waves, and the colours of the head and breaſt are rather brown than rufous. We have been aſſured that this Sarcelle was found in Egypt.

The MADAGASCAR SARCELLE.

FIFTH SPECIES.

Anas Madagascariensis. Gmel.
The Madagascar Teal. Lath.

THIS Sarcelle is nearly the size of the common teal; but its head and bill are smaller. The character which distinguishes it best is a broad spot of pale-green or water-green, placed behind the ear, and inclosed with black, which covers the back of the head and the neck; the face and the throat are white; the lower part of the neck, as far as the breast, is handsomely worked with little brown fringes in rufous and white; this last colour covers the fore side of the body; the back and the tail are tinged and glossed with green on a black or blackish ground. This Sarcelle was sent to us from Madagascar.

The COROMANDEL SARCELLE.

SIXTH SPECIES.

Anas Coromandeliana. Gmel.
The Coromandel Teal. Lath.

THIS bird is ſmaller than the garganey. The plumage conſiſts of white and dark brown: white predominates on the fore ſide of the body; it is pure in the male, and mixed with gray in the female: the dark brown forms a cowl on the head, ſtains all the mantle, and marks the neck of the male with ſpots and ſpeckles, and the lower part of the neck of the female with little tranſverſe waves; alſo the wing of the male ſhines, on its blackiſh tint, with a green and reddiſh reflection.

The JAVA SARCELLE.

SEVENTH SPECIES.

Anas Falcaria, var. Gmel.
The Falcated Duck, var. Penn. and Lath.

THE plumage of this Sarcelle, on the fore ſide of the body, on the top of the back, and on the tail, is richly worked with black and white feſtoons; the mantle is brown; the throat is white; the head is enveloped in a fine purple violet, with a green reflection on the feathers of the occiput, which extend to the nape, and ſeem parted in ſhape of a bunch: the violet tint re-commences under this little tuft, and forms a broad ſpot on the ſides of the neck; it marks a ſimilar one, accompanied with two white ſpots, on the feathers of the wing next the body. This Sarcelle was brought to us from the iſland of Java; it is as large as the garganey.

No. 250

THE CHINESE GARGANEY

The CHINESE SARCELLE.

EIGHTH SPECIES.

Anas Galericulata. Linn. and Gmel.
Querquedula Sinensis. Briss.
Querquedula Indica. Aldrov.
Anas Sinensis. Klein.
The Chinese Teal. Lath.

THIS beautiful Teal is very remarkable for the richness and the singularity of its plumage: it is painted with the most vivid colours, and adorned on the head with a magnificent green and purple bunch, which extends beyond the nape; the neck and the sides of the face are enriched with narrow and pointed feathers of an orange red; the throat is white, and also the part above the eyes; the breast is of a purple or wine rufous; the flanks are pleasantly worked with little black fringes, and the quills of the wings are elegantly bordered with white streaks: to these beauties, add a remarkable singularity, that two feathers, one on each side, between those of the wing next the body, have on the outside of their shaft webs of an uncommon length, of a beautiful orange rufous, fringed with white and black on the edge, which form, as it were,

were, two fans or two broad papilionaceous wings raiſed on the back: theſe two ſingular feathers diſtinguiſh ſufficiently this Sarcelle from all the others, beſides the beautiful creſt which uſually floats on its head, but which it can erect. The beautiful colours of this bird have ſtruck the eyes of the Chineſe: they have painted it on their porcelain and their fineſt paper. The female, which they have alſo delineated, appears uniformly in a brown ſuit; and this is indeed its colour, with ſome mixture of white. In both ſexes, the bill and the legs are red.

This beautiful Sarcelle is found in Japan as well as in China; for we may perceive it to be the *kimnodſui**, of whoſe beauty Kæmpfer ſpeaks with admiration: and Aldrovandus relates, that the embaſſadors, who came in his time from Japan to Rome, brought, among other rarities of their country, figures of that bird.

* "There is (in Japan) a ſort of duck which I cannot help "ſpeaking of, becauſe of the remarkable beauty of the male, called "*kimnodſui*; it is ſo exquiſite, that when its picture was ſhown to "me, I could not believe it to be a faithful likeneſs, till I ſaw the "bird itſelf, which is pretty common. Its feathers form a ſhade of "the moſt beautiful colours imaginable; but red predominates "about the neck and the throat; its head is crowned with a magnificent tuft; its tail, which riſes obliquely, and its wings, which "are placed on the back in a ſingular faſhion, exhibit to the eye an "object as ſingular as it is extraordinary." *Natural Hiſtory of Japan.*

The FEROE SARCELLE*.

NINTH SPECIES.

Querquedula Ferroensis. Briss.

THIS Sarcelle, which is somewhat smaller than the garganey, has all its plumage of an uniform white gray on the fore side of the body, of the neck and of the head; only it is slightly spotted with blackish behind the eyes, and also on the throat and the sides of the breast: all the mantle, with the upper surface of the head and of the neck, is of a dull blackish, without any reflections.

ALL the preceding species of Sarcelles are inhabitants of the ancient continent; those which we are now to describe belong to the new: and though the same species of water-fowl are often common to both worlds, yet each of the species of Sarcelles seems to be appropriated to the one or other continent, except the garganey and the teal, which are found in both.

* Called *Oedel* in the island of Feroe, according to Brisson.

The SOUCROUROU SARCELLE.

TENTH SPECIES.

Anas Discors. Linn. and Gmel.
Querquedula Americana. Briss.
Querquedula Minor Varia. Barrere.
Anas Querquedula Americana Variegata. Klein.
The White-faced Teal. Catesby.
The White-faced Duck. Penn. and Lath.

THIS species is common in Cayenne, where it is called *Soucrourou.* It is nearly the size of the garganey: the male is richly festooned and waved on the back; the neck, the breast, and all the fore side of the body, are spotted with blackish on a rusty brown ground; on the top of the wing is a beautiful plate of light blue, below which is a white streak, and then a green spangle; there is also a broad streak of white on the cheeks; the upper side of the head is blackish, with green and purple reflections: the female is quite brown.

These birds are found in Carolina, and probably in many other parts of America. Their flesh is, according to Barrere, delicate and well tasted.

[A] Specific character of the *Anas Discors*: "The coverts of "its wings are blue; its secondary wing-quills are green on the "outside; there is a white bar on the front."

The SOUCROURETTE SARCELLF.

ELEVENTH SPECIES.

Anas Discors, var. Gmel.
Querquedula Virginiana (fœmina). Briss.
Anas Quacula. Klein.
The Blue-winged Teal. Catesby and Lath.

THOUGH the Cayenne Sarcelle represented in our *Planches Enluminées* is smaller-sized than Catesby's *Blue-winged Teal*, the great resemblance in their colours induces us to regard them as the same species; and we are much inclined to class both with the preceding, and have therefore adopted a similar name. The *Soucrourette* has on the shoulder a blue plate with a white zone below, and then a green spangle, exactly as in the *soucrourou*: the rest of the body, and the head, are covered with spots of brown-gray, waved with white-gray. Catesby's figure does not show this mixture, but presents a brown colour, spread too uniformly, that would suit the female, which, according to him, is entirely brown. He adds, that these birds come in great numbers to Carolina in the month of August, and remain there till the middle of October, at which time they gather rice in the fields, being very fond of that grain. In Virginia, he says, where there is no rice, they eat a sort of wild oats that grow in the

the ſwamps. When fed in either of theſe ways, they become extremely fat, and their fleſh acquires an exquiſite reliſh *.

The SPINOUS-TAILED SARCELLE.

TWELFTH SPECIES.

Anas Spinoſa. Gmel.
The Spinous-tailed Teal. Lath.

THIS ſpecies of Sarcelle, which is a native of Guiana, is diſtinguiſhed from all the others by the tail-feathers, which are longer, and terminated by a little ſtiff filament like a ſpine, formed by the point of the ſhaft, produced a line or two beyond the webs of theſe feathers, which are blackiſh brown. The plumage of the body is unvaried, conſiſting of waves or blackiſh ſpots, deeper on the upper ſurface, lighter on the under, and feſtooned with white-gray in a ruſty or yellowiſh ground: the top of the head is blackiſh, and two ſtreaks of the ſame colour, parted by two white ſtreaks, paſs, the one as high as the eye, the other lower on the cheek; the quills of the wing are alſo blackiſh. This Sarcelle is ſcarcely eleven or twelve inches long.

* Mr. Latham, after Briſſon, reckons this to be the female of the preceding.—*T.*

The LONG-TAILED RUFOUS SARCELLE.

THIRTEENTH SPECIES.

Anas Dominica. Gmel.
Querquedula Dominicenſis. Briſſ.
Chilcanauhtli. Fernandez.
Colcanauhtli (fem). Id.
The St. Domingo Teal. Lath.

THIS is ſomewhat larger than the preceding, and differs much in its colours: it has however the character of the long tail, with the quills terminating in a point, though the unwebbed ſhaft is not ſo nicely defined. We will not venture to claſs theſe two ſpecies together, but we conceive them to be related. The upper ſide of the head, the face, and the tail, are blackiſh; the wing is of the ſame colour, with ſome blue and green reflections, and has a white ſpot; the neck is of a fine cheſnut-rufous; the flanks are of the ſame colour; and the upper ſurface of the body is waved with it on blackiſh.

This Sarcelle was ſent to us from Guadaloupe. Briſſon received one from St. Domingo, and refers it, with the utmoſt probability, to the *chilcanauhtli* of New Spain, deſcribed by Fernandez, who ſeems to denominate the female of the ſame ſpecies *colcanauhtli.*

The WHITE AND BLACK SARCELLE; *or*, the NUN.

FOURTEENTH SPECIES.

Anas Albeola. Linn. and Gmel.
Querquedula Ludoviciana. Briss.
The Little Black and White Duck. Edw.
The Spirit Duck. Penn.

A WHITE robe, a white band with a black cap and mantle, have procured this Louisiane Sarcelle the name of Nun (*Religieuse*). It is nearly as large as the garganey. The black of its head is decorated with green and purple lustre, and the white band encircles it behind from the eyes. " The Newfoundland fishers," says Edwards, " call this bird *Spirit*, I know not " for what reason, unless because it is a very " nimble diver: the instant after it has plunged, " it appears again at a very great distance; a power " which might recall to the imagination of the " vulgar the fantastic ideas of apparitions."

[A] Specific character of the Spirit, *Anas Albeola*: " It is white; " its back and wing-quills are black; its head blueish, and the back of " the head white." It extends over the whole of North America. It nestles in trees, near fresh water.

The MEXICAN SARCELLE.

FIFTEENTH SPECIES.

Anas Novæ Hiſpaniæ. Gmel.
Querquedula Mexicana. Briſſ.
Toltecoloctli, ſeu Metzcanauhtli. Fernandez.
The Mexican Duck. Lath.

FERNANDEZ gives this Sarcelle the Mexican name *Metzcanahachtli*, or *Metzcanauhtli*, which ſignifies, he ſays, *moon-bird*; becauſe it is hunted by moon-light. He adds, that it is one of the moſt beautiful ſpecies of the genus: almoſt its whole plumage is white, dotted with black, eſpecially on the breaſt; the wings exhibit a mixture of blue, of green, of fulvous, of black and white; the head is blackiſh brown, with varying colours; the tail is blue below, blackiſh above, and terminated with white: there is a black ſpot between the eyes and the bill, which is black below and blue above.

The female, as in all the ſpecies of this genus, differs from the male by its colours, which are not ſo diſtinct and vivid. The epithet which Fernandez gives it, *avis ſtertrix junceti*, ſeems to imply that it clears away or cuts the ruſhes, to form or place its neſt.

The CAROLINA SARCELLE.

SIXTEENTH SPECIES.

Anas Rustica. Linn.
Querquedula Carolinensis. Briss.
The Little Brown Duck. Lath.

THIS Sarcelle is found in Carolina, near the mouths of rivers, where the water begins to taste saltish. The plumage of the male is broken with black and white, like a magpie. The female, which Catesby describes at greater length, has its breast and belly of a light gray; all the upper side of the body and of the wings is deep brown; there is a white spot on each side of the head behind the eye, and another on the lower part of the wing. It is evident that Catesby gave it the appellation of the *little brown duck* from the garb of the female: he had better called it *the magpie-teal*, or *the black and white teal.*

The BROWN AND WHITE SARCELLE.

SEVENTEENTH SPECIES.

Anas Minuta. Linn.
Querquedula Freti Hudſonis. Briſſ.
The Little Brown and White Duck. Edw.

THIS bird, though called a duck by Edwards, ought to be ranged among the Sarcelles, ſince it has nearly the ſize and figure of the firſt ſpecies, the garganey: but the colour of its plumage is different; it is entirely of dark brown on the head, the neck, and the quills of the wing. The deep brown dilutes into whitiſh on the fore part of the body, which is beſides ſtriped acroſs with brown lines: there is a white ſpot on the ſides of the head, and a ſimilar one on the corner of the bill. This Sarcelle dreads not the moſt intenſe cold, ſince it is one of thoſe which inhabit the bottom of Hudſon's Bay *.

* Teals are reckoned among the number of birds that are ſeen to paſs in the ſpring at Hudſon's Bay, on their way to breed in the north. *Hiſt. Gen. des Voy.* *tom.* xv. *p.* 267.

SPECIES WHICH ARE RELATED TO THE DUCKS AND SARCELLES.

After the defcription and hiftory of the fpecies well known and difcriminated, it remains to indicate thofe to which the following accounts feem to refer; in order that obfervers and travellers may difcover to what preceding fpecies each belongs; or, if different, to delineate the new fpecies.

I. We muft mention the ducks commonly called *four wings*, of which the Collection of the Academy fpeaks in thefe terms. "About 1680, "appeared in the Boulonois a kind of ducks, "which had their wings turned differently from "others, the great feathers parting from the "body and projecting out: which has occafioned "the people to fay, and believe, that they have "four wings." (*Collect. Acad. Part. Etr. tom.* i. *p.* 304). We conceive that this character might be accidental, from the bare comparifon of the preceding paffage with the following. "M. "l'Abbé Nollet faw in Italy a flock of geefe, "among which were many that feemed to have "four wings: but this appearance, which took "place only when the bird flew, was caufed by "the inverfion of the laft portion of the wing, "which kept the great feathers elevated inftead "of lying flat along the body. Thefe ducks "came

"came from the ſame hatch with the reſt which "carried their wings as uſual; and neither of "their parents had its pinions folded back." *Hiſtoire de l'Academie*, 1750, *p.* 7.

Thus theſe ducks, like the four-winged geeſe, muſt not be conſidered as peculiar ſpecies, but as accidental and even individual varieties, which may occur in any kind of birds.

II. The duck, or rather the very little teal, mentioned by Rzaczynſki in the following paſſage: *Lithuania Poleſia alit anates innumeras, inter quas . . . ſunt . . . in cavis arborum natæ, molem ſturni non excedentes (Hiſt. p.* 269.) [Poliſh Lithuania maintains ducks innumerable, among which...are...that breed in the hollows of trees, and exceed not the bulk of a ſtare.] If this author is accurate with regard to the ſize, which he makes to be ſo diminutive, we muſt confeſs that the ſpecies is unknown to us.

III. The white-headed Barbary duck of Dr. Shaw, which is not the ſame with the muſk duck, but ought rather to be claſſed with the Sarcelles, ſince it is only, he ſays, of the *ſize of the lapwing:* it has a broad, thick, blue bill; its head entirely white, and its body flame-coloured.

IV. The *anas platyrinchos* of the ſame author, who calls it the *Barbary pelican*, improperly, ſince nothing can be further from a pelican than a duck. This is ſmaller than the preceding: its legs are red; its bill flat, broad, black, and indented; its

 breaſt,

breaſt, belly, and head are flame-coloured; its back is of a deeper caſt; and there are ſpots on the wing, a blue, a white, and a green.

V. The ſpecies which this traveller, with equal inaccuracy, denominates *the little-billed Barbary pelican*. "It is," ſays he, "ſomewhat "larger than the preceding; its neck is reddiſh, "and its head is adorned with a little tuft of "tawny feathers; its bill is entirely white, and "its back variegated with a number of white "and black ſtripes; the feathers of the tail are "pointed, and the wings are each marked with "two contiguous ſpots, the one black and the "other white; the extremity of the bill is black, "and the legs are of a deeper blue than thoſe of "the lapwing." This ſpecies appears to us much a-kin to the foregoing.

VI. The *turpan*, or *tourpan*, a Siberian duck, found by Gmelin in the vicinity of Selinginſki, of which he has given an account too ſhort for recognizing the bird *. It appears, however, that this ſame duck is found in Kamtſchatka, and is even common to Ochotſk, where, at the mouth of the river Ochotſka, multitudes are

* "In the neighbourhood of Selinginſk, we found a ſmall lake, "whoſe ſides were covered with ſwans, *geeſe*, tourpans, and ſnipes: "I cannot expreſs the ſatisfaction which the ſight of theſe birds be-"gat; their ſong, inſpired by nature, was as pleaſing as the imita-"tion with inſtruments would be diſagreeable. The tones of the "tourpans reſemble much thoſe of an hautbois; and, in this concert "of birds, they performed nearly the part of the baſs. This bird is "a kind of duck; its plumage is fox-red, except the tail and the "wings, which have a great mixture of black." *Gmelin.*

 caught

caught in boats, as deſcribed by Kracheninikoff. We ſhall remark, with regard to this traveller, that he mentions his meeting with eleven ſpecies of ducks and ſarcelles at Kamtſchatka; in which we can only aſcertain the turpan, and the long-tailed duck of Newfoundland: the nine remaining are called, according to him, *ſeloſni, tchirki, krohali, gogoli, lutki, tcherneti, pulonoſi, ſuaſi, and the mountain duck.* "The four firſt," ſays he, "paſs the winter near fountains; the "reſt arrive in ſpring, and retire in autumn, like "the geeſe." We may preſume, that many of theſe ſpecies might be referred to thoſe which we have deſcribed, had this obſerver told us any thing more than their names.

VII. The little duck of the Philippines, called at Luçon the *ſaloyazir*, and which, according to Camel, being not larger than the hand, ſhould be regarded as a ſarcelle.

VIII. The *woures-feique*, or *hatchet-bird* of Madagaſcar, a ſort of duck, ſo called by the iſlanders, ſays Francis Cauche, becauſe it has on its front an excreſcence of black fleſh, which is round, and extends, bending back a little on their bill, like their hatchets. This traveller adds, that this ſpecies is of the ſize of our goſlings, and of the plumage of our ducks. We will add, that it is perhaps only a variety *.

IX.

* Flacourt names three or four kinds of teals, or *ſirvire*, which, he ſays, occur in Madagaſcar:—*Tahie*; its cry ſeems to articulate this name;

IX. The two ſpecies of ducks, and the two ſarcelles, ſeen by Bougainville at the Malouine or Falkland Iſlands, of which he ſays, that the firſt differ not much from thoſe of our countries, adding, however, that he killed ſome which were entirely black, and others entirely white. With regard to the ſarcelles, " the one," ſays he, " is of *the ſize of the duck*, with its bill blue; " the other is much ſmaller, and of the latter " are ſome whoſe ventral feathers are *tinged with* " *carnation*." Theſe birds are very plentiful in theſe iſlands, and are well taſted.

X. The ducks of the Straits of Magellan*, which,

name; its wings, its bill, and its legs, are red: *Halive*, has its bill and legs red: *Hach*, has its plumage gray, and its wings ſtriped with green and white: *Tatach*, is a kind of halive, but ſmaller.

* " The ducks (at Magellan's Straits) are conſiderably dif- " ferent from ours, and much inferior; they are pretty numerous, " and poſſeſs a particular diſtrict in the iſland, upon the lofty rocks, " out of the reach of muſket-ſhot. I never in my life ſaw ſo much " art and induſtry in animals void of reaſon; they are ſo arranged " on the heights, that the greateſt geometer could not diſtribute the " ſpace to better advantage; all the diſtricts are divided by little " paths, no broader than to allow a bird to walk; the ground on " which the neſts are placed is ſmoothed, as if it were levelled by " the hand of man; the neſts are formed of kneaded earth, and " ſeem as if they were caſt upon the ſame mould; the ducks carry " water in their bill, with which they make a mortar of clay, and " faſhion it into a round ſhape, as well as with a pair of compaſſes; " the bottom is a foot broad, the mouth eight inches wide, and of " an equal height; they are all alike with reſpect to form and pro- " portions: theſe neſts ſerve them more than a year, and their eggs " are hatched, I believe, in the ſun. We could not find, in the whole " place, a ſingle ſtalk of graſs, or ſtraw, or feathers, or birds' dung; " the whole was as clean and neat, both in the neſts and the paths, " as if it had been newly waſhed and ſwept." *Hiſt. des Navigations aux Terres Auſtrales, tom.* i. *p.* 243.

which, according to ſome navigators, conſtruct their neſts after a ſingular faſhion, with kneaded mud, and plaſtered with the utmoſt neatneſs; if this account be true, which from ſeveral circumſtances ſeems ſuſpicious, and little to be depended on.

XI. The *painted duck* of New Zealand, ſo named in Captain Cook's Second Voyage, and thus deſcribed: "The largeſt is as big as a "Muſcovy duck, with a very beautiful varie-"gated plumage, on which account we called it "the painted duck; both male and female have a "large white ſpot on each wing; the head and "neck of the latter is white, but all the other "feathers, as well as thoſe on the head and neck "of the drake, are of a dark variegated colour." *Vol.* i. *pp.* 96 & 97.

XII. *The ſoft-billed whiſtling duck*, otherwiſe called the *blue-gray duck* of New Zealand; remarkable for this property, that its bill is ſoft and almoſt cartilaginous, inſomuch that it cannot ſubſiſt but by gathering, or, ſo to ſpeak, by ſucking the worms which the tide leaves on the beach.

XIII. The red-combed duck, alſo of New Zealand, which was found only on the river at the bottom of Duſky Bay: this duck, which is only a little larger than the ſarcelle, is of a very gloſſy dark gray on the upper ſide of the back, and of a deep gray ſoot-colour on the belly; the bill

bill and the legs are lead-coloured; it has a golden iris, and a red creſt on its head.

XIV. Laſtly, Fernandez gives ten ſpecies as belonging to the genus of ducks. We ſhall throw into notes the Mexican names *, and the deſcriptions,

* "*Xalcuani*, or ſand-ſwallower.—It is a kind of wild duck, "ſomewhat ſmaller than the tame; its bill moderately broad; the "feathers on the under ſide of the body white, and thoſe on the breaſt "and on the upper ſide fulvous, but others bright white run acroſs; "the wings and the tail are greeniſh, variegated above with bright "white, with black, and with brown, and below with white and ci- "nereous; a green band runs from the back of the head to the eyes; "the reſt of the head is white, inclining to cinereous, and mixed "with ruſſet and blackiſh; the legs are longer in proportion than "the reſt of the body, of a ruſſet-colour: This bird viſits the lake." *Cap.* 121. *p.* 39.

"*Yacatexotli*, or blue-billed bird.—It is almoſt as big as the tame "duck; its bill is ſky-blue above, and reddiſh white below; the up- "per ſide of the body is fulvous, and the under ſide ſilvery black; "the upper part of the wing black." *Cap.* 70. *p.* 29.

"*Yztactzonyayauhqui* (different from that of p. 28.)—It is a kind "of ſmall wild duck; its bill is blue, and marked near the tip with "a white ſpot; the legs incline alſo to blue; and the reſt of the body "is variegated with white and fulvous." *Cap.* 156. *p.* 45.

"*Colcanauhtlicioubt*.—It is a wild duck; the greater part of its "upper ſide brown, and a ſmall part whitiſh; its under ſide is white "and partly brown, except the wings, which below are en irely of "a bright white. The head is black and cinereous at its upper part, "but inclining to a deep black, and below to cinereous." *Cap.* 64. *p.* 28.

"*Atapalcatl*, or water-pot.—It would be exactly like the teal, if "it had not its bill twice as broad; its colour whitiſh and fulvous; "it bites the hand angrily, but without hurting it.

"*Tzonyayauhqui*, or variegated-head (*male*).—It is a wild duck "that lives about the lake, and is almoſt as large as a tame duck: "its bill is broad, above ſky-blue, only marked with two ſpots, and "having

deſcriptions, which are for the moſt part incomplete; waiting till new obſervations, or the inſpection of the ſubjects themſelves, enable us to complete and arrange them.

" having a ſmall ſlender projection with which it bites; the under " part blackiſh blue; the legs ſhort and blue, ſometimes mixed how- " ever with a pale colour; its head and neck thick, and of a peacock- " colour at the ſides, the top ſometimes blacker, however; the breaſt " is black: the ſides of the belly and of the body are whitiſh, al- " though black lines, running tranſverſely, decorate the tail; a black " tawny bar, three inches broad, and extending to the end of the " tail, marks the back; finally, the wings are tinged promiſcuouſly " with black, fulvous, bright white, and cinereous." *Cap.* 108. *p.* 36.

" *Nepapantototl.*—It is a wild duck, frequent in the Mexican lake, " its bill ending ſomewhat ſquare; in other reſpects ſimilar, except " that there is no ſort of colour which uſually decorates the wild " ducks, but falls to the ſhare of this, and beſtows on it ornament " and beauty, whence is derived its name." *Cap.* 127. *p.* 40.

" *Opipixcan.*—It is a wild duck with a reddiſh bill; its thighs and " its legs variegated with rufous and whitiſh; the reſt of its body " cinereous and black. *Cap.* 147. *p.* 44.

" *Perutototl.*—A Peruvian duck; which being already known in " our world, I ſhall not take the trouble to deſcribe." *Cap.* 16. " *p.* 47.

" *Concanauhtli.*—A kind of large duck, like our *lavancos*, and " which, for that reaſon, we have deemed it unneceſſary to deli- " neate."

The PETRELS.

Procellariæ of Linnæus.

Of all the marine birds, thoſe which the moſt conſtantly live on the great ſeas, are the Petrels; the moſt ſtrangers to the land, the moſt adventurous in roving on the vaſt ocean: they commit themſelves with equal confidence and audacity to the rolling billows, the impetuous winds, and ſeem to brave the fury of the tempeſts. In the remoteſt portions of the globe, in every zone which navigators have viſited, theſe birds ſeemed to expect their arrival, and even to have ſtretched beyond them into more diſtant and more ſtormy latitudes. Every where they have been ſeen to ſport with ſecurity, and even gaiety, on that element, ſo terrible in its fury, which unnerves the moſt intrepid man: as if nature meant to demonſtrate, that the inſtincts and faculties which ſhe has allotted to the inferior creatures, excel the combined powers of our reaſon and our art.

Furniſhed with long wings, accommodated with palmated feet, the Petrels add to the eaſe and nimbleneſs of flying, and to the facility of ſwimming,

ſwimming, the ſingular power of running and walking on the water, razing the waves in a rapid paſſage, their body being ſupported horizontally, and balanced by the wings, and their feet ſtriking alternately and precipitately on the ſurface. Hence is derived the Engliſh name *Petrel*, or *Peterel*, which alludes to St. Peter's walking on the ſea.

The ſpecies of the Petrels are numerous: they have all large and ſtrong wings; yet they riſe not to a great height, and commonly they raze the water in their flight. They have three toes connected by a membrane; their two lateral toes have a ledge on their outer part; their fourth toe is only a little ſpur that riſes immediately from the heel, without joint or phalanx *.

The bill, like that of the albatroſs, is articulated, and ſeems compoſed of four pieces, two of which, as if they were added portions, form the extremities of the mandibles. There are alſo, along the upper mandible, near the head, two little tubes or flat rolls, in which the noſtrils are perforated. From its general conformation, the bill would ſeem to be that of a ravenous bird, for it is thick, ſharp, and hooked at its extremity: but this figure of the bill is not exactly the ſame in all the Petrels, and the difference is even ſuch as to afford a character for the ſubdiviſion of the genus. In fact, the point of the upper

* Willughby calls this ſpur, *a little hind-toe*, not imagining that it proceeds immediately from the heel.

upper mandible alone is in many ſpecies bent into a hook; the point of the lower, on the contrary, is channelled and truncated like a ſpoon.—Theſe ſpecies are the ſimple Petrels. In others, the points of both mandibles are ſharp, reflected, and form together the hook. This difference of character has been remarked by Briſſon; and we think that it ought not to be omitted or rejected, as done by Forſter.—We ſhall denominate theſe ſpecies *puffin-petrels*.

All theſe birds, the puffins as well as the Petrels, ſeem to have the ſame inſtinct and common habits in hatching. They inhabit the land only during that time, which is pretty ſhort; and, as if they were ſenſible of the incongruity of that reſidence, they hide or rather bury themſelves in holes under the rocks by the ſea-ſhore. From the bottom of theſe holes is heard their diſagreeable voice, which would generally be taken for the croaking of a reptile *. They lay few eggs: they feed and fatten their young by diſgorging into their bill the half-digeſted, oily ſubſtance of fiſh, which are their chief and almoſt only ſupport. But they have a ſingular property, of which perſons who ſeek their neſts ought to be

* The Petrels bury themſelves by thouſands in holes under ground; there they rear their young, and lodge every night. *Forſter's Obſervations.*—The woods (at New Zealand) reſound with the noiſe of the Petrels, concealed in holes under ground, which croak like frogs, or cluck like hens. It would ſeem that all the Petrels make their neſts uſually in ſubterraneous cavities; for we ſaw the blue kind in ſuch lodgments at Duſky Bay. *Idem.*

well aware: when attacked, they, whether from fear, or the hope of defending themſelves, diſcharge the oil with which their ſtomach is filled; they ſpout it in the face of the fowler; and as their neſts are uſually lodged on rocky ſhores, in the clifts of lofty precipices, ignorance of this fact has coſt ſome obſervers their lives *.

Forſter remarks, that Linnæus knew little of the Petrels, ſince he reckons only *ſix ſpecies*; whereas Forſter diſcovered himſelf *twelve new ſpecies* in the South Sea. It is to be wiſhed, that this learned voyager would deſcribe all theſe ſpecies: meanwhile, we can only give thoſe which we know from other ſources.

* In the General Advertiſer, for June 1761, is the following remarkable account from the Iſle of Mull: "A gentleman of the name "of Campbell, being fowling among the rocks, and having mounted "a ladder to take ſome birds out of their holes, was ſo ſurprized, by "one of this ſpecies ſpurting a quantity of oil in his face, that he "quitted his hold, fell down, and periſhed."—Smith, in his Hiſtory of Kerry, mentions the ſame property of the ſtormy petrel.

The CINEREOUS PETREL.

FIRST SPECIES.

Procellaria Glacialis. Linn. and Gmel.
Procellaria Cinerea. Briſſ.
The Wagel of the Corniſh. Ray.
The Fulmar Petrel. Penn. and Lath.

THIS Petrel inhabits the northern ſeas. Cluſius compares its ſize to that of a middling hen: Rolandſon Martin, a Swediſh obſerver, ſays that it is equal in bulk to a crow. The firſt of theſe authors finds a reſemblance in its port and figure to a falcon: indeed its bill, ſtrongly jointed and much hooked, is formed for rapine; the hook of the upper mandible, and the truncated channel which terminates the under, are of a yellowiſh colour; and the reſt of the bill, with the two tubulated noſtrils, are blackiſh in the dead ſubject which we deſcribe; but we are aſſured that the bill is entirely red as well as the legs in the living bird: the plumage of the body is a cinereous white; the mantle is blue cinereous, and the quills of the wing are of a deeper blue, and almoſt black: the feathers are very cloſe and full, clothed below with a thick and fine down, with which the ſkin of the body is completely inveſted.

Obſervers

Nº 251

THE CINEREOUS PETREL.

Obſervers agree to give this Petrel the name of *haff-hert* or *hav-heſt*, that is, ſea-horſe; "be-"cauſe," ſays Pontoppidan, "it utters a ſound "like the neighing of a horſe, and the noiſe "which it makes in ſwimming is like the trot "of that animal." But it is difficult to conceive how a bird ſwimming can occaſion a noiſe like a horſe's trot. Was not the name impoſed becauſe of the Petrel's running on the water? The ſame author adds, that theſe birds invariably follow the boats employed in fiſhing for ſea-dogs, in expectation of the entrails that are thrown out. He ſays, that they faſten ſo keenly on the dead whales, or ſuch as are wounded and riſen to the ſurface, that the fiſhermen knock them down with ſticks, and yet cannot diſperſe the reſt of the flock. Hence Rolandſon Martin applies to them the name *mallemucke*; which, as we have formerly remarked, belongs properly to a gull.

Theſe cinereous petrels are found from the ſixty-ſecond degree of north latitude to the eightieth. They fly among the ice of thoſe regions, and when they are ſeen on the main, making towards land for ſhelter, it is, as in the *tempeſt-bird* or *little petrel*, a ſign to navigators of an approaching ſtorm.

[A] Specific character of the Fulmar, *Procellaria Glacialis*: "It "is whitiſh, and its back ſomewhat hoary." This bird inhabits the iſland of St. Kilda, on the weſt of Scotland, the whole year, except during the months of September and October. It breeds about the middle of June, laying but a ſingle egg, which is large, white, and very brittle. The iſlanders feed on its fleſh, ſtuff their beds with its

its down, and chear their tedious winter-nights with lamps ſupplied with its oil.

The dead ſubject deſcribed in the text, was perhaps a variety of the ſhearwater, as Gmelin and Latham ſtate; but the hiſtorical part of the article belongs undoubtedly to the Fulmar. See alſo *Species the Eighth.*

The WHITE and BLACK PETREL; or the CHECKER.*

SECOND SPECIES.

Procellaria Capenſis. Linn. and Gmel.
Procellaria Nævia. Briſſ.
Pardela. Ulloa.
The White and Black ſpotted Peteril. Edw.
The Pintado Petrel. Lath.

THE plumage of this Petrel, marked with white and black, regularly interſected and checkered, has procured it the name *damier (cheſs-board)* from our navigators. For the ſame reaſon the Spaniſh have termed it *pardelas*, and the Portugueſe *pintado*, which the Engliſh have adopted. It is nearly the ſize of a common pigeon, and, as it has in its flight the air and port of that bird, the ſhort neck, the round head, its length fourteen or fifteen inches, and

* Damier, *i. e.* Cheſs-board: I have adopted the word *checker*, for the ſake of ſhortneſs.—*T.*

its

No. 252

THE PINTADO.

its alar extent thirty-two or thirty-three, navigators have often ſtiled it the *ſea pigeon.*

The Checker has its bill and legs black; the outer toe is compoſed of four joints, the middle one of three, and the inner of two only; inſtead of a little toe it has a pointed ſpur, hard, a line and a half long, and the point turned outwards; the bill has over it the two little tubes or rolls in which the noſtrils are perforated; the point of the upper mandible is curved, that of the lower is channeled, and, as it were, truncated: this character places the Checker among the family of Petrels, and excludes it from that of the puffins. The upper ſide of its head is black, the great quills of its wings are of the ſame colour, with white ſpots; the tail is fringed with white and black, and when ſpread it *reſembles,* ſays Frezier, *a mourning ſcarf*; its belly is white, and its mantle is regularly interſperſed with black and white ſpots. This deſcription correſponds perfectly with what Dampier has given of the *pintado* *. The male and female ſcarce

* The *pintadoes* are admirably ſpeckled with white and black; their head is almoſt black, as well as the end of the wings and the tail: but in this black of the wings there appear white ſpots about the ſize of half-a-crown when it flies, and the ſpots are then beſt ſeen. The wings are alſo bordered entirely round with a ſlender black edging, which gradually becomes more dilute, and approaches to a dull gray on the back of the bird: the inner edge of the wings, and the back itſelf, from the head to the end of the tail, are enamelled with an infinite number of handſome round ſpots, white and black, of the ſize of a half-penny; the belly, the thighs, the flanks, and the under ſurface of the wings, are light gray. *Dampier.*

 differ

differ ſenſibly from each other in bulk or in plumage.

The Checker, as well as many other Petrels, receives birth on the antarctic ſeas; and if Dampier conſidered them as belonging to the ſouthern temperate zone *, it was becauſe that voyager did not ſufficiently penetrate into that cold, gloomy ocean: for Captain Cook aſſures us, "that theſe Petrels, and alſo the blue Petrels, "frequent every portion of the South Sea in the "higheſt latitudes." The beſt obſervers agree likewiſe, that they are very rarely met with before paſſing the tropic †; and it appears from many relations ‡, that the firſt latitudes where theſe

* We ſaw pintadoes when about two hundred leagues from the coaſt of Brazil, and thence till we approached nearly the ſame diſtance from New Holland. The pintado is a native of the ſouthern hemiſphere, and of the temperate part of it; at leaſt I hardly ever ſaw any to the north of the thirty-firſt degree of ſouth latitude. *Dampier.*

† The Checker is an inhabitant of the temperate and frigid zones of the ſouthern hemiſphere; and if a few pairs of theſe birds follow veſſels beyond the tropic, they halt but a ſhort time: and hence the Checker and the tropic-bird are ſeldom ſeen at once. *Obſervations communicated by the Viſcount de Querhoënt.*—On the 4th of October, in 25° 29′ ſouth latitude, a great number of ſmall common Petrels, of a ſooty-brown with a white rump (*procellaria pelagica*) flew about us; the air was cold and piercing: next day the albatroſſes and the pintadoes (*procellaria capenſis*) appeared for the firſt time. *Cook.*

‡ The following days we ſaw the ſame birds in greater numbers, nor did they leave us till we were very far beyond the Cape: ſome were black on the back and white under the belly, having the upper ſide of the wings variegated with theſe two colours, nearly like a cheſs-board: they are ſomewhat larger than a pigeon. There are others ſtill bigger than the former, blackiſh above and entirely

theſe birds begin to be found in numbers, are in the ſeas near the Cape of Good Hope; they occur alſo on the ſame parallel about the coaſts of America *. Admiral Anſon ſought for them unſucceſsfully at the iſland of Juan Fernandez; yet he perceived many of their holes, and he concluded that the wild dogs which were ſpread through this iſland had chaſed them away or deſtroyed them. But in another ſeaſon he might have there found theſe birds, ſuppoſing that the time he before made the ſearch was not that of their hatching; for, as we have already ſaid, they never reſide on land, except when detained by incubation, but ſpend their days in open ſea, reſting on the water in calm, and even dwelling on it when it rolls in commotion; they ſeat themſelves in the hollow between two waves,

entirely white below, except the extremity of their wings, which appears of a velvet black, and which the Portugueſe call *mangas de velado. Tachard.*—Dampier was, according to his reckoning, 1,200 leagues eaſt of the Cape. Nothing occurred remarkable on this run, except that he was accompanied by numbers of birds, eſpecially pintadoes. *Hiſt. Gen. des Voy. tom.* xi. *p.* 217.

* "In the paſſage from Rio de Janeiro to Port Deſire, and "about the latitude of 36° ſouth, we began to ſee a great number "of birds about the ſhip, many of them very large, of which ſome "were brown and white, and ſome black: there were among them "large flocks of pintadoes, which are ſomewhat larger than a pi- "geon, and ſpotted with black and white." *Byron's Voyage, p.* 9. —In this latitude (43° 30′ ſouth, on the coaſts of Brazil) and in that of Cape Blanc, which is in 46°, we ſaw numbers of whales and new birds like pigeons, their plumage regularly mottled with black and white; which has made the French give them the name *damier*, and the Spaniards, *pardela. Frezier.*

with their wings expanded, and are borne up by the wind.

Since they are almoſt continually in motion, their ſleep muſt be much interrupted. They are accordingly heard flying about veſſels at all hours of the night *: in the evening they often aſſemble under the poop, ſwimming at eaſe, and approaching the ſhip with a familiar air, and at the ſame time emitting their grating, hoarſe voice, which cloſes in ſomething like the cry of a gull †.

In their flight they glance the ſurface of the water, and, at intervals, dip their feet, which they hold pendent. It appears that they live on the fiſh ſpawn which floats on the ſea ‡: however, the Checker is ſeen, with the crowd of other ſea-fowls, to faſten greedily on the carcaſes of whales §. They are caught by a hook baited with a bit of fleſh ‖: ſometimes alſo they are entangled by the wings in the lines that drag at the ſhip's ſtern. When taken and car-

* Obſervation of the Viſcount de Querhoënt.

† *Idem.*

‡ In the ſtomach of thoſe which I opened I found a thick white mucilage, which I believe to be fiſh-ſpawn.

§ Dampier.

‖ *Lettres Edifiantes*, xv. *Recueil, p.* 341. Approaching the iſland of St. Helena, two hundred leagues from the land of Nativity, a number of birds came to the ſides of our veſſel: we took them in plenty with bits of fleſh with which we covered our hooks: they are as large as a pigeon, their feathers checquered with black and white, which was the reaſon that we called them *damiers*; their tail is broad, and their foot is like that of a duck. *Cauche.*

ried

ried aſhore, or ſet on the deck, they will jump, but cannot walk, or riſe on wing. This alſo is the caſe with moſt ſea-birds, which inceſſantly fly and ſwim at large: they cannot walk on the firm ground, and it is equally impoſſible for them to commence their flight. It is remarked even that, on the water, they wait till, raiſed by the ſwelling wave, they catch the wind, and are ſprung through the air.

Tho' the Checkers appear uſually in flocks * on the vaſt ſeas which they inhabit, and where a ſort of ſocial inſtinct holds them together, we are aſſured that a more particular and a very marked attachment binds the male and female, and that ſcarce has the one alighted on the water, than the other haſtens to join it; that they mutually invite each other to partake of the food which chance has thrown in their way; and laſtly, that if one of the pair is killed, the whole flock give ſigns of regret, by alighting and ſtaying ſome minutes beſide the dead body, but that the ſurviving mate ſhows evident marks of tenderneſs and ſorrow; that it pecks its inanimate companion, as if to recall it to life; and after the reſt of the troop has retired, it long continues to mourn over the corpſe †.

* All the pintadoes go generally in flocks, and almoſt ſweep the water as they fly. *Dampier.*

† Cloſe of the obſervations which the Viſcount de Querhoënt made at ſea, and which he obligingly communicated.

[A] Specific character of the *Procellaria Capensis*: "It is variegated with white and brown." It lays an egg of the size of a hen's in the month of December, which corresponds to June in our hemisphere. It is said to chatter like a parrot, if taken and confined.

The ANTARCTIC PETREL; or BROWN CHECKER.

THIRD SPECIES.

Procellaria Antarctica. Gmel.

THIS Petrel resembles the Checker, except the colour of its plumage, of which the spots, instead of black, are brown on a white ground. The denomination of Antarctic Petrel, given to it by Captain Cook, seems to suit it perfectly, since it occurs only in the highest southern latitudes *; while many species of Petrels, common in the lower latitudes, particularly that of the black checker, appear not in those dismal regions.

In the second voyage of that great navigator, he gives the following account of this new spe-

* In 62° 10′ south latitude, and 172° longitude, we saw the first island of ice, and at the same time we perceived an Antarctic Petrel, some gray albatrosses, pintadoes, and blue petrels. *Cook.*—In latitude 66°, Captain Cook saw some Antarctic Petrels in the air. —In 67° 8′, he was visited by a small number of Antarctic Petrels.

cies

cies of Petrels. "In 67° 15′ ſouth latitude, we "ſaw numbers of whales playing about the iſlands "of ice. Two days after, we remarked many "flocks of *pintados*, brown and white, which I "called *Antarctic Petrels*, becauſe they ſeemed "peculiar to thoſe regions. They are in every "reſpect ſhaped like the *pintadoes*, from which "they differ only in colour; the head and the "fore ſide of their body are brown, and the hind "part of their back, their tail, and the extremities "of their wings, are white." In another part, he ſays, "While we were collecting ice, we "caught two *Antarctic Petrels*, and upon exa-"mining them, we were ſtill diſpoſed to believe "that they belonged to the family of the Pe-"trels. They are nearly of the ſize of a large "pigeon; the feathers of the head, the back, "and a part of the upper ſide of the wings, are "of a light brown; the belly, and the under ſide "of the wings, are white; the feathers of the "tail are white alſo, but brown at the tips. I "remarked that theſe birds had more plumage "than thoſe we had ſeen; ſo careful is nature "to accommodate the cloathing to the climate. "We found theſe Petrels among the ſnow."

Yet theſe Petrels, ſo common among the floating iſlands of ice, diſappear, as well as all the other birds, when the firm ice is approached, whoſe formidable bed extends very far into the polar regions of the ſouthern continent. Of this fact we are informed by that great navigator,

navigator, the firſt and the laſt perhaps of mortals, that has dared to viſit the frozen barriers which nature gradually forms and enlarges in proportion as our globe cools. "After our arrival amidſt the ice," he ſays, "no Antarctic Petrel any more called our attention."

[A] Specific character of the *Procellaria Antarctica*: "It is brown; below blueiſh white; its tail white tipt with black; its legs lead-coloured."

The WHITE PETREL, or SNOWY PETREL.

FOURTH SPECIES.

Procellaria Nivea. Gmel.

This Petrel is very juſtly denominated the *Snowy Petrel*, not only on account of the whiteneſs of its plumage, but becauſe it is always met with in the vicinity of the frozen regions, and announces to the navigator in the South Sea his approach to the ice-iſlands. Captain Cook, when he firſt ſaw them at a diſtance, termed them *white birds* *; but afterwards he diſcovered

* "At noon we were in the latitude of 51° 50′ ſouth, and longitude 21° 3′ eaſt, where we ſaw ſome white birds about the ſize of "pigeons,

diſcovered from the ſtructure of their bill that they belonged to the genus of Petrels. They are as large as a pigeon; their bill is blueiſh-black; their legs are blue, and their plumage ſeems to be entirely white.

"When we approached a broad ridge of ſolid "ice," ſays Forſter, the learned and laborious companion of the illuſtrious Cook, "we ob-"ſerved at the horizon, what the Greenland-"men call an *ice-twinkle*; inſomuch that, from "the appearance of this phænomenon, we were "ſure of meeting ice at a few leagues diſtance. "Then it was that we commonly ſaw flights of "White Petrels of the ſize of pigeons, which "we called *Snowy Petrels*, and which are the "fore-runners of the ice."

Theſe White Petrels, intermingled with the antarctic petrels, ſeem to have conſtantly accompanied theſe adventurous navigators in all their traverſes amidſt the iſlands of ice, as far as the vicinity of the immenſe *glaciere* of the ſouthern pole. The flight of theſe birds on the waves, and the motion of ſome whales in the icy flood, are the laſt, and the only objects that preſerve the remains of animation in thoſe frightful regions, the ſcene of expiring nature.

"pigeons, with blackiſh bills and feet. I never ſaw any ſuch be-"fore; and Mr. Forſter had no knowledge of them. I believe "them to be of the Peterel tribe, and natives of theſe icy ſeas. At "this time we paſſed between two ice iſlands, which lay at a little "diſtance from each other." *Cook's ſecond Voyage, vol.* i. *pp.* 22 *and* 23.

[A] Specific

[A] Specific character of the *Procellaria Nivea*: "It is snowy; "the shafts of its feathers, and its bill, are black; its legs are dull "blue."

The BLUE PETREL.

FIFTH SPECIES.

Procellaria Vittata. } Gmel.
Procellaria Cærulea. }
The Vittated Petrel. Forster.
The Blue-billed Petrel. Lath.

THE Blue Petrel, so called because its plumage is blue gray, as well as its bill and legs, occurs only in the South Seas, from the twenty-eighth to the thirtieth degree of latitude, and thence towards the pole. Captain Cook was accompanied from the Cape of Good Hope as far as the forty-first degree by flocks of these Blue Petrels, and flocks of Checkers, whose numbers the rough sea and boisterous winds seem to augment. He again saw the Blue Petrels in the fifty-fifth degree to the fifty-eighth; and, no doubt, they inhabit all the intermediate points of these southern latitudes.

It is remarked as a peculiarity in these Blue Petrels, that their bill is exceeding broad, and their tongue very thick: they are somewhat larger than the snowy petrels *. In the blue gray

* The Blue Petrel has nearly the size of a little pigeon. *Cook.*

gray tint that covers the upper ſide of the body, we perceive a deeper band, cutting tranſverſely the wings and the lower part of the back: the end of the tail is alſo of the ſame deep blue or blackiſh caſt: the belly and the under ſide of the wings are of a blueiſh-white. Their plumage is thick and abundant. "The Blue Petrels, which are ſeen in this immenſe ſea," (between America and New Zealand) ſays Mr. Forſter, "are no leſs provided againſt the cold than the penguins. Two feathers, inſtead of one, grow from each root; they are laid one upon another, and form a very warm covering. As they are continually in the air, their wings are very ſtrong and long. We found them between New Zealand and America, more than ſeven hundred leagues from land; a ſpace which it would be impoſſible for them to traverſe, were not their bones and muſcles prodigiouſly firm, and were they not aided by long wings."

"Theſe ſailor-birds," continues Mr. Forſter, "live perhaps a conſiderable time without food. . . . Our experience demonſtrates and corroborates in ſome reſpects this ſuppoſition: when we wounded ſome of theſe Petrels, they inſtantly diſcharged a quantity of viſcous aliments, newly digeſted, which the others ſwallowed with an avidity that betrayed a long faſting. It is probable, that in thoſe frozen ſeas there are many ſpecies of *molluſca*, which

"riſe

" rise to the surface in fine weather, and serve " to support these birds."

The same observer again found these Petrels in vast numbers assembled to nestle in New Zealand. " Some were flying, others were in " the middle of the woods, under the roots of " trees, in the crevices of rocks where they could " not be caught, but where they undoubtedly " hatch their young. The noise which they " made resembled the croaking of frogs. None " appeared in the day, but they flew much dur- " ing the night."

These Blue Petrels were of the broad-billed species which we have just described; but Captain Cook seems to point out another in the following passage: " We killed Petrels; many " were of the blue kind, but they had not a " broad bill, as those of which I have spoken " above; and the end of their tail was tinged " with white, instead of deep blue. Our natu- " ralists could not agree, whether this form of " the bill, and this shade of colour, distinguished " only the male from the female *." It is not probable that such a difference in the fashion of the bill could take place between the male and female of the same species; and it would seem, that we ought to admit two species of Blue Petrels, the first with a broad bill, and the second with a narrow bill, and the tip of the tail white.

* " We were in the fifty-eighth degree of south latitude." *Cook.*

[A] Specific

[A] Specific character of the *Procellaria Vittata*: "It is blueish "cinereous; below white; its legs black." The other Blue Petrel is termed *Procellaria Cærulea*, and is thus characterized: "It is "blueish cinereous; below white; the bill and legs blue."

The GREATEST PETREL;

QUEBRANTAHUESSOS of the Spaniards.

SIXTH SPECIES.

Procellària Gigantea. Gmel.
The Osprey Petrel. Forst. Obs.
Glupisha. Hist. Kamtsch.
Ossifraga, or *Break-bones.* Ulloa.
The Giant Petrel. Penn. and Lath.

QUEBRANTHUESSOS signifies *bone-breaker*; and this denomination refers no doubt to the force of the bill of this great bird, which is said to approach the bulk of the albatross. We have not seen it; but Forster, a learned and accurate naturalist, describes its magnitude, and ranges it among the Petrels. In another place, he says, "We found at Staten-"land gray petrels, of the size of the albatross, "and of the species which the Spaniards term "*Quebrantahuessos*, or bone-breaker." Our sailors called this bird *Mother Carey's Goose*; they ate it, and found it pretty good. A circumstance which

which the more aſſimilates it to the Petrels, is, that it ſeldom appears near veſſels but on the approach of ſtormy weather. This is related in the *Hiſtoire Generale des Voyages*: ſome deſcriptive details are there added, which appear however too uncertain to be adopted, and which we ſhall therefore be contented to throw into a note*.

* The pilots in the South Sea have long remarked, that a day or two before a north-wind blows, a ſort of birds, which they ſee at no other time, then advance to the coaſt, and hover about veſſels: they are called *quebranthueſſos* (that is, *bone-breakers*); and they are obſerved to alight and float on the waves beſide the ſhip till the weather calms. It is pretty ſtrange that, except at this time, they never appear either on water or on land, and that we know not their retreats, which they ſo punctually leave when their inſtinct forewarns them of danger. This bird is ſomewhat larger than a duck; its neck is thick, ſhort, and a little curved; its head large, its bill broad, and not long; its tail ſmall, its back raiſed, its wings ſpacious, its thighs ſmall: ſome have the plumage whitiſh, in others it is ſpotted with dull brown; in others the whole craw, the inner part of the wings, the lower part of the neck, and the whole of the head, are perfectly white; but the back, and the upper part of the wings and of the neck, are brown verging on black; hence they are called *lomos-prietos* (blackiſh-backs): they are reckoned the ſureſt forerunners of foul weather. *Hiſt. Gen. des Voy. tom.* xiii. *p.* 498.

[A] Specific character of the Giant Petrel, *Procellaria Gigantea*: "It is browniſh ſpotted with white; below white; its ſhoulders, "its wings, and its tail, are brown; its bill and legs yellow." It is forty inches long. It is nimble, and lives on fiſh and the carcaſes of ſeals. Its fleſh is palatable food.

The PUFFIN-PETREL.

SEVENTH SPECIES.

Procellaria Puffinus. Linn. and Gmel.
Puffinus. Briff.
Puffinus Anglorum. Will. Ray, and Sibb.
Avis Diomedea. Gefn. Aldrov. Johnft. and Charlt.
Larus Piger Cunicularis. Klein.
Sterna Medica. Brown.
The Manks' Puffin, or *Puffin of the Isle of Man.* Johnft. Will. and Edw.
The Shear-water Petrel. Will. Penn. and Lath. *

THE character of the branch of *Puffins* in the genus of Petrels confifts, as we have faid, in both mandibles being hooked and bent downwards; a ftructure undoubtedly of very little advantage to the bird, and which, in the ufe of its bill and in the act of feizing, allows the upper mandible to exert fmall force on the reflected part of the lower. The noftrils are of a tubulated form, as in all the Petrels; the ftructure of its feet with the fpur at the heel, as well as the general fhape of its body, are the fame. It is fifteen inches long; its breaft and belly are white; a gray tint is fpread over the whole

* In Norway it is called *Skraap:* in the Feroe iflands *Skrabe*; and the young *Liere.*

 upper

upper side of the body, pretty clear on the head, and which becomes deeper and blueish on the wings and the tail, in such manner however that each feather appears fringed or festooned with a lighter tint.

These birds reside in our seas, and seem to have their rendezvous in the Scilly islands, but more especially on the Calf of Man: they resort there in multitudes during the spring, and begin by making war on the rabbits, the only inhabitants of that rock; they drive these from their burrows, of which they take possession. They lay two eggs, one of which, it is said, usually never hatches: but Willughby positively asserts, that they have only a single egg. As soon as the chick is hatched, the mother leaves it early in the morning, and returns not till evening. During the night she feeds it, disgorging at intervals the substance of the fish which she caught in the course of the day at sea. The aliment, half digested in her stomach, turns into a sort of oil, which she gives to her young one. This nourishment makes it extremely fat; and, at this time, some fowlers land on the rocky islet, where they lodge in huts, and catch multitudes of the young birds in their burrows. But to render this game palatable, it must be cured with salt, in order to temper in part the rankness of its excessive fat. Willughby, from whom we borrow these facts, adds, that as the fowlers have a custom of cutting away a foot from each of theſe birds, for the ſake of reckoning the number caught, the people entertain a notion that they are hatched with a ſingle foot.

Klein pretends, that the name *Puffin* or *Pupin* is formed from the cry of the bird. He remarks, that this ſpecies has its times of appearance and diſappearance; which muſt indeed be the caſe with birds that never come on land but to neſtle, and that dwell on the ſea ſometimes in one latitude, ſometimes in another, always attending the ſhoals of little migratory fiſh, or their collections of ſpawn, on both which they feed.

Though the obſervations above related were all made in the northern ſea, it appears that this ſpecies is not excluſively attached to that part of our globe. It is common on all ſeas, for it is the ſame with the *Jamaica ſhear-water* of Brown, and the *artenna* of Aldrovandus. In ſhort, it ſeems to frequent equally the different portions of the ocean, and even to advance into the Mediterranean, as far as the Gulf of Venice and the *Tremiti* iſles, anciently called *the iſles of Diomede*. All that Aldrovandus ſays, whether of the figure or of the natural habits of his *artenna*, correſponds with thoſe of the ſhear-water. He aſſures us, that the cry of theſe birds reſembles exactly the wailing of a new-born infant. Finally, he is diſpoſed to believe that they are *the birds of Diomede**, famous in antiquity from an affecting fable. It was of those Greeks, who, with their valiant leader, pursued by the wrath of the gods, were found in those islands metamorphosed into birds, which still retaining something human, and a tender remembrance of their ancient country, flocked to the shore when the Greeks disembarked, and seemed, by their tender accents, to express their melancholy regret. But this interesting mythology, whose fictions, too much censured by persons of cold temper, diffused to the apprehension of sensible minds so much grace, life, and charms in nature, appears really to allude, in this instance, to a point in natural history, and to have been imagined from the moaning voice of these birds.

* Ovid, speaking of these birds of Diomede, says:

Si volucrum quæ sit dubiarum forma requiris,
Ut non cygnorum, sic albis proxima cygnis.

This does not come very near to the Petrel; but poetry and mythology are here so blended, that we cannot expect to find exact traces of nature. Linnæus was not very happy in applying his erudition, when he gave the name of *Diomedea* to the albatross; since this large bird occurs only in the seas of the east and south, and was therefore unknown to the Greeks.

[A] Specific character of the Shear-water, *Procellaria Puffinus*: "Its body is black above, and white below; its legs are rufous."

Nº. 253

THE FULMAR, FROM THE ISLAND OF Sᵗ. KILDA.

The FULMAR; or, WHITE-GRAY PUFFIN-PETREL OF THE ISLAND OF ST. KILDA.

EIGHTH SPECIES.

FULMAR is the name which this bird has at the island of St. Kilda. It seems to us a species closely related to the preceding; the only difference being this, that the plumage of the under side of the body is white-gray in the Fulmar, and blueish-gray in the shear-water.

"The Fulmar," says Dr. Martin *, "feeds "on the back of living whales; its spur serves "to hold it firm on their slippery skin, without "which precaution they would be blown off by "the wind, always violent in those stormy seas. "... If one attempts to seize or even touch the "young Fulmar in its nest, it spurts from its bill "a quantity of the oil in the person's face."

This eighth species is the same with the first, which was not so distinctly described as usual.

* Voyage to St. Kilda, *London*, 1698, *p.* 55.

The BROWN PUFFIN-PETREL.

NINTH SPECIES.

Procellaria Æquinoctialis. Linn. and Gmel.
Puffinus Capitis Bonæ Spei. Briss.
Plautus Albatross Spurius Major. Klein.
Avis Diomedea. Redi Dissert.
The Great Black Petrel. Edw. and Lath.

EDWARDS, though he gives this bird under the name of the *great black Peteril*, remarks, that the uniform colour of its plumage is rather blackish brown than jet black. He compares its size to that of a raven, and describes very well the conformation of its bill, which character places it among the Puffins. "The nostrils," says he, "seem to have been two tubes joined "together, which rising from the fore part of "the head, advance about a third of the length "of the bill, of which both points bent down-"wards into a hook, look like two pieces added "and soldered."

Edwards reckons this species a native of the seas adjacent to the Cape of Good Hope; but this is merely conjecture.

[A] Specific character of the *Procellaria Æquinoctialis*: "It is "brown and spotless; its bill bright yellow; its legs brown."

N°. 254

THE STORMY PETREL.

The STORMY PETREL.

L'OISEAU DE TEMPETE. *Buff.*

TENTH SPECIES.

Procellaria Pelagica. Linn. and Gmel.
Procellaria. Briff.
Plautus Minimus, Procellarius. Klein.
The Storm-finch. Will. and Penn.
The Small Petrel. Edw. and Borlafe.
The Gourder. Smith's Hift. Kerry.
The Affilag. Martin's St. Kilda, and Sibbald's Hift. Fife *.

THOUGH the epithet *ftormy* is applicable more or lefs to all the Petrels, yet navigators have agreed to appropriate it to this fpecies. The Stormy Petrel is the laft in the order of fize, not exceeding that of a finch; whence it has fometimes received its name. It is the fmalleft of all the palmipede birds; and one might be furprifed that fo little a bird fhould expofe itfelf on the ocean at an immenfe diftance from land. But amidft its audacity, it ftill feems confcious of its weaknefs, and it is the firft that feeks fhelter from the impending ftorm.

* In Swedifh *Stormwaders Vogel:* in Norwegian *St. Peder's Fugl, Soren Peder, Veften Vinds Are, Sonden Vinds Fugl*; and in the Feroo iflands, it is called *Strunk Vit.*

By

By force of inſtinct, it perceives thoſe indications which eſcape our ſenſes; and its motions and its approach warn the ſailors to be prepared for the tempeſt *.

When, in calm weather, theſe little Petrels are ſeen to flock behind a veſſel, flying on the wake, and ſheltering themſelves under the ſtern, the mariners haſten to furl the ſails, and prepare for the ſtorm, which infallibly comes on a few hours after †. Thus, the appearance of theſe birds at ſea is at once diſmal and ſalutary; and nature would ſeem to have diſperſed them over the wide ocean to convey the friendly intelligence. The ſpecies of the Stormy Petrel is univerſally diffuſed: " It is found," ſays Forſter, " equally in the northern and the ſouthern ſeas, " and almoſt in all latitudes." Many ſailors have

* Cluſius.

† More than ſix hours before the ſtorm, it foreſees its approach, and haſtens to ſhelter itſelf beſide the veſſels which it deſcries at ſea. *Linnæus, in the Stockholm Memoirs.*—On the 14th of May, between the iſland of Corſica and that of Monte Chriſto, we ſaw behind the veſſel a flock of *Petrels,* known by the name of *Storm-birds.* When theſe birds arrived, it was three o'clock in the afternoon; the weather was fine, the wind ſouth-eaſt, and almoſt calm: but at ſeven o'clock the wind turned into the ſouth-weſt with much violence, the ſky thickened and grew ſtormy, the night was very dark, and repeated flaſhes of lightning augmented the horror, the ſea ſwelled prodigiouſly, and we were obliged to paſs the whole night under a reefed main-ſail. *Extract from the Journal of a Navigator.*—It would ſeem that many navigators apply the name of *alcyon* to the Stormy Petrel, or ſome other ſpecies, which follows their veſſels, but is very different from the kingfiſher, or the alcyon of the ancients.

averred,

averred, that they met with these birds in every track of their voyages *. But they have not n that account been the easier to catch; they have long even escaped the search of observers, because, when shot, they were almost always lost in the eddy of the ship's wake, which swallowed up their little body †.

The Stormy Petrel flies with amazing swiftness by means of its long wings, which are pretty much like those of the swallow ‡. It can rest amidst the tumbling billows; it shelters itself in the hollow between two high waves, where it remains a few seconds, though the swell rolls on with extreme rapidity. In these watery undulating furrows it runs, like the lark in the furrows of ploughed land, it supports and moves itself not by flying but by running, in which, balanced on its wings, it with astonishing swiftness razes and strikes the surface of the water with its feet §.

The

* These birds fly on all the coasts of the Atlantic Ocean, and are seen on the shores of America, as well as those of Europe, several hundred leagues from land; sea-faring people generally reckon their appearance as the prognostic of a storm. *Catesby.*—I have seen many of these birds together in the broadest and most northern parts of the German Ocean, when they must have been upwards of a thousand English miles from land. *Edwards.*

† "One of these birds," says Linnæus, "was fired at on wing, "but missed; yet it was not intimidated by the report; and perceiv-"ing the wad, it alighted, mistaking it for food, and was caught by "the hand."

‡ Salerne.

§ "You would say it was Pegasus, if you saw it running like light-"ning on the water." *Clusius.*—Though their feet are formed for swimming,

The colour of the plumage of this bird is a blackiſh brown or a ſmoky black, with purple reflections on the fore ſide of the neck, and on the coverts of the wings, and other blueiſh reflections on their great quills: the rump is white; the point of its folded and croſſed wings proiects beyond the tail; its legs are pretty tall, and, like the other Petrels, it has a ſpur inſtead of a hind toe; and as the two mandibles are bent downwards, it belongs to the family of *Puffins*.

It appears that there is a variety in this ſpecies: the little Petrel of Kamtſchatka has the tips of its wings white*; that of the Italian ſeas, which Salerne deſcribes minutely, and at the ſame time diſcriminates from the ſtormy Petrel †, is, according to that ornithologiſt,

ſwimming, they are alſo calculated for running; indeed they moſt commonly uſe them in the latter, for they are ſeen very often running ſwiftly on the ſurface of the waves, when thrown into the greateſt commotion. *Cateſby*.

* The *procellariæ*, or the birds that foretell ſtorms, are about the bulk of a ſwallow; they are entirely black, except the wings, whoſe tips are white. *Hiſtory of Kamtſchatka*.

† "It is not," ſays he, "larger than the *ſea-finch*; its head is almoſt wholly blue, as well as its craw and its ſides, with reflections of violet and of black; the upper ſide of its neck is green and purple, changing like that of the pigeon; the top of its wings and its rump are ſpeckled with white; all the reſt is black: it has a very quick, confident look. This bird ſeems to be a ſtranger to land, at leaſt no perſon can ſay that he ever ſaw it on the coaſt: its preſence is a ſure ſign of an approaching ſtorm, though the ſky, the air, and the ſea, betray no indication of it, but are calm and ſerene: at this time they do not fly one by one, but they all direct their flight to ſome veſſel which they deſcry from a diſtance, and at which they meet." *Salerne*.

tinged with blue, violet, and purple. But we think that theſe colours are nothing elſe than the reflections with which the dull ground of its plumage is gloſſed. And with reſpect to the white or whitiſh feathers on the coverts of the wing, which Linnæus mentions in his deſcription of the little Swediſh Petrel, which is the ſame with ours, the difference ariſes undoubtedly from the age. [A]

To this little Petrel we ſhall refer the *rotje* of Greenland and Spitzbergen, which the Dutch navigators ſpeak of; for though their accounts are in ſome reſpects incongruous, they are ſufficient to ſhew the identity of the *rotje* and our Stormy Petrel. "The *rotje*," according to theſe voyagers, "has a hooked bill . . . it has only "three toes, which are connected by a mem- "brane . . . it is almoſt black over all the body, "except on the belly, which is white: ſome alſo "have their wings ſpotted with black and white. "... In other reſpects, it much reſembles a ſwal- "low." Anderſon ſays, that *rojet* ſignifies *little rat*, and that "this bird has, in fact, the black colour, "the diminutive ſize, and the cry of a rat *." It ſeems

[A] Specific character of the Stormy Petrel, *Procellaria Pelagica*: "It is black; its rump white." Its length is ſix inches; its alar extent thirteen. This bird is particularly frequent on the Atlantic Ocean: it is ſilent in the day, and clamorous during the night. The ſailors call it the *Witch*.

* They cry *rottet, tet, tet, tet, tet,* at firſt very high, and afterwards lowering the tone gradually; perhaps this cry has occaſioned

ſeems that theſe birds never come aſhore in Spitzbergen and Greenland, but to breed their young: they place their neſt, like all the Petrels, in narrow deep holes, under the ruins of fallen rocks, on the coaſts, and cloſe on the water's edge. As ſoon as the young are able to come out of the neſt, the parents accompany them, and ſlip out of their holes into the ſea, and return not to land.

With regard to the *little diving Petrel* of Cook and Forſter, we ſhould have alſo given it the ſame arrangement, had not theſe voyagers indicated, by that epithet, a habit which we know not in our Stormy Petrel, that of diving *

Finally,

ſioned their receiving the name of *rotje:* they make more noiſe than any other bird, becauſe their voice is ſhriller and more piercing. They build their neſts with moſs, and ſome on the mountains, where we killed a great number of the young ones with ſticks: they feed on certain gray worms reſembling crabs ... they alſo eat red ſhrimps and lobſters. We killed ſome of theſe birds, for the firſt time, on the ice, on the 29th of May; but afterwards we took many at Spitzbergen. Theſe birds are very good to eat, and the beſt next to thoſe which are called *ſtrand copers runers* (ſhore-runners); they are fleſhy and fat. *Recueil des Voyages du Nord*; *Rouen*, 1716, *tom.* ii. *p.* 93.

* In Queen Charlotte's Sound (at New Zealand), we ſaw great flocks of little diving Petrels *(procellaria tridactyla)* flying or ſitting on the ſurface or the ſea, or ſwimming under water to a conſiderable diſtance with aſtoniſhing agility. They appeared to be exactly the ſame with thoſe which we had met with in our ſearch for Kerguelin's land, in the 48th degree of latitude. *Cook.*—In latitude 56° 46′, longitude 139° 45′, the weather became fair, and the wind veered to the S. W. About this time we ſaw a few ſmall divers (as we called them) of the Peterel tribe, which we judged to be ſuch as are uſually ſeen near land, eſpecially in the

Finally, we ſhall refer, not indeed to the Stormy Petrel, but to the tribe of Petrels in general, the ſpecies hinted at in the following notices.

I. The Petrel, which Captain Carteret's ſailors called *Mother Carey's Chicken*, "which appeared," he ſays, "to walk on the water, and "of which we ſaw many from the time we "cleared the Straits of Magellan, along the coaſts "of Chili *." This Petrel is probably one of theſe which we have deſcribed; perhaps the *quebranta-bueſſos*, called *Mother Carey* by Cook's people †.—A word on the ſize of this bird would have decided the queſtion.

the bays and on the coaſt of New Zealand. I cannot tell what to think of theſe birds. Had there been none of them, I ſhould have been ready enough to believe that we were, at this time, not very far from land; as I never ſaw one ſo far from land before. Probably theſe few had been drawn thus far by ſome ſhoal of fiſh; for ſuch were certainly about us, by the vaſt number of blue Peterels, albatroſſes, and ſuch other birds as are uſually ſeen in the great ocean: all, or moſt of them, left us before night. *Cook's Second Voyage*, *vol.* i. *pp.* 260 & 261.

[The bird mentioned in theſe extracts is the diving Petrel of Latham, and the *Procellaria Urinatrix* of Gmelin, which is thus characterized: "It is brown and deep black; its under ſide white; its "bill and chin black; its feet blue green, and having three toes." It is eight inches and a half long.—*T.*]

* It is alſo the ſame probably which Wafer mentions in the following terms. "The gray birds (of the iſland of Juan Fernandez) "are nearly of the bulk of a ſmall pullet, and make holes in the "ground like rabbits; in theſe they lodge night and day; they go "a-fiſhing."

† Our author's conjecture is right; it is the giant Petrel.—*T.*

II. The

II. The *devil birds* of Father Labat, of which we can hardly determine the ſpecies, notwithſtanding all that this prolix author ſpeaks of it. We ſhall give his account, much abridged. "The devils, or *diablotins*, begin," ſays he, "to "appear at Guadaloupe and St. Domingo about "the end of the month of September. They "are then found two and two in each hole. "They diſappear in November, and appear "again in March; at which time the mother is "found in her hole with two young ones, which "are covered with a thick and yellow down, and "are lumps of fat: they are now called *cottons*. "They are able to fly, and they depart about the "end of May. During this month many are "caught, and the negroes live on nothing elſe. "... The great ſulphur mountain (*ſoufrière*) "in Guadaloupe is all bored, like a warren, with "the holes which theſe devils excavate: but as "they ſelect the ſteepeſt parts, it is very dan-"gerous to catch them ... All the night we "ſpent on that mountain, we heard the great "noiſe made by them going out and in, and call-"ing and anſwering each other... By our mutual "aſſiſtance, dragging each other with cords, we "reached places ſtocked with theſe birds. In "three hours our four negroes took thirty-eight "devils out of their burrows, and I ſeventeen ... "A young devil newly roaſted is a delicious "food... The old devil is nearly of the ſize of "a pullet ready to lay; its plumage is black; "its

"its wings are broad and ſtrong; its legs are "pretty ſhort; its toes are furniſhed with ſtout "and long claws; the bill is hard, and very "hooked, pointed, and an inch and an half long; "it has large eyes level with its head, which "ſerve admirably for ſeeing in the night, but are "ſo uſeleſs during the day, that it cannot bear "the light, or diſcern objects; inſomuch, that if "it be overtaken by day, while out of its retreat, "it daſhes againſt every thing it meets, and at "laſt tumbles to the ground . . . and hence it "never goes to ſea but in the night."

What Father Dutertre ſays of the *devil-bird* does not aſſiſt us to diſcover this. He ſpeaks only from the reports of fowlers; and all that we can infer from the natural habits of the bird is, that it is a Petrel.

III. The *Alma de Maeſtro* of the Spaniards, which appears to be a Petrel, and might even be referred to the checker, if the account given of it were a little more preciſe, and did not begin with an error, by applying the name *pardela*, which conſtantly applies to the checker, to two Petrels, a gray and a black, with which it does not correſpond *.

IV. The

* In the paſſage from Chili to Peru, at a great diſtance from land, we ſaw birds remarkable for roving on the ocean; they are called *pardelas:* they are nearly of the ſize of a pigeon; their body is long, their neck very ſhort, their tail in proportion, their wings long and thin. They are diſtinguiſhed into two kinds, the one gray, the other black; and their only difference conſiſts in the colour. We ſaw

IV. The *Majagué* of the Brazilians *, which Piso describes as follows: " It is," says he, " of the size of a goose, but its hooked bill enables it to catch fish; its head is round, its eye " brilliant; its neck bends gracefully like that of " the swan; the feathers on the fore side of the " neck are yellowish; the rest of the plumage is " of a blackish brown. This bird swims and " dives swiftly, and easily eludes ambushes: it " is seen on the sea *near the mouths of rivers.*" This last circumstance, were it constant, would incline us to doubt whether this bird belonged to the Petrels, which all affect to live remote from the shores.

saw also, but at a less distance on sea, another bird, which the Spaniards call *Alma de Maestro*, black and white; it has a long tail, and is not so common as the pardelas; it seldom appears but in rough weather, and hence its name. *Run of the Frigates le Veles & la Rosa from Callao to Juan Fernandez*; *Hist. Gen. des Voy. tom.* xiii. *p.* 497.

* The *Procellaria Brasiliana* of Linnæus.

Nº 255

THE WANDERING ALBATROSS.

The WANDERING ALBATROSS.

L'Albatros. *Buff.*

Diomedea Exulans. Linn. and Gmel.
Albatrus. Briſſ.
Plautus Albatrus. Klein.
Tchaiki. Pallas, and Hiſt. Kamtſch.
The Man-of-War Bird. Alb. and Grew.

This is the largeſt of the water-fowl, not excepting even the ſwan; and though inferior in bulk to the pelican or flamingo, its body is much thicker, its neck and legs ſhorter and better proportioned. Beſides its lofty ſtature, the Albatroſs is remarkable for many other attributes, that diſtinguiſh it from all the other ſpecies of birds. It inhabits only the South Sea, and is found in the whole extent, from the promontory of Africa to thoſe of America and New Holland. It never has been ſeemin the ſeas of the northern hemiſphere, no more than the manchots, and ſome others which ſeem to be attached to that portion of our globe, where they can ſcarce be diſturbed by man, and where they have long remained unknown. It is ſouthwards, beyond the Cape of Good Hope, that the firſt

Albatroſſes were ſeen; nor before our own times were they examined with attention ſufficient to diſcriminate the varieties, which, in this large ſpecies, ſeem to be more numerous than in other large ſpecies of birds or quadrupeds.

The very great corpulence of the Albatroſs has procured it the appellation of *Cape Sheep* *. The ground of its plumage is a dun white on the mantle, with little black hatches on the back and on the wings, where theſe hatches multiply and thicken into ſpeckles: a part of the great quills of the wing, and the extremity of the tail, are black: the head is thick, and of a round form: the bill is of a ſtructure ſimilar to that of the bill of the frigat, the booby, and the cormorant; it is compoſed in the ſame manner of ſeveral pieces that ſeem articulated and joined by ſutures, with a hook ſuperadded, and the end of the lower part hollowed with a channel, and, as it were, truncated; this very large and ſtrong bill reſembles that of the petrels, in the remarkable property that its noſtrils are open in ſhape of little rolls or ſheaths, laid near the root of the bill in a groove which, on each ſide, runs the whole length; it is yellowiſh white, at leaſt in the dead bird: the legs, which are thick and ſtout, have only three toes connected by a broad membrane, that edges alſo the outſide of each exterior toe: the length of the body is near three feet; the alar extent at

* *Mouton du Cap.*

leaſt

leaſt ten *; and, according to Edwards, the firſt bone of the wing is as long as the whole body.

With this force of body, and theſe arms, the Albatroſs might ſeem to be a warrior bird. Yet we are not told that it aſſails the other fowl, which alſo croſs thoſe vaſt ſeas: it ſeems even to act on the defenſive againſt the gulls, which, ever quarrelſome and voracious, harraſs and annoy it †. It attacks not even the great fiſh; and, according to Forſter, it ſubſiſts almoſt wholly on little marine animals and mucilaginous zoophytes, which float in abundance on the South Sea ‡. It feeds alſo on the ſpawn and fry of fiſh, which the currents bear along, and which ſometimes cover a great extent. The Viſcount de Querhoënt, an accurate and judicious obſerver, aſſures us that he invariably found their ſtomachs to contain only a thick mucilage, and no veſtiges of fiſh.

* Our latitude was 60° 10 ſouth, our longitude 64° 30′ . . . As the weather was very calm, Mr. Banks went into a ſmall boat to ſhoot birds, and he brought ſome Albatroſſes: we remarked, that theſe were larger than ſuch as we had taken on the north of the Strait Lemaire; one of them, which we meaſured, was ten feet two inches in alar extent. *Cook*.—The Albatroſſes, the frigats, the flying fiſh, the dolphins, and the ſharks, played about the ſhip: our gentlemen had killed Albatroſſes that were ten feet acroſs the wings. *Idem*.

† Several large gray gulls, that were purſuing a white Albatroſs, afforded us a diverting ſpectacle; they overtook it, notwithſtanding the length of its wings, and they tried to attack it under the belly, that part being probably defenceleſs; the Albatroſs had now no means of eſcaping, but by dipping its body into the water; its formidable bill ſeemed then to repel them. *Cook*.

‡ *Idem*.

Captain Cook's people caught the Albatroſſes, which often appeared about the ſhip, with hook and line *. The capture was the more agreeable † to theſe navigators, as they were in the midſt of the ocean, far from any land ‡; for theſe large birds were met with on the whole extent of the South Sea, at leaſt in the high latitudes §. They frequent alſo the iſlands ſcattered in

* "We were in latitude 35° 25′ ſouth, and 29′ weſt of the Cape, "and had abundance of Albatroſſes about us, ſeveral of which "we caught with hook and line; and were very well reliſhed by "many of the people, notwithſtanding they were at this time ſerved "with freſh mutton." *Cook, vol.* i. *p.* 20.—[I have here corrected an error in our author's text, occaſioned by a very extraordinary inaccuracy in a French tranſlation of Cook's Voyage, to which he refers; where it is ſaid, that *they caught the Albatroſſes with a line and hook baited with a bit of ſheep's-ſkin.—T.*]

† "We ſkinned the Albatroſſes, and after ſoaking them till next "morning in ſalt water, we boiled them, and ſeaſoned them with a "rich ſauce; every body found it thus dreſſed to be very palatable, "and we ate it when there was freſh pork on the table." *Cook's Firſt Voyage.*—"In 40° 40′ ſouth latitude, and 23° 47′ eaſt longitude . . . "we killed Albatroſſes and petrels, which we were then glad to "eat." *Idem.*

‡ "We had another opportunity of examining two different kinds "of Albatroſſes . . . We had now been nine weeks without ſeeing any "land." *Cook's Second Voyage.*—"On the 8th, being in the latitude "of 41° 30′ S. longitude 26° 51′ E . . . We daily ſaw Albatroſſes, "peterels, and other oceanic birds, but no ſign of land." *Idem. vol.* ii. *p.* 245.

§ "We were now in the latitude of 32° 30′, longitude 133° 40′ "weſt . . . This day was remarkable, by our not ſeeing a ſingle bird. "Not one had paſſed ſince we left the land, without ſeeing ſome "of the following birds, viz. Albatroſſes, ſheerwaters, pintadoes, "blue peterels, and Port Egmont hens. But theſe frequent every "part in the antarctic ocean *, as well as the extremity of America † and that of Africa ‡.

" These birds, like most of those of the South " Sea," says the Viscount de Querhoënt, " glance " on the surface, and never mount higher, except " in rough weather, when they are borne up by " the wind." Since they are found at such distances from land, they must rest on the water §: in fact, Albatrosses even sleep on the surface; and Le Maire and Schooten are the only || voyagers who assert their having seen them alight on their ships ¶.

" part of the Southern Ocean in the higher latitudes." *Cook, vol.* i. *pp.* 135 & 136.—" In latitude 42° 32′ south, longitude 161° west, " we often saw Albatrosses and petrels." *Idem.*—" In 45° 20′ south " latitude, and 134° west longitude, we saw Albatrosses." *Idem*— " On the 10th of January, observed at noon, in latitude 54° 35′ S. " longitude 47° 56′ west, a great many Albatrosses and blue peterels " about the ship." *Vol.* ii. *p.* 209.—On the 11th of July, in 34° 56′ south latitude, and 4° 41′ longitude, M. de Querhoënt saw some *croiseurs* and an Albatross.

* In general, no part of New Zealand contains so many birds as Dusky Bay; we have found there Albatrosses, penguins, &c. *Forster.*—There were likewise Albatrosses in New Georgia. *Cook.*

† From our clearing the Strait of Magellan, and during our run along the coast of Chili, we saw a great number of sea-birds, and particularly Albatrosses. *Carteret.*

‡ Mr. Edwards had not seen the narratives of the illustrious navigators just cited, when he said, " These birds are brought from " the Cape of Good Hope, where they are numerous. I have never " heard that they were frequent in any other part of the world."

§ *Voyage d'un Officier du Roi aux Isles-de-France & de Bourbon, page* 68.

|| See the quotation from Forster, in the Discourse on the Water Fowl.

¶ We saw *jeans-de-genten* of an extraordinary bulk; these are seagulls with a body as large as that of a swan, and each wing extend-

The celebrated Cook met with Albatroſſes differing ſo much from each other *, that he regarded them as diſtinct ſpecies. But from the deſcriptions which he gives we are diſpoſed to reckon them only mere varieties. He diſtinguiſhes three; *the gray Albatroſs* †, which appears to be the great ſpecies we have juſt delineated; the *dark brown*, or *chocolate Albatroſs* ‡; and

ing not leſs than a fathom. They alighted on the ſhip, and ſuffered the ſailors to catch them (in the Strait of Lemaire). *Relation de Le Maire & Schooten.*—The following extract alſo refers to an Albatroſs. At ſome diſtance from the Cape of Good Hope, as it was a perfect calm, we ſaw ſomething floating on the water; we let down the yawl into the water, and found this to be two large gulls, which could not riſe by reaſon of their unwieldineſs and the want of the aſſiſtance of the wind; ſo they were taken. They were as white as ſnow; but their wings were gray, and longer than the whole extent of a man's arms; their bill was hooked, and a quarter of a Dutch ell in length *(this appears to be exaggerated)*; they bit fiercely with it. Their feet were like thoſe of the ſwan, and were a ſpan in breadth. They taſted tolerably; we ſaw alſo two great whales. *Voyage de Hagenar aux Indes Orientales, dans le Recueil des Voyages qui ont ſervi à l'Etabliſſement de la Compagnie*; *Amſterdam*, 1702, *tom.* v. *page* 161.

* In 53° 35′ ſouth latitude, there was a great number of Albatroſſes of different kinds about the ſhip. *Cook.*

† "In latitude 67° 5′ ſouth, the fog being ſomewhat diſſipated, "we reſumed our courſe. The ice iſlands we met with in the morn-"ing were very high and rugged, forming at their tops many peaks; "whereas moſt of thoſe we had ſeen before were flat at top, and not "ſo high; though many of them were between two and three "hundred feet in height, and between two and three miles in cir-"cuit, with perpendicular cliffs or ſides, aſtoniſhing to behold. Moſt "of our winged companions had now left us; the gray Albatroſſes "only remained; and, inſtead of the other birds, we were viſited by "a few antarctic peterels." *Cook, vol.* i. *p.* 256.

‡ The *Diomedea Spadicea* of Gmelin: "It is chocolate; its front, "its orbits, its chin, its throat, the lower coverts of its wings, its "belly,

and the *ſooty* or *brown Albatroſs*, which the ſailors, on account of its ſober garb, ſtyled the *quaker-bird* *. The laſt appears to be the ſame with the *Chineſe Albatroſs* repreſented in the *Planches Enluminées:* it is ſomewhat larger than the firſt; its bill ſeems not to have its ſutures ſo ſtrongly marked. Perhaps it is only a young bird, that had not yet attained its proper form or colours. In the ſame manner, the ſpotted gray might be the male, and the brown one the female. We are the more diſpoſed to entertain theſe views, as the large animals, whether quadrupeds or birds, exiſt generally detached, and ſeldom include contiguous ſpecies. In ſhort, we ſhall only admit one ſpecies of Albatroſs, until we are better informed.

Theſe birds are no where more plentiful than among the iſlands of ice in the South Sea †, from

" belly, and its legs, are white; its bill ochry white." Captain Cook met with it in latitude 37° ſouth: it is larger than the ſooty Albatroſs.

* We alſo ſaw, from time to time, two ſpecies of Albatroſſes, of which we have already ſpoken, and alſo a third ſmaller than theſe, which we called the *ſooty*; our ſailors named it the *quaker-bird*, becauſe of its dingy colour. *Cook*.—[This is the *Diomedea Fuliginoſa* of Gmelin: " It is brown; its head, its bill, its tail, its wing-quills " and its tail, are brown and deep black; the ſpace about its eyes is " white." It is about the bulk of a gooſe, being near three feet long: it occurs in the latitude of 47°, and in the whole of the antarctic circle.—*T*.]

† " We began to ſee theſe birds about the time of our firſt falling " in with the ice iſlands; and ſome had accompanied us ever ſince. " Theſe, and the dark brown ſort with a yellow bill, were the only " Albatroſſes that had not now forſaken us." *Cook, vol.* i. *p*. 38.

the fortieth degree of latitude to the frozen barriers under the ſixty-fifth and ſixty-ſixth degrees. Forſter killed an Albatroſs with brown plumage in latitude 64° 11′ *; and from the fifty-third degree this ſame navigator ſaw ſeveral of different colours; he found them even in latitude 48°. Other voyagers have met with them at ſome diſtance from the Cape of Good Hope †. It ſeems even that theſe birds advance ſometimes nearer the ſouthern tropic ‡, which appears to be their limit in the Atlantic Ocean: but they have paſſed it, and have even traverſed the torrid zone in the weſt part of the Pacific Ocean, if the account of Captain Cook's third voyage may be relied on. The veſſels purſued a tract from Japan ſouthwards: "We approach-" ed," ſays this relater, "the latitudes where "occur the Albatroſſes, the bonitoes, the dol-"phins, and the flying fiſh."

* The head and the upper ſide of the wings were ſomewhat blackiſh, and the eye-lids white. *Forſter.*

† There are ſeveral other ſigns of approach to the Cape of Good Hope; for inſtance, the ſea-fowl met with, and eſpecially the *algatros* birds with very long wings. *Dampier.*

‡ After the boobies had left us, we ſaw no more birds till we came up with Madagaſcar . . . we then ſaw an Albatroſs, and daily afterwards we met with more. *Cook.*—We ſaw an Albatroſs (*Diomedea Exulans*) in 25° 29′ ſouth latitude, and 24° 54′ longitude, on the 5th of October, the air being ſharp and cold. *Idem.*

[A] Specific character of the Wandering Albatroſs, *Diomedea Exulans*: "It is white, its back and wings lineated with black, its "bill yellow, its legs carnation, its wing-quills black, its tail lead-"coloured and rounded." The bulk of the Albatroſs is between that

that of a goofe and of a fwan; its weight varying from twelve to eighteen pounds. It is not confined to the antarctic feas; numbers refort every fummer to the northern fhores in queft of the fhoals of falmon, and it is fo voracious as fometimes to be taken while it dozes furfeited on the water. It brays like an afs. The flefh of thefe birds is tough and dry; but the Kamtfchadales feek them for the fake of their entrails, which they blow and ufe as buoys for their nets: their method is to faften a cord to a large hook baited with a whole fifh, which the Albatroffes greedily feize. The bones of the wing ferve thefe people for tobacco-pipes. Such as frequent the feas near the tropics fubfift chiefly on flying-fifh. Thofe of the fouthern hemifphere repair to the fhore in the month of October, and build their neft with fedges, like a rick three feet high, leaving a fmall hole in the top for receiving their egg, which is four inches and a half long, white, with dull fpots near the large end. They are much annoyed with hawks.

The GUILLEMOT.

Colymbus Troile. Linn. and Gmel.
Uria Troile. Lath. Ind.
Uria. Geſn. Aldrov. and Briſſ.
Lomwia. Cluſ. Nieremb. Johnſt. Charlet. Sibb. and Will.
Lomben. Klein.
The Lavy. Martin's St. Kilda.
The Guillem *of Wales, the* Sea-hen *of Northumberland, the* Skout *of Yorkſhire, and the* Kiddaw *of Cornwall.* Will. Ray, and Edw.
The Fooliſh Guillemot. Lath. Syn. *

THE Guillemot exhibits the ſtrokes by which nature prepares to cloſe the numerous ſeries of the varied forms of birds. Its wings are ſo narrow and ſhort, that it ſcarce can fly above the ſurſace of the ſea †; and to reach its neſt, which is placed on the rocks, it is obliged to flutter, or rather to leap from cliff to reſting a moment at each throw ‡. This habit, or rather this neceſſity, is common to it with the puffin, the penguin, and other ſhort-winged birds;

* In the Feroe iſlands, the Guillemot is called *Lomwier* or *Lomwia:* in Norway *Lomvie, Longivie, Langvire, Lumbe,* and *Storfugl:* in Denmark *Aalge:* in Lapland *Doppau:* in Greenland *Tuglok.*—The name *Uria* is given by Geſner, from a ſtrained application of the Greek ουρια, or *diver:* the Greeks could never have known the Guillemot, which is confined to the northern ſeas.

† They fly very low on the ſea, and their flight reſembles that of the partridges. *Recueil des Voyages du Nord, tom.* ii. *p.* 89.

‡ Edwards.

of

Nº 256

THE FOOLISH GUILLEMOT.

of which the ſpecies, almoſt baniſhed from the temperate countries of Europe, have ſettled on the extremity of Scotland, and on the coaſts of Norway and Iceland, and on the Feroe iſlands, the laſt inhabited tracts of our northern world, where theſe birds ſeem to ſtruggle againſt the progreſs and incroachment of the ice. It is even impoſſible for them to inhabit thoſe latitudes in the winter: they are much accuſtomed indeed to the utmoſt ſeverity of cold, and remain on the floating ice *; but they cannot ſubſiſt except in an open ſea, and muſt leave it when frozen over.

It is in this migration, or rather in this diſperſion during the winter, and after having quitted their abodes in the region of the north, that they deſcend along the coaſts of England †, where ſome pairs remain even, and ſettle on the ſhelves and deſert iſlets, particularly in a little iſland, uninhabited for want of ſprings, and facing Angleſey ‡. There they breed on the projecting crags, as near as they can reach the ſummit of the rocks §: their eggs are of a blueiſh colour, more or leſs clouded with black ſtains, they are pointed at the end, and very large in proportion to the ſize of the bird ||, which is

* It was the 3d of May and on the ice, I ſhot for the firſt time one of theſe birds; I afterwards killed ſeveral at Spitzbergen, where they are very numerous. *Recueil des Voyages du Nord, tom.* ii. *p.* 89.

† Britiſh Zoology. ‡ Willughby. § Cluſius.

|| Willughby.

nearly

nearly that of the morillon: their body is short, round, and compact: their bill straight, pointed, three fingers long, and black throughout; the upper mandible has at its point two little productions, which on each side jut over the lower. This bill is in a great measure covered with a velvet down, of the same brown cinereous or smoky black that covers all the head, the neck, the back, and the wings: all the fore side of the body is of a snowy white: the feet have only three toes, and are placed quite behind the body, a position which makes the bird as agile in swimming and diving, as tardy in walking, and feeble in flying. Its only retreat, when pursued or wounded, is under the water, or even under the ice *; the danger must be urgent however to rouse it; for it is not a shy bird, but suffers a person to approach and catch it with great ease †. This appearance of stupidity has given origin to the English name *Guillemot*.

* They swim under water as fast as we could row the boat; when pursued or fired at, they plunge, and continue very long concealed under water; so that as they pass often under the ice, they must then be undoubtedly suffocated. *Recueil des Voyages du Nord, tom.* ii. *p.* 89.

† Ray.

[A] Specific character of the Foolish Guillemot, *Colymbus Troile*: "Its body is black; its breast and belly snowy; its secondary wing-quills tipt with white." Its length is seventeen inches; its alar extent twenty-seven and a half; its weight twenty ounces. It lays a large egg, three inches long, and of a various colour. It winters on the coast of Italy.—Gmelin and Latham make the Guillemot to be the lumme of the northern nations.

The LITTLE GUILLEMOT,

IMPROPERLY CALLED

The GREENLAND DOVE.

Colymbus Grylle. Linn. and Gmel.
Uria Grylle. Lath. Ind.
Uria Minor Nigra. Briſſ.
Columba Gröenlandica dicta. Will. Ray, Sibb.
Columbus Gröenlandicus. } Klein.
Plautus Columbarius. }
Turtur Maritimus Insulæ Baſs. Sibb. Hiſt. Fife.
Kaiaver, vel *Kaior.* Hiſt. Kamtſch.
The Scraber. Martin's St. Kilda.
The Greenland Dove, or *Sea Turtle.* Alb. and Will.
The Black Guillemot. Penn. and Lath. *

IN thoſe frozen countries, where ſtern Boreas reigns alone, and where the gentle zephyrs never ſport, the ſweet murmurs of the tender dove are no more heard. The charming votary of love ſhuns ſuch chilling ſcenes; and the pretended dove of Greenland is a melancholy water-fowl, which can only ſwim and dive, ſcreaming inceſſantly, in a dry re-iterated tone, *rottetet, tet, tet, tet* †. It bears no reſemblance to our pigeon, except in bulk, which is nearly the ſame

* In Swediſh *Sjoe-orre, Griſla*: in the iſland of Oëland *Alle*; in that of Gothland *Grylle*; and in the Feroe iſles *Fuldkoppe*: in Iceland *Teiſta*: in Norway *Teiſte*: in Greenland *Sarpak*.

† Klein.

in both *. It is a Guillemot ſmaller than the preceding, and its wings alſo ſhorter in proportion. Its legs are placed in the ſame manner in the abdomen: its walk is as feeble and tottering †. Its bill only is ſhorter, more inflated, and not ſo much pointed. Its feathers are all unwebbed, and reſemble ſilky hair ‡. The colours are only ſmoky black, with a white ſpot on each wing, and more or leſs of white on the fore ſide of the neck and of the body: this laſt character varies to ſuch degree, that ſome individuals are entirely black, and others almoſt entirely white §. "It is in winter," ſays Willughby, "that they are found completely white; "and as, in the tranſition from one of theſe "garbs to the other, they muſt neceſſarily be "more or leſs mixed or variegated with black "and white, we may reckon *the ſpotted Greenland dove* of Edwards to be the ſame ſpecies "with the two *little Greenland doves* repreſented "in his ninety-firſt plate; becauſe they differ "not from each other, or from the preceding, "unleſs in the greater or leſs mixture of black "and white in their plumage."

Theſe fly commonly in pairs, razing the ſurface of the ſea, like the great guillemot, with a

* Ray.—According to Martens, the ſailors gave it this name, becauſe it pules like young doves; yet there is little reſemblance between puling and the cry which Klein expreſſes.

† Linnæus. ‡ Klein. § Willughby and Klein.

brisk

brisk flapping of their narrow wings *. They place their nests in the crevices of the low rocks †, from which the young can throw themselves into the sea, and avoid becoming the prey of the foxes ‡, that incessantly watch them. These birds lay only two eggs: some of their nests are found on the coasts of Wales and of Scotland §, and also in Sweden, in the province of Gothland ||. But the far greater number breed in much more northern countries, in Spitzbergen and in Greenland, the principal abode of both the great and the little Guillemot ¶.

To the little Guillemot we shall refer the *kaiover* or *kaior* of Kamtschatka, since Kracheneninikow applies to it, after Steller, the denomination of *the Greenland pigeon of the Dutch.* "It has," says he, "its bill and legs red; it "builds its nest on the top of rocks, whose bot-"tom is washed by the sea, and screams or "whistles very loud, whence the Cossacs have "stiled it *ivoskik*, or the postilion."

* Ray. † Linnæus. ‡ Anderson.
§ Klein. || Linnæus. ¶ Ray.

[A] Specific character of the Black Guillemot, *Colymbus Grylle*: "Its body is deep black, the coverts of its wings white." Its length is fourteen inches, and its alar extent twenty-two. For the most part, these birds fly in pairs: they nestle under ground, and lay an egg as large as a hen's, and of an ash-colour. They occur in St. Kilda, on the Bass isle in the Firth of Forth, in the Farn islands off the Northumbrian coast, and on the Llandidno in Caernarvonshire.

The PUFFIN.

LE MACAREUX. *Buff.*

Alca Arctica. Linn. and Gmel.
Fratercula. Briss.
Anas Arctica. Sibb. Will. and Ray.
Plautus Arcticus. Klein.
Lunda. Clusius, Nierem. and Johnst.
Puphinus Anglicus. Gesner and Aldrovandus.
The Bowger. Martin's St. Kilda.
In North Wales, Puffin; *in South Wales,* Golden-head, Bottle-nose, *and* Helegug: *in Yorkshire, near Scarborough,* Mullet: *in Durham, at the mouth of the Tees,* Coulterneb *.

THE bill is the principal organ of birds, the instrument by which their powers and faculties are exercised; it serves as a mouth, as a hand, as an arm. It is that part of their body whose structure the most determines their instincts, and directs their habits of life: and if the winged tribes disperse through the air, on the sea, and on the land, if they engage in an endless variety of pursuits, it is because nature has bestowed on their bill an infinite diversity of form. A sharp, lacerating hook arms the head of the fierce birds of prey; their appetite for flesh and their thirst for blood, joined to the

* Anderson calls the Puffin *the Greenland parrot*; and in collections of voyages it is often named *the diver parrot*, *the ducker parrot*, and *the thick-billed sea-magpie*. In the Kamtschadale language it is termed *Ypatka*: in the Norwegian and in the Feroe islands *Lunde*, *Soë-Papegay*; the chicken *Lund-toëller*: in Greenland *Killengak*.

means

N.o 257

THE PUFFIN.

means of ſatisfying theſe, precipitate them from their towering heights upon all other birds, and even upon all the weak and timorous animals, which are equally their victims. A bill ſhaped like a broad and flat ſpoon, induces another genus of birds to gather their ſubſiſtence at the bottom of the water: while a conical bill, ſhort and truncated, enables the gallinaceous kind to pick up the ſeeds on the ground, diſpoſes them to aſſemble round us, and ſeems to invite them to receive their food from our hands. A bill, faſhioned like a ſlender pliant probe, which lengthens out the face of the curlews, of the woodcock, of the ſnipe, and of moſt other waders, conſtrains them to inhabit marſhy grounds, there to dig in the ſoft mud and the wet ſlime. The ſharp taper form of the woodpecker's bill condemns it to bore the bark of trees. And finally, the little awl-ſhaped bill of moſt of the field-birds permits them only to catch gnats and other minute inſects, and forbids every other ſort of food. Thus the different form of the bill modifies the inſtincts, and gives riſe to moſt of the habits of birds *; and this ſtructure varies infi-

* It is proper to put the reader on his guard againſt this ſpecious ſort of declamation, in which the materialiſts have ſo much indulged. If an animal were directed by its organization to follow its particular mode of life, it muſt be ſuppoſed to make trial of every poſſible ſituation, and to adopt that which, on due experience, is found to be the beſt ſuited to its nature. But this hypotheſis is completely abſurd. Prior to all reflection, inſtinct leads irreſiſtibly to a certain courſe of action, to which the corporeal ſtructure is in general admirably adapted.—*T.*

 nitely,

nitely, not only by ſhades, as in all Nature's productions, but even by ſteps, and ſudden leaps. The enormous ſize of the bill of the toucan, the monſtrous ſwelling of that of the calao, the deformity of that of the flamingo, the ſtrange ſhape of the bill of the ſpoonbill, the reverſed arch of that of the avoſet, &c. demonſtrate ſufficiently that all the poſſible figures have been traced, and every form moulded. That for completing this ſeries nothing may be imagined wanting, the extreme of all the faſhions is exhibited in the vertical blade of the Puffin's bill. It exactly reſembles two very ſhort blades of a knife applied one againſt the other by the edge: the tip is red, and channelled tranſverſely with three or four little furrows, while the ſpace near the head is ſmooth and tinged with blue. The two mandibles being joined, are almoſt as high as they are long, and form a triangle very nearly iſoceles: the circuit of the upper mandible is edged near the head, and as it were hemmed with a ledge of a membranous or callous ſubſtance, interſperſed with little holes, and whoſe expanſion forms a roſe on each corner of the bill *.

This

* M. Geoffroy de Valognes, who appears to me to be a good obſerver, has been ſo obliging as to ſend me the following note on the ſubject of the Puffin:

"I received," ſays he, "a Puffin that had been taken the beginning of this month (of May) in its paſſage on our coaſts: this "bird was viewed with aſtoniſhment, even by perſons who ofteneſt "frequent the ſea-ſhore; which makes me think that it is a ſtranger "to this country.

"The

This imperfect analogy to the bill of the parrot, which is alſo edged with a membrane at its baſe, and the no leſs diſtant analogy to the ſhort neck and the round ſhape, have procured the

"The poſition of the legs of the Puffin near the anus leads me "to preſume that it walks with difficulty, and that it is more formed for ſwimming on the water: cinereous, black, and white, are "ſenſibly contraſted on its plumage; the firſt of theſe colours marks "the cheeks, the ſides of the head, the under part of the throat, "where it takes a deeper ſhade; the ſecond prevails on the head, "the neck, the back, the wings, the tail, and extends to the throat, "where it forms a broad collar, that divides at this place the gray "from the pure white, which alone appears on the under ſide of "the body, where the feathers conceal from view a thick gray "down which clothes the belly: the black on the upper ſide of the "head grows a little dilute near the origin of the neck, on the "quills of the wings, and at the termination of the feathers which "cover the back; on the tip of the wings there is a white border, "which is not very apparent unleſs they are ſpread.

"The bill is longer than it is broad, if we meaſure from its "origin; its form is almoſt triangular, the two mandibles are moveable; the iron-gray, which partly paints it, is ſeparated as it "were by a white ſemi-circle from a bright red that covers the "point, and completes the decoration: the upper mandible preſents four ſtreaks, the lower three, which correſpond to the three "laſt of the upper; all theſe ſtreaks form a ſort of ſemi-circles: the "upper mandible has at its baſe a little roll, on which there are "ſmall holes diſpoſed regularly; from ſome of theſe holes very "ſmall feathers grow; the noſtrils are placed on the edges of the "upper mandible, and extend three lines in the length of the bill: "I perceived on the palate of the bird ſeveral rows of fleſhy points "directed towards the opening of the throat, of which the tranſparent and gloſſy extremity ſeemed to be ſomewhat harder than the "reſt; the eyes, edged with vermillion, have this peculiarity, that "they occupy the centre of a gray triangular excreſcence: the legs "are ſhort, and of a bright orange like the feet; the nails are black "and ſhining, that of the hind toe is the longeſt and broadeſt." *Extract of a letter from M. Geoffroy, to M. le Comte de Buffon, dated from Valognes, the 8th of May, 1782.*

 Puffin

Puffin the name of *sea parrot*; a denomination as improper as that of *sea dove* for the little guillemot.

The Puffin has not more of wings than this guillemot, and in its short, skimming flutters, it assists itself by the rapid motion of its feet, with which it only razes the surface *: and hence to support itself it has been said to strike the water continually with its wings †. The quills are very short, as well as those of the tail ‡; and the plumage of the whole body is rather down than real feathers. With respect to its colours, "imagine," says Gesner, "a bird clothed in a "white robe, with a black frock or mantle, and "a cowl of the same, and you will have a picture "of the Puffin, which, for that reason, I call "the little monk, *fratercula*."

This little monk lives on prawns, shrimps, star-fish, and sea-spiders, and several other sorts of fish, which it catches by diving in the water, beneath which it willingly retires § and shelters itself from danger. It is said even to drag its enemy, the raven, under the flood ‖: such exertions of force or dexterity seem to exceed

* Gesner.

† Willughby.

‡ Twelve are reckoned to be the number, though Edwards counted sixteen in a subject of this species.

§ *Recueil des Voyages du Nord, tom.* iii. p. 102.

‖ The bill of the sea parrot is an inch broad, and so sharp, that it is able to master its enemy, the raven, and to drag it under water. *Hist. Gen. des Voy. tom.* xix. *p.* 46.

the

the ſtrength of its body, which is not larger than that of a pigeon *; they muſt therefore be aſcribed to the power of its weapons, and the bill is indeed formidable by its ſharp blades and its terminating hook.

The noſtrils are pretty near the edge of the bill, and appear like two oblong ſlits: the eye-lids are red; on the upper one is a little excreſcence of a triangular ſhape, and on the lower is a ſimilar excreſcence, but of an oblong form: the feet are orange, furniſhed with a membrane between the toes; the Puffin, like the guillemot, wants the hind toe; the nails are very ſtrong and hooked: as its thighs are ſhort, and concealed under the abdomen, it is obliged to keep quite erect, and ſeems to totter and rock in its walk †. It is accordingly never found on land, except retired in caverns or in holes excavated under the ſhores ‡, and always in ſuch ſituations, that it can throw itſelf into the water, as ſoon as the calm invites its return: for it has been remarked, that theſe birds cannot remain on the ſea, or fiſh, except when it is ſmooth; and that if they be overtaken by a ſtorm, either on their departure in autumn or on their return in ſpring, numbers periſh. The winds caſt theſe dead

* A foot from the point of the bill to the end of the tail; thirteen inches from the bill to the nails.

† "It walks turning every moment from ſide to ſide." *Voyage du Nord.*

‡ Geſner.

Puffins ashore *, sometimes even on our coasts † where these birds are seldom seen.

They constantly inhabit the most northern islands ‡ and promontories of Europe and Asia, and probably also those of America, since they are found in Greenland as well as in Kamtschatka §. They leave the Orknies and other islands near Scotland regularly in the month of August; and it is said, that in the first days of April a few come to reconnoitre the places, and in two or three days after retire to inform the main body, which they lead back in the beginning of May ||.

* Willughby.

† "The north wind has sent us this winter thousands of dead "and drowned Puffins. These every year take a sea voyage, about "the end of February or the beginning of March; when it is "stormy, many are drowned, and at all times the ravenous birds "devour great numbers of them. Probably this passage is laborious, for all the bodies of these drowned birds are constantly "very lean. These birds are found on the coasts of Picardy also "in the month of August, but are then few in number. The male "differs not from the female, except that his colours are deeper: "the old ones have their bill broader." *Letter of M. Baillon, dated Montreuil-sur-mer, 10th of April*, 1781.—"The Puffin is known on "this coast (of Croisic) under the name of *gode*, and occurs at all "seasons; it seldom comes to land, and then only on the nearest shore: it nestles in the holes of craggy rocks, especially near "Belle-isle, at the place called *the Old Castle*; it there lays on the "bare ground three eggs. It is found in the whole of the gulf of "Gascogny." *Letter from the Viscount de Querhoënt, 29th of June*, 1781.

‡ In the islands Anglesey, Bardsey, Caldey, Priestholm, Farn, Godreve, the Scillys, and others. *Willughby*.

§ The Kamtschadales call the sea-diver *yatka*: it occurs on all the coasts of that peninsula. *Hist. Gen. des Voy. tom.* xviii. *p.* 270.

|| Willughby.

These

Theſe birds build no neſt; the female lays on the naked ground and in holes, which they excavate and enlarge: they have only one egg, it is ſaid, which is very large, much pointed at the end, and of a gray or grayiſh colour *. The young that are unable to follow the troop in their autumnal retreat are abandoned †, and perhaps periſh. On their return in ſpring, theſe birds do not all occupy the moſt northern ſpots; ſmall flocks halt on different iſlets along the Engliſh coaſts, and they are found with the guillemots and the penguins on the *Needles*, which lie on the weſt ſide of the Iſle of Wight. Edwards paſſed ſeveral days among theſe rocks to obſerve and deſcribe the birds ‡.

* Willughby.

† *Idem.*

‡ He repreſents it as one of the moſt aſtoniſhing works of nature. "I have ſometimes admired," ſays he, "the palaces of kings; "the antique majeſty of our old cathedrals have often inſpired me "with religious fear: but when from the ocean I ſaw diſplayed "this vaſt, ſtupendous work of nature, how little and diminutive "appeared all the monuments of human power! Imagine a maſs "of rocks ſix hundred feet in height, and ſtretching about four miles "in length, flanked with obeliſks and ſhapeleſs columns, which "ſeemed to riſe out of the ſea, and which were indented by the dark "mouths of caverns formed by the billows: if from this gloomy "depth the affrighted eye meaſures the broken perpendicular ſides "of theſe rocks, whoſe projecting cliffs ſeem to threaten every moment to plunge the ſpectator into the abyſs: if retiring a quarter "of a mile to enjoy a full view of this immenſe rock, we fire a cannon, the air will be darkened with a black cloud formed by the "riſing of thouſands of birds from all the crags and ledges, and "which, with ſome ſheep, are the only inhabitants of this rock."

[A] Specific character of the Puffin, *Alca Arctica*: "Its bill is "compressed, channelled on each side with four furrows; its or- "bits and its temples white; its upper eye-lid pointed." Its length is twelve inches, its alar extent twenty-one inches, its weight twelve ounces. They arrive on several of the coasts of Great Britain and Ireland in April, and take possession of the rabbit-burrows, where they lay a single egg, white, and as large as a hen's. They bite very hard when disturbed; their voice is disagreeable, and seems as if it cost them an effort. They retire in August.

The PUFFIN of KAMTSCHATKA.

Alca Cirrhata. Gmelin and Pallas.
Igilma. Hist. Kamtsch.
The Tufted Auk. Penn. and Lath.

"THE Kamtschadale women," says Steller, "make themselves a head-dress of a "glutton's skin, fashioned like a crescent, with "two white ears or beards, and say, that in this "ornament they resemble the *mitchagatchi* *, "which is a bird quite black, and hooded with "two pendulous crests or tufts of white fila- "ments, which look like tresses on the sides of "the neck." It is easy to perceive, that the bird alluded to is the Kamtschadale Puffin; and the *kallingak* of the Greenlanders appears to be

* Or *Monichagatka*, for so it is written in page 270 of the nineteenth vol. of the *Hist. Gen. des Voy.* while in page 253 of the same volume it is written *Mitchagatchi*.

the ſame *. Like this it has the two white treſſes and cheeks, and the reſt of the plumage black or blackiſh, with a deep blue tint on the back, and dull brown on the belly: its bill is furrowed on the upper blade, and the noſtrils are ſituated near the edge: laſtly, it has little roſes on the corners of the bill, as in the common puffin; only the ſize of the *kallingak* or Greenland Puffin is ſomewhat ſmaller than that of the Kamtſchadale Puffin.

* The Greenlanders know a ſea-parrot, which they call *kallingak*, and which is entirely black, and as large as a pigeon. *Idem*, *p.* 46.

[A] Specific character of the *Alca Cirrhata*: "It is entirely "black, has four furrows in its bill; the ſides of its head, the ſpace "about its eyes, and the corner of its throat, are white; a yellow-"iſh longitudinal tuft from the eye-brows to the nape." Its fleſh is hard and inſipid, but the Kamtſchadales uſe its eggs. The bills, mixed with thoſe of the common puffin and the hairs of the ſeal, were formerly regarded by theſe rude people as a powerful amulet.

The PENGUINS and the MANCHOTS; or, The BIRDS WITHOUT WINGS.

IT is difficult to ſeparate in imagination the idea of bird from that of wings: yet is the faculty of flying not eſſential to the feathered race. Some quadrupeds are provided with wings, and ſome birds are deſtitute of them. A wingleſs bird would ſeem a monſter produced by the neglect or overſight of nature; but what is apparently a derangement, an interruption of her plan, does really fill up the order of ſucceſſion, and connects the chain of exiſtence. As ſhe has deprived the quadruped of feet, ſhe has alſo deprived the bird of wings; and it is remarkable that the ſame defect begins with the land birds, and ends in the water fowl. The oſtrich may be ſaid to have no wings, the caſſowary is abſolutely deſtitute of them; it is covered with hair inſtead of feathers. Theſe two great birds ſeem in many reſpects to approach the land animals; while the Penguins and Manchots appear to form the ſhade between birds and fiſh. Inſtead of wings they have little pinions, which might be ſaid to be covered with ſcales rather than feathers,

thers, and which ſerve as fins *; their body is large, compact, and cylindrical, behind which are attached two broad oars, rather than two legs: the impoſſibility of advancing far into the land, the fatigue even of remaining there, otherwiſe than by lying; the neceſſity, the habit of being almoſt always at ſea, their whole œconomy of life, mark the analogy between the aquatic animals and theſe ſhapeleſs birds, ſtrangers to the regions of air, and almoſt equally exiled from thoſe of the land.

Thus between each of the great families, between the quadrupeds, the birds, and the fiſhes, nature has placed connecting links that bind together the whole: ſhe has ſent forth the bat to flutter among the birds, while ſhe has impriſoned the armadillo in a cruſtaceous ſhell. She has moulded the whale-kind after the quadruped, whoſe form ſhe has only truncated in the walrus: the ſeal, from the land, the place of his birth, plunges into the flood, and joins the cetaceous herd, to demonſtrate the univerſal conſanguinity of all the generations that ſpring from the boſom of the common mother: finally, ſhe

* They ſeem to form a middle ſpecies between the birds and the fiſhes; for the feathers, eſpecially thoſe of their wings, differ little from ſcales, and theſe wings, or rather pinions, muſt be regarded as fins. *Cook.*—The wings of theſe animals are without feathers, and ſerve only as fins; they live moſt of their time in the water. *De Gennes.*—Theſe ſtumps ſerve as fins when they are in the water. *Dampier.*

has

has produced birds partaking of the inſtincts and œconomy of fiſhes. Such are the two families of Penguins and Manchots, which ought however to be diſtinguiſhed, as they are actually in nature, not only by conformation, but by difference of climates.

The name of Penguin has been given indiſcriminately to all the ſpecies of theſe two families, which has introduced confuſion. We may ſee in Ray's Synopſis what difficulties ornithologiſts have met with to accommodate the characters aſcribed by Cluſius to his Magellanic Penguins, with the characters obſerved in the arctic Penguins. Edwards is the firſt who reconciled theſe contradictions: he juſtly remarks, that far from thinking, with Willughby, that the northern Penguin was the ſame ſpecies as the ſouthern, one ſhould rather be diſpoſed to range them in two different claſſes; the latter having four toes, and the former having the traces only of the hind toe, and *having its wings covered with nothing that can be called feathers*; whereas the northern Penguin has very ſmall wings, covered with real feathers.

To theſe differences we ſhall add another, ſtill more eſſential, that, in the ſpecies of the north, the bill is furrowed with channels on the ſides, and raiſed with a vertical blade; while, in thoſe of the ſouth, it is cylindrical and pointed. Thus all the *Penguins* of the ſouthern voyages are

are *Manchots* *, which are diſtinguiſhed from the real arctic *Penguins*, by eſſential differences in the ſtructure, as well as by the diſtance of the climates.

We proceed to prove this poſition by a compariſon of the relations of voyagers, and by an examination of the paſſages in which our Manchots are mentioned under the name of *Penguins*. All the navigators of the South Sea, from Narborough to Admiral Anſon, Commodore Byron, M. de Bougainville, Meſſieurs Cook and Forſter, agree in aſcribing to theſe Manchots the ſame characters, and all different from thoſe of the arctic Penguins †.

" The genus of the *Penguins* (Manchots)" ſays Forſter, " have been improperly confounded " with that of the *diomedea* (albatroſs) and that " of the *phaëton* (tropic bird). Though the

* *Manchot*, in French, ſignifies *maimed*. I have, for the ſake of perſpicuity, adopted the term.—*T.*

† The moſt ſingular birds that are ſeen on the coaſt of Patagonia have, inſtead of wings, two ſtumps, which can be of no ſervice but in ſwimming; their bill is *ſtraight* like that of an *albatroſs* (which points out the elongated cylindrical form). *Anſon.*—The Penguin, inſtead of wings, has two flat ſtumps, like the fins of fiſh; and its plumage is only a kind of ſhort down . . . its neck is thick, its head and bill *like that of a crow*, except that the point turns a little downwards. *Narborough.*—In this country (Lobos-del-mar, in the Pacific Ocean) there are many birds, ſuch as *boobies*, but eſpecially Penguins, of which I have ſeen prodigious numbers in all the South Seas, on the coaſt of the country lately diſcovered, and at the Cape of Good Hope. The Penguin is a ſea-bird, about as large as a duck, having its feet ſhaped the ſame, but its *bill pointed*; they do not fly, having ſtumps rather than wings. *Dampier.*

" thickneſs

"thickneſs of the bill varies, it has the ſame "character in all (cylindrical and pointed); ex"cept that in ſome ſpecies the end of the lower "mandible is truncated: their noſtrils are always "linear ſlits, which again proves them to be "diſtinguiſhed from the *albatroſſes*: they all "have exactly the ſame form of feet (three toes "before, without any trace of a hind toe): the "ſtumps of the wings are ſpread into fins by a "membrane, and covered with plumules laid ſo "near each other as to reſemble ſcales; this "character, as well as the ſhape of their bill and "feet, diſcriminates them from the *alcæ* (the auks "or true penguins) which are unable to fly, not "becauſe their wings abſolutely want feathers, "but becauſe theſe feathers are too ſhort."

It is the Manchot, therefore, that we may particularly ſtile the *wingleſs bird*; and at firſt ſight we might alſo call it the *featherleſs bird*. In fact, not only the hanging pinions ſeem covered with ſcales, but all the body is inveſted with a compreſſed down, exhibiting all the appearance of a thick, ſhaved beard, ſprouting in ſhort pencils of little gloſſy tubes, and which form a coat of mail impenetrable by water.

Yet, on a cloſe inſpection, we perceive in theſe *plumules*, and even in the ſcales of the pinions, the ſtructure of a feather, that is, a ſhaft and webs *. Wherefore Feuillée has reaſon to find

* Edwards.

fault

fault with Frezier, for aſſerting, without modification, that " the Manchots were covered with " hair exactly like that of ſea wolves."

On the contrary, the northern Penguin is clothed with real feathers, ſhort indeed, eſpecially on the wings, but which preſent unequivocally the appearance of feathers, and not that of hair, or down, or ſcales.

Here then is a diſtinction well eſtabliſhed, and founded on eſſential differences in the exterior conformation of the bill, and in the plumage. The *Penguins* alſo inhabit the moſt northern ſeas, and advance only a ſhort way into the temperate zone: but the *Manchots* fill the vaſt Pacific, and occur in moſt of the iſlets that are ſcattered through that immenſe ocean; they occupy, as their laſt aſylum, the formidable range of ice, which incruſts the whole region of the ſouth pole, and advances as far as the ſixtieth and fiftieth degrees of latitude.

" The body of the Penguins (Manchots)" ſays Forſter, " is entirely covered with oblong " *plumules*, thick, hard, and ſhining . . . laid as " near each other as the ſcales of fiſh . . . this " cuiraſs is neceſſary to them, as well as the " thickneſs of fat with which they are lined, and " enables them to reſiſt the cold; for they live " continually in the ſea, and are confined eſpe- " cially to the frigid and temperate zones, at " leaſt I have never known them between the " tropics."

According

According o this obſerver, and the illuſtrious Captain Cook, amidſt the ſouthern ice, where they penétrated with more intrepidity, and farther than any navigator before them, Manchots were every where found, and the more numerous, the higher the latitude and the colder the climate *, as far as the antarctic circle; on the borders of the icy mountains †, on the floating ſhoals ‡, at Statenland §, at the Sandwich iſlands, countries

* *Penguins* ſeen in latitude 51° 50′ ſouth. *Cook.*—In the latitude of 55° 16′ ſouth, we ſaw many whales, *Penguins*, and ſome of the white birds. *Id. vol.* i. *p.* 26.—In 55° 31′ ſouth latitude, we ſaw ſome *Penguins*. *Id.*—In 63° 25′, we ſaw a *Penguin* and a guillemot. *Id.*—In 58° ſouth latitude, we killed a ſecond *Penguin*, and ſome petrels. *Id.*

† On approaching the ice iſlands (under the antarctic circle) we heard *Penguins*. *Cook.*—Being in 55° 51′, we ſaw ſeveral *Penguins* and a ſnowy petrel, which we took to be the forerunners of the ice. *Id.*—On the 24th of January, our latitude was 53° 56′, and our longitude 39° 24′; we had round us a great number of blue petrels and *Penguins*. *Id.*

‡ Upon our getting among the ice iſlands, the albatroſſes left us; that is, we ſaw but one now and then. Nor did our other companions, the pintadoes, ſheerwaters, ſmall gray birds, fulmars, &c. appear in ſuch numbers; on the other hand, *Penguins* began to make their appearance. Two of theſe were ſeen to-day . . . we paſſed no leſs than eighteen ice iſlands, and ſaw more *Penguins* . . . we ſaw many whales, Penguins, ſome white birds, pintadoes, &c. *Cook. vol.* i. *pp.* 23 & 24.—The ſea was ſtrewed (latitude 60° 4′ ſouth, longitude 29° 23′ weſt) with large and ſmall ice; ſeveral *Penguins*, ſnow peterels, and other birds were ſeen, and ſome whales. *Id. vol.* ii. *p.* 223.—In 66° latitude, we ſaw many *Penguins* on the ice iſlands, and ſome antarctic petrels in the air. *Id.*—A number of *Penguins*, ſitting on pieces of ice, paſſed near us (in latitude 61°, and longitude 31°). *Id.*

§ *Cook's Second Voyage.*—The cold was intenſe, the two iſlands

were

countries desolate, deserted, without verdure, buried beneath eternal snow. "We saw them, "with the petrels, inhabit regions now inaccessible to all other species of animals, where "these birds alone seemed to resist destruction "and annihilation, in places where animated "nature has already sunk into its tomb. *Pars* "*mundi damnata a rerum naturâ, æterna mersa* "*caligine* *."

When the shoals of ice on which the Manchots settle are drifted, they remain on them, and are thus transported to immense distance from land †. "We saw," says Captain Cook, "on "the summit of the ice island, which passed near "us, eighty-six *Penguins* (Manchots). This "shoal was about half a mile in circumference, "and upwards of an hundred feet high, for it "withheld the wind some minutes from our "sails. The side which these Penguins occupied rose sloping from the sea, so that they "climbed with a gradual ascent." Hence this great navigator justly concludes, that the occurrence of the Manchots at sea is no certain token

were covered with hoar-frost and snow, and no trees or shrubs appeared; we saw no living creature, except the shags and the *Penguins*; the last were so numerous, that they seemed to incrust the rock. *Third Voyage.*

* i. e. *A part of the world condemned by nature, plunged in eternal darkness.* Pliny.

† We found *Penguins*, petrels, and albatrosses, six or seven hundred leagues in the middle of the South Sea.

 of

of the proximity of land, unleſs in latitudes where there is no floating ice.

It appears alſo, that they can perform diſtant excurſions by ſwimming, and thus paſs nights as well as days at ſea *. The element of the water agrees better than that of the land with their diſpoſitions and their ſtructure: on ſhore their pace is ſlow and heavy; as their legs are ſhort and placed quite behind their belly, they are obliged to maintain an erect poſture, and their large body extends in the ſame perpendicular with their neck and head; "in this atti-"tude," ſays Sir John Narborough, "they "would be taken at a diſtance for young chil-"dren with white bibs †."

But if they are heavy and aukward on land, as much are they lively and alert in the water: "They dive, and continue a long time under "the water," ſays Forſter, "and when they riſe "again, they dart ſtraight up to the ſurface, "with ſuch prodigious ſwiftneſs, that they are

* "The preceding evening, three Port Egmont hens were ſeen; "this morning another appeared. In the evening, and ſeveral "times in the night, *Penguins* were heard . . . Our latitude now was "49° 53′ ſouth, and longitude 63° 39′ eaſt." *Cook, Vol.* i. *p.* 50.— "In latitude 57° 8′ ſouth, longitude 80° 59′ eaſt, we ſaw one *Pen-"guin,* which appeared to be of the ſame ſort which we had for-"merly ſeen near the ice. But we had now been ſo often deceived "by theſe birds, that we could no longer look upon them, nor in-"deed upon any other oceanic birds, which frequent high latitudes, "as ſure ſigns of the vicinity of land." *Ibid. p.* 53.

† They walked *erect*, letting their fins hang like arms; ſo that at a diſtance they might be taken for pygmies. *Dampier.*

"difficult

" difficult to ſhoot." The ſort of cuiraſs alſo, or coat of mail, hard, ſhining, and ſcaly, with which they are clothed, and their very firm ſkin, reſiſt often the lead *.

Though the Manchots lay but two eggs, or three at moſt, or even only one †, yet as they are never diſturbed on the deſert lands where they aſſemble, and of which they are the ſole and peaceful poſſeſſors, they are very numerous. " We went aſhore on an iſland ‡," ſays Narborough, " where we caught three hundred *Penguins* (Manchots) in the ſpace of a quarter of " an hour. We could as eaſily have taken " three thouſand, had the boat been capable of " holding them. We drove them before us " in flocks, and knocked them on the head " with a ſtick."

" Theſe *Penguins*, (Manchots)" ſays Wood, " which are improperly ranked among the " birds, ſince they have neither feathers nor " wings, hatch their eggs, as I have been " aſſured, about the end of September or the " beginning of October; in that ſeaſon, as " many might be taken as would victual a fleet. " . . . On our return to Port Deſire, we gathered about an hundred thouſand of theſe

* We wounded one, and following cloſe, we fired at it more than ten times with ſmall ſhot, and though they took effect, it was neceſſary to make a diſcharge with ball. *Forſter*.

† Forſter.

‡ In ſight of Port Deſire, on the coaſt of Patagonia.

" eggs, ſome of which were kept on board near " four months without ſpoiling."

" On the 15th of January," ſays the compiler of the Voyages to the South Sea, " the " veſſel bore towards the *great iſle of Penguins,* " for the purpoſe of catching theſe birds. In " fact, we found there ſuch prodigious numbers, " that they might have ſupplied five-and-twenty " ſhips, and we took nine hundred in two " hours."

No navigator neglects an opportunity of providing himſelf with theſe eggs, which are ſaid to be very good *, and with the fleſh even of theſe birds †, which cannot indeed be excellent, but

* Their fleſh is but indifferent food, but their eggs are excellent. *Dampier.*

† On the 18th, we caſt anchor in the ſecond bay of Magellan's Straits, oppoſite to the *iſle of Penguins,* where the boats were ſoon loaded with theſe birds, which are larger than ducks. *Adams.*—We returned about the middle of September to Port Deſire, to procure new ſtore of ſeals, of *Penguins*, and of the eggs of theſe birds. *Narborough.*—A little iſland in the entrance of the Bay of Saldana is ſtocked ſo plentifully with ſeals and *Penguins,* as to afford refreſhment to the moſt numerous fleet. *Hiſt. Gen. des Voy. tom.* i. *p.* 384.—The *Penguin* is better than the diver of the Scilly iſlands; it has a fiſhy taſte. To render it palatable, it is ſkinned, becauſe of its exceſſive fat; upon the whole it is tolerable food, when roaſted, boiled, or baked, eſpecially roaſted. We ſalted twelve or ſixteen barrels of them, to ſerve us inſtead of cured beef. The taking of them afforded much diverſion; indeed nothing could be more amuſing, whether purſuing them, intercepting them as they want to gain their burrows, when they often tumble into the holes, or ſurrounding them and knocking them on the head with ſticks, for blows on the reſt of the body will not kill them, and beſides will blemiſh the fleſh, which is to be preſerved ſalted. . . . Theſe miſerable

but ſerves as a reſource on coaſts deſtitute of every other refreſhment *. The meat is ſaid not to taſte of fiſh, though in all probability the Penguins ſubſiſt on fiſh †: and if they are ſeen to frequent the tufts of coarſe graſs, the laſt veſtiges of vegetation that remain in thoſe frozen lands, they are induced leſs, it is ſuppoſed, for the ſake of food ‡ than for that of ſhelter.

Forſter has deſcribed their ſettlement in this ſort of aſylum, which they ſhare with the ſeals. " To neſtle §," ſays he, " they form holes or " burrows,

able Penguins, hunted on all ſides, threw themſelves one upon another, and were eaſily ſhot by thouſands; the reſt fell from the top of the rocks to the ground, and inſtantly expired . . . the more fortunate reached the ſea, where they were ſafe. *Hiſt. des Navig. aux Terres Auſtrales, tom.* i. *p.* 240.

* There are prodigious quantities of theſe amphibious birds (on ſome iſlets near Staten-land, ſo that we felled as many as we pleaſed with a ſtick; I cannot ſay that they are good eating; but in want of freſh proviſions, we often found them excellent. They do not lay here, or it was not the ſeaſon (in January) for we ſaw neither eggs nor young. *Cook.*—Spilberg and Wood found the fleſh of the Penguins to be very good; but this depends
of the ſailors, and their want of better food. much upon the hunger

† Cluſius.

‡ The Penguin iſlands (in Magellan's Strait) are three in number . . . they yield only a little graſs, which maintains the Penguins. *Spilberg.*

§ On New Year's iſland, near Statenland, and at New Georgia, a graſs of the ſpecies called *Dactylis Glomerata* takes a remarkable growth: it is perennial, and endures the coldeſt winters; it ſhoots always in tufts at ſome diſtance from one another; every year the buds riſe to a new head, and enlarge the tuft, till it is four or five feet high, and twice or thrice broader at the bottom than at the top; the leaves and ſtalks of this graſs are ſtrong, and often three or four feet long. The ſeals and the Penguins ſhelter themſelves under

 theſe

"burrows, and choose, for this purpose, a down "or sandy plain. The ground is every where "so much bored, that in walking a person often "sinks up to the knees, and if the Penguin "chance to be in her hole, she revenges herself "on the passenger, by fastening on his legs, "which she bites very close *."

The Manchots occur not only in all the southern tract of the great Pacific Ocean, and on all the islands scattered in it †, but also in those of the

these tufts, and as they come out of the sea quite drenched, the paths between these plants are rendered so dirty and slimy, that a person cannot walk without stepping from one tuft to another. *Forster.*—The most advanced and the largest of these islands (on the north-east of Spiring bay, in sight of Port Desire, in Magellan's Strait) is that named *the island of Penguins,* about three quarters of a mile in length. This island consists only of craggy rocks, except near the middle, where it is gravelly, and bears a little green herbage: it is the retreat of a prodigious number of Penguins and seals. *Narborough.*

* Voyage of five vessels to the Straits of Magellan.—They make holes in the ground, like our rabbits, and there lay eggs; but they live on fish, and cannot fly, having no feathers on their wings, which hang at their sides like bits of leather. *Noort.*—All the shore, near the sea, is strewed with burrows, where these birds hatch their eggs: the island of Detroit is full of these holes, except a beautiful vale clothed with fine green herbage, which we imagine these birds had reserved for their pasturage. *Hist. des Navig. tom.* i. *p.* 240.—In a bay on the coast of Brazil is an immense number of the birds which the English call *Penguins*; these birds have no wings, are larger than geese, and make holes or burrows in the ground, into which they creep; which has made the French call them *toads. Drake.*

† In general, no part of New Zealand contains so many birds as Dusky Bay; besides those just mentioned, there are also cormorants, albatrosses, gulls, and *Penguins* (Manchots). *Forster.*—We cannot reckon

the Atlantic, and, it would appear, at lower latitudes. There are vaſt flocks of them near the Cape of Good Hope, and even farther north *. We are of opinion, that the *divers*, which the ſhips *Eagle* and *Mary* met with in lat. 48° 50′ † ſouth, among the firſt floating ice, were Manchots. They muſt have advanced even into the Indian ſeas, if Pyrard is exact in placing them in the *Atollons* of the Maldives ‡, and if Sonnerat really found them in New Guinea §. But theſe places

reckon parrots and Penguins among the domeſtic animals; for though the natives of the Friendly and Society Iſlands tame a few individuals, theſe have never bred. *Drake.*

* Twenty leagues north from the Cape of Good Hope, there is a multitude of birds, and, among others, a prodigious number called *Penguins*; ſo that we could ſcarce turn ourſelves among them: they are not accuſtomed to ſee men, as ſeldom any veſſel touches at this iſland, unleſs it meets with ſome accident at ſea, as was our caſe. *Spilberg.*

† In the ſeventh degree of longitude.

‡ Many little iſlands, the Atollons of the Maldives, have no verdure, and are mere drifted ſand, of which a part is overflowed at ſtream-tides: they contain, at all times, plenty of ſea-crabs, and ſuch a prodigious number of *Penguins*, that one cannot ſtir a foot without cruſhing their eggs or their young. *Pyrard.*

§ This voyager ſpeaks of them as an enlightened naturaliſt:— "All the ſpecies of Manchots," ſays he, "are deprived of the power "of flying; they walk with difficulty, and carry their body erect and "perpendicular, their legs are entirely behind, and ſo ſhort that the "bird can only take very ſmall ſteps; the wings are only appendices "in the place where the true wings ſhould be attached, and their only "uſe is to balance the bird in its tottering pace. They come on ſhore "to paſs the night and to breed; the impoſſibility of their flying, and "the difficulty of their running, expoſe them to the mercy of thoſe "who chance to land on their retreats, and they are run down; the "defect places excepted, we may ſay with Forſter, that in general the tropic is the limit which the Manchots have ſeldom paſſed, and that the bulk of them affect the high and cold latitudes of the South Sea.

The true Penguins alſo, thoſe of the north, ſeem to prefer the icy ſea, though they ſometimes deſcend as far as the Iſle of Wight to breed: however, the Feroe Iſlands and the coaſts of Norway, ſeem to be their native territory in the ancient continent; and Greenland, Labrador, and Newfoundland, that in the new. Like the Manchots, they are entirely deſtitute of the power of flying, having only ſmall ends of wings, covered indeed with feathers, but theſe ſo ſhort as to be fit only for fluttering.

The Penguins, like the Manchots, remain almoſt conſtantly on ſea, and ſeldom come to land but to neſtle or reſt; they lie ſquat, it being equally painful for them to walk or to ſtand erect, though their legs are rather taller, and placed not quite ſo much behind the body as in the Manchots.

In fine, the analogy in their inſtinct, their mode of life, and their mutilated truncated ſhape,

" defect of their ſtructure, which incapacitates them from avoiding " their enemies, has made them be regarded as ſtupid creatures, in- " attentive even to ſelf-preſervation: they are never found in places " inhabited, and they never can; for, being incapable of reſiſtance " or eſcape, they muſt quickly diſappear, wherever deſtructive man " ſhall fix his abode, who permits nothing to ſubſiſt that he can ex- " tirpate."

is ſuch between theſe two families, notwithſtanding the characteriſtic differences which diſcriminate them, that in producing them nature ſeems evidently to have baniſhed to the extremities of the globe theſe extremes of the feathered kind; in the ſame manner as ſhe has baniſhed to thoſe retreats the great amphibious animals, the extremes of the quadrupeds, the ſeals and the walruſſes; unfiniſhed, mutilated forms, incapable of figuring in the animated ſcene among the more perfect models, and exiled into the remote confines of the world.

We proceed to enumerate and deſcribe the ſpecies of theſe two genera of wingleſs birds, the *Penguins* and *Manchots* *.

* Mr. Pennant, and after him Mr. Latham, gives the name *auk* to the northern ſpecies and appropriates that of *Pinguin* or *Penguin* to the ſouthern ſpecies — *T*.

The PENGUIN.

FIRST SPECIES.

Alca Torda. Linn. and Gmel.
Alca. Briſſ.
Plautus Tonſor. Klein.
Alka. Cluſius, Nieremb. and Johnſt.
Alka Hoieri. Sibb. Will. and Ray.
The Falk. Martin's Voy. St. Kilda.
The Marrot. Sibb. Hiſt. Fife.
The Auk. Penn. and Lath. *

THOUGH this firſt Penguin is furniſhed with wings of ſome length, and with ſeveral little feathers, we are aſſured that it cannot fly, nor even riſe from the water †. The head, the neck, and the whole of the upper ſide of the body, are black; but the under ſide, which is immerſed in the water when it ſwims, is entirely white. A little ſtreak of white runs from the bill to the eye, and a ſimilar ſtreak croſſes the wing obliquely.

We have ſaid that the feet of the Penguin has only three toes, and that this conformation, as

* In the north of England the *Auk*: in the weſt of England *the Razorbill*: in Cornwall *the Murre*: in Scotland *the Scout*: in Norway, and in the Feroe iſlands, *the Alke, Klub Klympæ*: in Gothland *Tord*, and in Angermania *Tordmulé*: in Iceland *Aulka, Klumbr, Klumbernevia*: in Greenland *Awarſak*.

† Edwards.

well

No. 258

THE RAZOR BILL.

No. 259

THE RAZOR-BILL THE FEMALE.

well as that of the bill, diſtinguiſhes it very manifeſtly from the Manchot. The bill of this firſt Penguin is black, ſharp at the edges, very flat on the ſides, which are channelled with three furrows, of which the middle one is white: juſt at its aperture, and under the down that covers the baſe of the bill, the noſtrils appear in long ſlits. The female wants the little white ſtreak between the bill and the eye, but its throat is white.

"This Penguin," ſays Edwards, "occurs "equally in the northern parts of America and "of Europe. It comes to breed on the Feroe "Iſlands *, along the weſt of England †, and on "the Iſle of Wight ‡, where it augments the "multitude of ſea-fowl that inhabit the great "rocks, called *the Needles*." We are aſſured, that it lays only one egg ||, which is very large in proportion to the ſize of the bird §.

It is ſtill uncertain in what aſylum the Penguins, eſpecially the preſent, paſs the winter ¶. As they cannot hold out on the ſea in the depth of that ſeaſon, and never appear then on ſhore, nor retire to ſouthern climates, Edwards ſuppoſes that they paſs the winter in the caverns of rocks, which open under water, but riſe internally as much above the level of the flood as to admit a receſs, where the Penguins remain torpid, and live upon their abundant fat.

* Hoierus. † Ray. ‡ Edwards.
|| Linnæus. § Ray. ¶ *Idem.*

We

We ſhould add, from Pontoppidan, ſome particulars concerning this ſpecies; that it is a great catcher of herrings, that it bites hooks baited with theſe fiſh, &c. if the account given by that writer did not betray the ſame inconſiſtencies that appear in his other narrations; for inſtance, he ſays, "that when theſe birds iſſue from the "caverns where they ſhelter themſelves and "neſtle, they darken the ſun by their number, "and make with their wings a noiſe like that of "a tempeſt." This aſſertion applies not to the Penguins, which at moſt can only flutter.

We recognize the Penguin in the *eſarokitſok*, or *little wing* of the Greenlanders; "a kind of "diver," ſays the narrator, "which has wings "at moſt only half a foot long, and ſo ſcantily "feathered that it cannot fly; its legs too are "placed ſo far back, that one cannot conceive "how it is able to ſtand erect and walk." In fact, the erect attitude is painful to the Penguin; its pace is heavy and ſluggiſh, and its ordinary poſture is that of ſwimming or floating on the water, or lying ſtretched on the rocks or on the ice.

[A] Specific character of the Auk, *Alca Torda*: "Its bill is "marked with four furrows, a white line on either ſide between the "bill and the eyes." The length is eighteen inches; the alar extent twenty-ſeven; the weight twenty-three ounces. The Auk lays her egg on the naked rock, to which it is faſtened by the concretion of viſcous moiſture that bedews the ſurface upon its excluſion: if this cement chance to be broken, it rolls down the precipice.

No. 260

THE GREAT AUK.

The GREAT PENGUIN.

SECOND SPECIES.

Alca Impennis. Linn. and Gmel.
Alca Major. Briſſ.
Mergus Americanus. Cluſius.
Goirfugel. Nieremb. Johnſt. and Cluſius.
The Penguin. Wormius, Will. Ray, Martin, &c.
The Northern Penguin. Edw.
The Gare. Sibb. Prodrom. Scotiæ.
The Great Auk. Penn. and Lath. *

WILLUGHBY ſays, that the ſize of this Penguin approaches that of the gooſe. He muſt mean the height of its head, and not the bulk of its body, which is much more ſlender than in the gooſe. The head, the neck, and the whole mantle, are of a fine black, with little ſhort feathers, ſoft and gloſſy like ſattin: a great oval white ſpot appears between the bill and the eye, and the margin of this ſpot riſes like a rim on each ſide of the top of the head, which is very flat: the bill, which, according to Edwards' compariſon, reſembles the end of a broad cutlaſs, has its ſides flat and hollowed with notches: the greateſt feathers of the wings exceed not three inches in length. We may eaſily judge, that

* In Iceland it is called *Goirfugl:* in Norway *Fiært, Anglemange, Penguin, Brillefugl:* In Swediſh *Penguin.*

plumage

plumage ſo ſcanty in proportion to the maſs of its body cannot raiſe it into the air *. It can ſcarce even walk, but continues always on the water, except in the time of breeding.

This ſpecies ſeems not to be numerous; at leaſt theſe great Penguins appear ſeldom on the coaſts of Norway †. They do not reſort every year to the Feroe Iſlands ‡; and they ſeldom deſcend more ſoutherly in our European ſeas §. That deſcribed by Edwards was caught by the fiſhers on the banks of Newfoundland. It is uncertain to what region they retire to neſtle ‖.

The *akpa* of the Greenlanders, a bird *as large as a duck, with the back black and the belly white, and which can neither run nor fly* ¶, appears to be the Great Penguin. With reſpect to the pretended Penguins, deſcribed in the voyage of Martiniere, they are evidently pelicans **.

* Hoierus. † Linnæus. ‡ Hoierus.

§ Edwards. ‖ Hoierus.

¶ The *akpa* of Greenland is as large as a duck; its back is black, its belly white: this ſpecies lives in flocks very far at ſea, and approaches not the land, except in the coldeſt weather; but it then repairs in ſuch numbers that the water round the iſlands ſeems covered with a thick dark fog: then the Greenlanders drive them upon the coaſt, and catch them with the hand, for theſe birds can neither run nor fly. They afford ſubſiſtence to the inhabitants during the months of February and March, at leaſt at the mouth of Ball River, for they do not reſort to all the ſhores indiſcriminately. They have the tendereſt and moſt nutritive fleſh of all the ſea-hens; and their down ſerves to line winter garments. *Hiſt. Gen. des Voy. tom.* xix. *p.* 46.

** Theſe birds, which our commander ſaid were called *Penguins*, are not taller than ſwans, but twice as large, and equally white; their

their neck as long as that of a goose, their head much larger; their eye red and sparkling, their bill tapered to a point, and yellowish-brown; their feet also are formed like those of a goose, and they have a sort of pouch, which begins under the bill, continuing along the neck to the breast, enlarging below, in which they store their provisions when they are satisfied, to feed as occasion requires . .. To prepare them for eating we were obliged to skin them, as their skin was very hard, and the feathers could not be plucked but with great difficulty. The flesh is very good, and of the same taste with that of *wild ducks*, and very fat.

[A] Specific character of the Great Auk, *Alca Impennis*: "Its "bill is compressed and channelled; an oval spot on either side be-"fore the eyes." Its length on both surfaces, to the end of the toes, is three feet: the tip of the longest wing-quills is only four inches and a quarter from the joint. Its egg is six inches long, white, and marked irregularly with ferruginous. It frequently visits St. Kilda, and breeds in June and July.

The LITTLE PENGUIN; or the SEA-DIVER of BELON.

Alca Pica. Linn. and Gmel.
Alca Minor. Briss.
Mergus Bellonii. Aldrov. Johnst. Will. and Ray.
Alca Unisulcata. Brunn. and Muller.
The Black-billed Auk. Penn. and Lath.

THIS bird is noticed by Belon under the name of *Sea-Diver*, and by Brisson under that of *Little Penguin*. Yet we much doubt the propriety of the latter denomination; for, upon examining the figure given by that ornithologist, we perceive a strong likeness between it and *the little guillemot* of our *Planches Enluminées*; and at any rate

rate its bill is different from that of the Penguin. The place, too, where Belon obſerved it, the Cretan Sea, throws in our way an additional doubt; ſince the Penguins never advance to the Mediterranean, and are all repreſented as peculiar to the northern ſeas. In ſhort, if we durſt in this inſtance ſuſpect the accuracy of an obſerver ſo well-informed, and ſo uniformly exact as Belon, we ſhould infer, notwithſtanding what he ſays concerning the ſtructure of the feet of his Cretan *uttamaria*, that it belongs rather to ſome ſpecies of diver or grebe, than to the family of Penguins. However, we cannot but tranſcribe the relation by our old and learned naturaliſt, who is the original author from whom Dapper and Aldrovandus have drawn their account of this bird.

"There is," ſays he, "in Crete, a particular "ſort of Sea-Diver, ſwimming beneath the ſur-"face, different from the cormorant and the "other divers called *mergi*, and which I con-"ceive to be what Ariſtotle has termed *æthia*. "The inhabitants on the Cretan ſhore call it "*uuttamaria* and *calicatczu*. It is of the ſize of "a garganey, white below the belly, and black "over the whole upper ſide of the body. It has "no ſpur behind, and it is likewiſe the only one "of all the flat-footed birds which has that pro-"perty: its bill is very ſharp at the edges, black "above, white below; hollow, and as it were "flat, and covered with down a good way

x "forward

" forward . . . which is occaſioned by a tuft of " feathers that grows upon ſomething over the " bill joining the head, raiſed like a half walnut " . . . The top of the head is broad, but the tail is " ſo ſhort, that it ſeems like a point; it is en- " tirely covered with fine down, which adheres " ſo cloſe to the ſkin, that it might juſtly be " looked upon as hair, and ſeems as delicate as " velvet; inſomuch that when flayed the ſkin is " found to be very thick, and, if curried, it re- " ſembles the ſkin of ſome land animal."

[A] Specific character of the Black-billed Auk, *Alca Pica*: " Its " bill is ſmooth and compreſſed; all the under ſide of the body, and " the tips of the poſterior wing-quills, are white; its legs red." Its length is eighteen inches and a half; its weight eighteen ounces. This ſpecies is very common in Greenland, where they breed on the cliffs. They feed on marine inſects, and grow very fat. In winter they paſs the day in the bays, but in the evening retire to the ſea. The Greenlanders eat their fleſh half putrid, ſuck their raw fat, and clothe themſelves with their ſkins. The bird, dreſſed with its entrails, is by theſe people eſteemed a great delicacy.

The GREAT MANCHOT.

FIRST SPECIES.

Aptenodytes Patachonica. Gmel.
Anser Magellanica. Clusius.
Plautus Pinguis. Klein.
The Patagonian Pinguin. Penn. and Lath.

CLUSIUS seems to attribute the discovery of the Manchots to the Dutch, who performed in 1598 a voyage to the South Sea. "These "navigators," says he, "having touched at cer-"tain islands near Port Desire, found them full "of a kind of unknown birds, which had come "there to nestle: they called these birds *pin-"guins*, on account of their fatness *(pinguedo)* *,

* This derivation is adopted by Dr. Grew; and Messrs. Pennant and Latham have gone so far, to favour that conjecture, as to alter the usual spelling into pinguin. But is it in the smallest degree probable, that illiterate sailors would think of bestowing a Latin name on a new object? And even admitting this, they would have called the bird *pingued*, not *pinguin*, surely, far less *penguin*, which is however the original orthography. A word of a similar sound signifies *white head* in Welch; and some authors have alledged this accidental coincidence as a further proof that a colony was carried from Wales to America. To this opinion Butler alludes in his Hudibras:

"British Indians named from penguins."

—But it appears that, in the northern languages, the great auk has the name of penguin, which the Dutch must have learnt in their frequent voyages to the Whale-fishery; when they met with a similar bird, therefore, on the coast of Patagonia, they would naturally bestow upon it the same appellation.—*T.*

"and

No. 261

THE PATAGONIAN PENGUIN.

"and named these iſlands *the iſlands of pin-*
"*guins.*

"Theſe ſingular birds," adds Cluſius, "have no wings, but in their ſtead two membranes that hang on each ſide like little arms; their neck is thick and ſhort; their ſkin is hard and thick like hog's leather. They were found three or four in a hole: the young ones weighed ten or twelve pounds, but the adults reached to ſixteen pounds, and, in general, they were of the bulk of the gooſe."

From theſe proportions, it is eaſy to recognize the Manchot repreſented in the *Planches Enluminées* under the name of *the Manchot of the Malouine iſlands*, and which occurs not only in the whole of the Straits of Magellan and the adjacent iſlands, but alſo at New Holland, from whence it has ſtretched to New Guinea *. It is indeed the largeſt of the Manchots; and the individual which we directed to be engraved, was twenty-three inches high: they attain to a much greater ſize; for Forſter found ſeveral that meaſured thirty-nine inches, and weighed thirty pounds †.

"Divers flocks of theſe penguins, the largeſt I ever ſaw, wandered on the coaſt (of New Georgia): their belly was of an enormous bulk, and covered with a large quantity of fat; they have on each ſide of the head a ſpot of bright yellow or orange-colour, edged with black; all

* Sonnerat.

† Forſter.

 the

"the back is of a blackiſh gray; the belly, the "under ſide of the pinions, and the fore part of "the body, are white. They were ſo ſtupid "that they made no effort to eſcape, and we "knocked them down with ſticks . . . Theſe are, "I think, what the Engliſh have termed at "the Falkland Iſlands, *yellow penguins* or *king* "*penguins*."

This deſcription of Forſter agrees exactly with our Great Manchot, obſerving only, that a bluiſh tint is ſpread on its cinereous mantle, and that the yellow of its throat is rather lemon or ſtraw-colour than orange. The French, indeed, found it in the Falkland or Malouine Iſlands; and Bougainville ſpeaks of it in the following terms. "It loves ſolitude and ſequeſtered re-"treats: its bill is longer and more ſlender "than in the other kinds of Manchots, and "its back is of a lighter blue; its belly is of a "dazzling whiteneſs; a jonquil tippet, which "riſing from the head interſects theſe white and "blue (gray-blue) ſpaces, and terminates on the "ſtomach, gives it a great air of magnificence: "when it ſcreams it ſtretches out its neck . . . "We hoped to be able to carry it to Europe: "at firſt it grew ſo tame as to diſtinguiſh and "follow the perſon who had the charge of "feeding it; and it ate indifferently bread, fleſh, "or fiſh. But this diet was not ſufficient; it "abſorbed its fat, became exceſſively emaciated "and died."

[A] Specific character of the Patagonian Penguin, *Aptenodytes Patachonica*: "Its bill and legs are black; a gold ſpot on the ears." This is rather a ſcarce ſpecies. They lay in the end of September or the beginning of October. They are very full of blood, ſo that in killing them their head muſt be ſevered, to allow it to flow.

The MIDDLE MANCHOT.

SECOND SPECIES.

Aptenodytes Demerſa. Gmel.
Diomedea Demerſa. Linn.
Spheniſcus. } Briſſ.
Spheniſcus Nævius. }
The Black-footed Pinguin. Edw.
The Leſſer Penguin. Philoſ. Tranſ. and Sparr.
The Cape Penguin. Lath.

OF all the characters which might be employed to denominate this ſecond ſpecies of Manchot, we have pitched on the ſize as the moſt conſtant and diſcriminating. It is what Edwards calls *the black-footed penguin*; but the feet of the great Manchot are black likewiſe. It appears in the *Planches Enluminées* under the name of *the Manchot of the Cape of Good Hope*, or of *the Hottentots*. But the ſpecies occurs in other places beſide the Cape, and is met with alſo on the South Seas. We had thought of calling it *the collared Manchot*; and in fact the black mantle of the back encircles the fore part

 of

of the neck by a collar, and ſends off upon the ſides two long bands after the manner of a ſcapulary: but this livery appears not to be conſtant except in the male, and the female has ſcarce ſome obſcure trace of a collar. In both, the bill is coloured near the tip by a little yellow band, which perhaps depends on the age. So that we can denominate it only from its ſize, which is about the average in this genus, ſeldom ever exceeding a foot and an half.

All the upper ſurface of the body is ſlaty, that is of a blackiſh aſh-colour; and the fore part, with the ſides of the body, are of a fine white, except the collar and the ſcapulary; the end of the lower mandible ſeems a little truncated, and the fourth toe, though free and not attached to the membrane, is turned more before than behind; the pinion is all flat, and looks as if covered with a ſhagreen, the pencils of feathers which clothe it are ſo little, ſtiff, and preſſed, the largeſt of theſe plumules is not half an inch long, and according to Edwards' remark, above an hundred may be counted in the firſt row of the wing.

Theſe Manchots are very numerous at the Cape of Good Hope *, and in the adjacent latitudes.

* There were at the Cape of Good Hope birds called *penguins* in great numbers, which are as large as a pretty ſmall *gooſe*; their body is covered with ſmall feathers; their wings are like thoſe of a duck after the feathers are plucked: they cannot fly, but they ſwim very

titudes. The Viſcount de Querhoënt obſerved them off the Cape, and communicated to me the

very well, and dive ſtill better; they are frightened at the ſight of men, and endeavour to eſcape, but they may be eaſily caught by running: each female lays two eggs as large as thoſe of a gooſe; they make their neſt among the brambles, ſcraping in the ſand and forming a hole, in which they lurk ſo cloſe, that, in paſſing along, one can hardly perceive them; they bite very ſtrong when they are near a perſon who is off his guard;—they are ſpotted with black and white. *Recueil des Voyages qui ont ſervi à l'etabliſſement de la Compagnie des Indes Orientales, tom.* iii. *p.* 581; *Amſterdam,* 1702.—The birds which are the moſt frequent in this bay (of Saldana) are the *penguins*; they do not fly, and their wings aſſiſt them only in ſwimming; they ſwim as faſt in the ſea as other birds fly in the air, *Flaccourt.*—We called a little iſland, which is four leagues beyond the Cape of Good Hope, *the iſland of birds,* on account of the great number and different ſpecies that were on it; there are *penguins* differing only from thoſe which occur in the Straits of Magellan, in that their bill is ſtraight like that of a *heron,* and not bent back as in the others; they are about the ſize of a gooſe, weighing ſixteen pounds; their back is covered with black feathers; their belly with white; their neck is ſhort and thick, with a white collar; their ſkin is very thick, and they have ſmall pinions like leather, which hang as ſmall arms covered with ſmall ſtiff feathers, white, and intermixed with black, which ſerve them to ſwim and not to fly; they ſeldom come on ſhore, unleſs it be to lay their eggs and hatch; their tail is ſhort, their feet black and flat; they conceal themſelves in holes which they make on the brink of the ſea, never more than two at once; they lay on the ground, and hatch only two eggs, which are about the bulk of thoſe of turkies. *Cauche.*—At Aguada de San Bras, twenty-five leagues from the Cape, is a ſmall iſland or a great rock, where is a multitude of birds called *penguins,* about the ſize of a goſling; they have no wings, or at leaſt theſe are ſo ſmall and ſo ſhort, as to reſemble more the ſhaggy ſkin of a beaſt than wings; but inſtead of wings, they have a feathered fin with which they ſwim; they ſuffer themſelves to be taken without making an effort to eſcape, a proof that they ſee few men or none at all; when one is killed, the ſkin is found to be ſo hard that a ſabre can ſcarce cut any part but the head. There were alſo on this rock

the following note: "The penguins (Manchots) "of the Cape are black and white, and of the "bulk of a duck; their eggs are white, two at "each hatch, and they defend their brood cou-"rageoufly: they neftle on the iflets along the "coaft; and an obferver of credit affured me, "that in one of thefe was a raifed knoll, which "thefe birds preferred, though more than half "a league from the fea. As they walk flowly, "he thinks it impoffible that they fhould every "day refort to the fea for food. He took fome "therefore to try how long they could live "without fuftenance; he kept them a fortnight "without any thing to eat or drink, and at the "end of that time they were ftill alive, and fo "ftout, that they bit keenly."

M. de Pages, in the manufcript relation of his voyage towards the South Pole, agrees with refpect to thefe facts. "The fize of the Cape "Manchots," fays he, "is equal to that of our "largeft ducks: they have two oblong cravats "of a black colour, the one on the ftomach, "the other on the neck. We found commonly "in the neft two eggs or two young ones, laid "head to tail, the one always a fourth at leaft "bigger than the other. The adults were as "eafy to take as the young: they could walk

rock many fea-dogs, which made refiftance to the failors: we killed fome of them, but neither the dogs nor the birds were good to eat. *Recueil des Voyages qui ont fervi à l'etabliffement de la Compagnie, tom.* i. *pp.* 213 & 214.

"only

"only ſlowly; and ſought to lie among the "rocks."

This voyager adds a curious fact, that the Manchots uſe their pinions from time to time as fore-feet, and then they go faſter, walking as it were on four. But in all probabllity this is a ſort of tumbling, and not a real walk.

This middle ſpecies ſeems to be the ſecond of thoſe deſcribed by Bougainville at the Malouine iſlands; for he ſays, that it is the ſame with that of Admiral Anſon *, which is alſo that of Narborough: but from the weight and colours which Narborough aſcribes to his penguin, we may regard it as the ſame with the ſpecies in queſtion †. It ſeems alſo to be that which Forſter deſcribes as the moſt common in the Straits of Magellan, and which he ſays is of the bulk of a little gooſe, and ſtiled by the Engliſh at the Falkland Iſlands the *jumping Jack*.

* On the eaſt coaſt of Patagonia, we found immenſe troops of ſeals, and a great variety of ſea-fowl, of which the moſt ſingular were the *penguins*; they are of the ſize and nearly of the figure of a gooſe; but inſtead of wings they have two ſtumps, which are of no uſe to them but in ſwimming; when they ſtand or walk, they hold their body erect, and not in a ſituation nearly horizontal like the other birds. This peculiarity joined to their having a white belly, ſuggeſted to Sir John Narborough the whimſical idea of comparing them to children ſtanding with white bibs. *Anſon.*

† It weighs about eight pounds; its head and back are black, its neck and belly white, and the reſt of its body blackiſh; its legs are as ſhort as thoſe of a gooſe; when there are many in flocks, and ſeen at a diſtance, one would ſuppoſe them to be children dreſſed in white; it bites very hard, but is not at all ſhy, for they came in whole flocks about our boats, where we eaſily killed them one after another, ſtriking them on the head. *Narborough.*

Forſter

Forſter obſerved theſe Manchots at Staten-land, where he had a little adventure with them: "They were," ſays he, "in a profound ſleep, "for Dr. Sparrmann lighted on one, which he "rolled ſeveral yards without waking it: to "rouſe them from their ſlumber, we were ob-"liged to jog them repeatedly. At length they "roſe in flocks, and when they ſaw that we "ſurrounded them, they took courage, darted "with violence upon us, and bit our legs and "our clothes. After leaving a great number "apparently dead on the field of battle, we "chaſed the reſt, but the firſt ſtarted ſuddenly, "and paced gravely behind us."

[A] Specific character of the Cape Penguin, *Aptenodytes Demerſa*: "Its bill and legs are black; its eye-brows, and the bar on "its breaſt, are white."

The HOPPING MANCHOT.

THIRD SPECIES.

Aptenodytes Chryſocome. Gmel.
The Hopping Penguin. Bougainville, and Phil. Tranſ.
The Creſted Pinguin. Penn. and Lath. W

THIS Manchot is ſcarcely a foot and half high from the bill to the feet, and nearly as much when, its head and body extended, it ſits on on its rump, which is neceſſarily its poſture on land: its bill is red, and ſo is its iris, over the eye there paſſes a white line tinged with yellow, which dilates and expands behind into two little tufts of briſtled filaments, that riſe from both ſides of the top of the head; this part is black, or of a very deep blackiſh aſh-colour, as well as the throat, the face, the upper ſide of the neck, of the back, and of the pinions; all the fore ſide of the body is of a ſnowy white.

In the *Planches Enluminées* this bird is indicated under the name of *Siberian Manchot:* we no longer retain that denomination, ſince nature ſeems to have marked the great diviſion of the northern penguins and the ſouthern Manchots; and as M. Bougainville has diſcovered it on the *Terra Magellanica*, we ſuſpect that it is not found in Siberia, but only in the iſlands of the South Sea, where the ſame navigator has deſcribed them under the name of *hopping penguin*. "The third ſpecies of theſe half birds," ſays he, "lives in families like the ſecond, on the high "rocks where they lay. The characters which "diſtinguiſh theſe from the two others are their "ſmallneſs, their fulvous colour, a tuft of gold-"coloured feathers ſhorter than thoſe of the "*egrets*, and which they erect when angry; and "laſtly, other little feathers of the ſame colour, "which ſerve as eye-brows. They are called "*hopping penguins:* in fact, they move by leaps "and

"and ſprings. This ſpecies has more livelineſs "in its mien than the two others."

It is, in all probability, the ſame creſted and red-billed Hopping Manchot that Captain Cook alludes to in the following paſſage: "Hi-"therto (in lat. 53° 57′ ſouth) we had continu-"ally round the ſhip a great number of *penguins*, "which ſeemed to be different from thoſe we "ſaw near the ice; they were ſmaller, with "reddiſh bills and brown heads. The meeting "with ſuch a multitude of theſe birds gave me "ſome hope of finding land." And in another place . . . "On the 2d of December, lat. 48° 23′ "ſouth, long. 179° 16′, we obſerved ſeveral red-"billed penguins which continued with us next "day."

[A] Specific character of the Creſted Penguin, *Aptenodytes Chryſocome*: "Its bill is rufous-brown; its legs yellowiſh; the creſt "on its front deep black and erect, a deflected tuft from the ear of "a ſulphur-colour." Its length is twenty-three inches. It is not quite ſo unwieldy as the other penguins.

No. 262

THE PENGUIN, WITH A MUTILATED BILL.

The MANCHOT WITH A TRUNCATED BILL.

FOURTH SPECIES.

Aptenodytes Catarractes. Gmel.
Phaeton Demersus. Linn.
Catarractes. Briss.
The Red-footed Penguin. Edw. and Lath.

THE bill of the Manchots usually terminates in a point. In this species the extremity of the lower mandible is truncated. This character seemed sufficient to Brisson for constituting a distinct genus under the denomination of *gorfou,* of which he was completely master according to the hypothetical and systematical order of his divisions: but it was not a matter equally arbitrary to apply to the same Manchot the name of *Catarractes* or *Catarracta,* by which Aristotle denoted an aquatic bird of prey *, which was certainly not a Manchot, with which Aristotle must have been totally unacquainted.

However, Edwards, to whom we owe our knowledge of this species, applies to it this passage of Sir Thomas Roe, in his voyage to India: " On the isle of Penguins (at the Cape of

* *Hist. Anim.* lib. ix. 12.

" Good

"Good Hope) is a ſort of fowl of that name "that goes upright; his wings without feathers, "hanging down like ſleeves faced with white; "they do not fly, but walk in companies, keep-"ing regularly their own quarters *."

Yet Edwards does not inform us if this Manchot be an inhabitant of the Cape, rather than of the Straits of Magellan. It was, he ſays, *as large as a gooſe*; its bill was open as far as the eyes, and red, as well as the feet; the face was of a dull brown; all the fore ſide of the body was white; the hind part of the head, the top of the neck, and the back, were of a dull purple, and covered with very little feathers ſtiff and cloſe: "Theſe feathers," adds Edwards, "reſemble "more the ſcales of a ſerpent than feathers; the "wings," he continues, "are ſmall and flat like "brown plates, and covered with feathers ſo lit-"tle and ſo ſtiff, that at ſome diſtance they "might be taken for ſhagreen: there is no ap-"pearance of tail, but ſome ſhort and black "briſtles at the rump."

* Churchill's Coll. of Voyages, *vol.* i. *p.* 767.

[A] Specific character of the Red-footed Penguin, *Aptenodytes Catarractes*: "Its bill and legs are red; its head brown."—Our reader will find a full and diſtinct deſcription of the penguins, with an excellent figure, by Mr. Pennant, in the Philoſophical Tranſactions for 1768.

SUCH

* * * *

SUCH are the four ſpecies of Manchots which we could exhibit as known and well deſcribed. If this genus is more numerous, as Forſter ſeems to inſinuate, each new ſpecies will naturally aſſume its place. Meanwhile we ſhall remark ſome that are mentioned, though imperfectly and confuſedly, in the following notes:

I. " Of the Maldive iſles," ſays one of our old voyagers, " a prodigious number are uninha-" bited . . . and others covered with large crabs, " and a croud of birds called *pingui*, which lay " and breed in theſe retreats. Their multitude " is ſo aſtoniſhing, that one cannot any where " ſot a foot without trampling on their eggs " and young, or the birds themſelves. The " iſlanders will not eat them, though they are " very palatable, *and are of the ſize of pigeons*, " with a white and black plumage *."

We are unacquainted with this ſpecies of Manchot as ſmall as a pigeon, and yet a ſimilar ſmall ſpecies of wingleſs bird, under the name of *calcamar*, occurs on the coaſt of Brazil. " The " calcamar is of the bulk of a pigeon; its wings " are of no aſſiſtance to it in flying, but it ſwims " very nimbly: it never leaves the water; the

* Voyage de François Pyrard de Laval; *Paris*, 1619, *tom*. i.

8 " Brazilians

" Brazilians aſſert even that it there depoſits its " eggs, but do not explain how it could hatch " them on the water *."

II. The *aponars* or *aponats* of Thevet †, " which," ſays he, " have little wings, by which " reaſon they cannot fly; their belly is white, " their back black, their bill ſimilar to that of a " cormorant or a raven, and when they cry, it " is like the grunting of hogs." Theſe are in all probability Manchots. Thevet found them on the iſland of Aſcenſion: but under the name of *aponar*, he makes the ſame confuſion with what has happened under that of penguin; for he ſpeaks of *aponars which ſhips meet with in ſailing from France to Canada.* Theſe laſt are penguins.

III. The bird of the South Seas, which Captain Wallis's people, and afterwards Captain Cook's, called the *race-horſe*, becauſe it ran on the water very ſwiftly, ſtriking the ſurface with its feet and wings, which are too ſmall for its flying. This bird ſeems from theſe characters to be a Manchot; yet Forſter denominates it the *logger-head duck* in the Philoſophical Tranſactions, Vol. lxvi. Part 1. He thus ſpeaks: " It reſembled a duck, except in the extreme " ſhortneſs of its wings, and in its bulk, which

* Hiſt. Gen. des Voy. *tom.* xiv. *p.* 303.

† Singularités de la France Antarctique, par André Thevet; *Paris*, 1558, *p.* 40.

" is

" is that of a goose; its plumage was gray, with " a few white feathers; its bill and legs yellow, " and two large scaly bumps of the same colour " at the joint of each wing. Our sailors called " it race-horse, on account of its swiftness; but " in the Falkland Islands the English have " given it the name of *logger-head-duck*."

IV. Lastly, according to other voyagers * there is found on the islands of the Chilian coast, beyond Chiloë, and towards the Straits of Magellan, a " species of goose which does not fly, " but runs on the water as nimbly as others fly. " This bird has a very fine down, which the " American women spin, and make it into co- " verlets, which they sell to the Spaniards." If these particulars are to be depended on, they indicate a species between the large feathered birds and the Manchots with scaly feathers, which bear little resemblance to down, and seem capable of being spun.

* Anson and Wager.

NOTES AND HINTS
OF CERTAIN SPECIES OF BIRDS THAT ARE UNCERTAIN OR UNKNOWN.

NOTWITHSTANDING the pains that we have taken, through the whole of this Work, to discuſs, elucidate, and refer to their true objects the imperfect or obſcure indications of voyagers or naturaliſts, on different ſpecies, real or nominal, of birds; notwithſtanding the extent and even the ſucceſs of our reſearches, we muſt confeſs, that there ſtill remains a certain number of ſpecies which we cannot recognize with certainty, becauſe they are mentioned under unknown names, or exhibited with obſcure or vague features, which quadrate not exactly with any real object. Theſe names and theſe features, however confuſed, we here collect, not only to omit nothing material, but to prevent theſe dubious hints from being admitted as certain; and, above all, to ſet obſervers in the way of verifying or elucidating them.

In this ſummary ſurvey we ſhall follow the order of the work, beginning with the Land Birds, paſſing to the Waders, and concluding with the Water Fowl.

I. The *great bird* at Port Deſire, on Magellan's Land, which is undoubtedly a bird of prey, and

and ſeems, from the ſtatement of Commodore Byron, to be a *vultur*. "The head," ſays he, "reſembled that of an eagle, except that it had "a comb upon it; round the neck there was a "white ruff, exactly reſembling a lady's tippet; "the feathers on the back were as black as jet, "and as bright as the fineſt poliſh could render "that mineral; the legs were remarkably ſtrong "and large; the talons were like thoſe of an ea-"gle, except that they were not ſo ſharp; and "the wings, when they were extended, mea-"ſured, from point to point, no leſs than twelve "feet."

II. The bird of New Caledonia, mentioned in Captain Cook's ſecond voyage, as *a ſpecies of raven*; though he ſays at the ſame time, that *it is only half as large as the raven, and its feathers ſhaded with blue*. This new-diſcovered iſland has preſented but few birds, and among theſe *beautiful turtles, and ſeveral unknown ſmall birds*.

III. The *avis venatica* of Belon, the only one perhaps which that judicious naturaliſt has not diſcriminated by his numerous obſervations. "We ſaw alſo (near Gaza) a bird which, in "our opinion, excels all the reſt by the charms "of its ſong; and we think it was denominated "by the ancients *avis venatica*. It is ſomewhat "larger than a ſtare; its plumage is white below "the belly, cinereous on the back, as in the "*molliceps* or groſbeak: the tail is black, and

" extends beyond the wings, as in the magpie; " it flies like the green woodpecker."

From the ſize, the colours, and the name *avis venatica (hunting bird)* we might take this bird to be a ſpecies of ſhrike; but *a pleaſant warble* is no attribute of this miſchievous and cruel ſpecies.

IV. The *ſea-ſparrow*, " which the inhabitants " of Newfoundland call the *ice-bird*, becauſe it " lives conſtantly among the ice; it is not larger " than a thruſh; it reſembles the ſparrow by its " bill, and its plumage is black and white."

Hiſt. Gen. des Voy. tom. xix. *p.* 46.

Notwithſtanding the name of *ſea-ſparrow*, the form of its bill indicates it to be a land bird, and it ſeems to be a-kin to the ſnow bunting.

V. The little *yellow-bird*, ſo called at the Cape of Good Hope, and which Captain Cook found in New Georgia. It is perhaps known to ornithologiſts, but not under that name. With reſpect to the *little birds with handſome plumage*, which this ſame navigator found at Tanna, one of the New Hebrides, we readily agree with him in opinion, that in land ſo remote and unconnected they are abſolutely new ſpecies.

VI. The bird which the naturaliſts that accompanied Captain Cook in his firſt voyage denominated *motacilla velificans*, who ſaw it alight on the ſhip's rigging at ſea, ten leagues from Cape Finifterre. We fhould certainly have found it to be a fhepherdefs, had not Linnæus, whofe nomenclature they follow, applied the term *motacilla*, as generic, to all birds that wag their tail.

VII. The *occolin* of Fernandez, which fhould have ranged among the woodpeckers; for he exprefsly fays, that *it is a woodpecker of the fize of a ftare, its plumage agreeably variegated with black and yellow.* Fernandez, *Hift. Avi. Nov. Hifp.* ccii. 54.

VIII. The birds feen by Dampier at Ceram, and which, from the form and bulk of their bill, feem to be *calaos*. He defcribes them as follows: "Their body was black, and their tail white; "they were as large as a crow; their neck was "pretty long, and faffron-coloured; their bill "was like a ram's horn; their legs were fhort "and ftrong; their feet refembled thofe of a "pigeon, and their wings were of an ordinary "fize, though they made great noife in flying: "they feed on wild berries, and perch upon the "largeft trees. Dampier found their flefh fo "good, that he feemed to regret his not having "feen thefe birds except at Ceram and New "Guinea." *Hift. Gen. des Voy. tom.* ii. *p.* 244.

IX. *The hoitzitzillin of Tepufcullula* of Fernandez, and the *nexhoitzillin* of the fame author, which muft be colibris; living, he fays,

 on

on the honey of flowers, which they ſuck with their little curved bill, almoſt as long as their body; and with its brilliant feathers ſkilful hands form precious little pictures. Fernandez, clxxiv. p. 47. & lxxxii. 31.

With reſpect to the *hoitzitzil-papalotl* of this Spaniſh naturaliſt, though he compares it to the *hoitzitzillin*, he ſays expreſsly that it is a ſort of butterfly.

X. The *quauchichil*, or *little red-headed bird*, alſo of Fernandez, cliv. p. 21. It is only ſomething larger, he ſays, than the *hoitzitzillin*, and yet appears not to be a colibri or fly-bird, *for it occurs likewiſe in cold countries, and lives and ſings in the cage.*

XI. The half-aquatic bird, deſcribed by Forſter, and which he ſays is of *a new genus:* "This "bird, which we met with in our excurſion, "was of the ſize of a pigeon, and perfectly white; "it belongs to the claſs of aquatic birds that "*wade*; its feet are ſemi-palmated, and its eyes, "and the baſe of its bill, are encircled with lit-"tle glands or warts: it exhaled ſo inſupport-"able a ſmell, that we could not eat its fleſh, "though at that time we were not eaſily diſguſted "with the moſt unpalatable food." (It was at Statenland). *Forſter's Voyage.*

XII. The *corbijeau* of Page Dupratz *(Hiſtory of Louiſiana, tom.* ii. *p.* 128) which is nothing

thing but the *curlew*; and we here inſert the name to complete the whole ſyſtem of the denominations relative to this bird, and to ornithology in general.

XIII. The *chochopitli* of Fernandez, *a bird*, ſays this naturaliſt, *of the kind of what the Spaniards call chorlito* (which is the curlew). It ſeems to be *the white and brown great curlew of Cayenne*. This bird, Fernandez adds, is migratory on the lake of Mexico, and its fleſh has a diſagreeable fiſhy taſte.

XIV. The *ayaca*, which, both from the ſimilarity of its name to *ayaia*, applied to the ſpoonbill in Brazil, and from the reſemblance of its characters, except the alterations which objects always undergo in paſſing through the hands of the compilers of voyages, appears to be a ſpoonbill. "This Brazilian bird *(ayaca)* is remark-"ably diligent in catching little fiſh; it never "darts without effect upon the water: it is of "the bulk of a magpie; its plumage is white "marked with red ſpots, and the bill is ſhaped "like a ſpoon." *Hiſt. Gen. des Voyages, tom.* iv. *p.* 303.

The *aboukerdan* of Montconys *(I. partie, page* 98.) is our ſpoonbill.

XV. The *acacahoactli*, or *the bird of the Mexican lake, with a raucous voice*, mentioned by Fernandez; which, he ſays, is a kind of *alcyon* or kingfiſher.

 But,

But, according to the remark of Adanſon, it is rather a ſpecies of heron or of bittern; ſince *it has a very long neck, which it often ſolds, bringing it between its ſhoulders.* It is ſomewhat ſmaller than the wild duck; its bill is three inches long, pointed, and ſharp; the ground of its plumage is white ſpotted with brown, browner above, and whiter below the body; the wings are of a bright and reddiſh fulvous, with the point black. According to Fernandez, we may tame this bird, feeding it with fiſh, and even fleſh; and, what is not very conſiſtent with its raucous voice, *its ſong*, he ſays, *is not diſagreeable. (Fernandez, vol.* ii. *p.* 16.) It is the ſame with the *avis aquatica raucum ſonans* of Nieremberg, *lib.* x. 236.

XVI. The *atototl*, a little bird, likewiſe of the Mexican lake, of the form and ſize of a ſparrow, with the plumage white on the under ſide of the body, varied above with white, fulvous, and black; which neſtles in the ruſhes, and which from morning to evening emits a feeble cry, like the ſhrill ſqueak of a rat: its fleſh is eaten. *Fernandez, cap.* viii. *p.* 15.

It is hard to ſay whether this *atototl* is really a ſhore-bird, or only an inhabitant of marſhes, like the reed thruſh or the ſedge warbler. At any rate, it is very different from another *atototl*, given by Faber, at the end of Hernandez' work, (*p.* 672.) and which is the *alcatraz*, or *Mexican pelican*.

XVII.

XVII. The *mentavaza* of Madagaſcar, "a "bird with a hooked bill, as large as a par- "tridge, which haunts the ſea-ſhore." The voyager Flaccourt ſays nothing more of it. *Voy. à Madagaſcar, Paris,* 1661, *p.* 165.

XVIII. The *chungar* of the Turks, and the *kratzhot* of the Ruſſians, of which we can only tranſcribe the relation given by the hiſtorian of the voyages, without adopting his conjectures. "The plains of Tartary," ſays he, "produce "numerous birds of rare beauty: that deſcribed "in Abulghazi-Khan, is ſeemingly a ſpecies of "heron, which frequents the part of the Mo- "gul's dominions which borders on China; it is "entirely white, except the bill, the wings, and "the tail, which are of a beautiful red; its fleſh "is delicate, and taſtes like that of the hazel "grous." But as the author ſays that it is very rare, we may ſuppoſe it to be the bittern, which is in fact very rare in Ruſſia, Siberia, and Great Tartary, but which occurs ſometimes in the territories of the Mogul, near China, and which is almoſt always white. Abulghazi-Khan ſays, that its eyes, its legs, and its bill, are red; and he adds, that the head is of the ſame colour. He tells us, that this bird is named *chungar* in the Turkiſh language, and *kratzhot* in the Ruſſian; which has led the Engliſh tranſlator to conjecture that it is the ſame with that denominated *chon-kui*, in the hiſtory of Timur-Bek, and which was preſented to Gengis-Khan

Khan by the ambaſſador of Kadjak *. *Hiſt. Gen. des Voy. tom.* vi. *p.* 604.

XIX. The *obeitſok*, or, the *ſhort-tongue*, which is ſaid " to be a ſea-fowl of Greenland, which " having ſcarce any tongue, preſerves an eternal " ſilence, but in compenſation, it has a long " bill and leg, ſo that it might be called the " ſea ſtork. This gluttonous bird devours an " incredible number of fiſh, which it brings up " from the depth of twenty or thirty fathoms, " and which it ſwallows whole, though they be " very large. It can be killed only when en- " gaged fiſhing, for it has large eyes, protu- " berant, and very vivid, crowned with a yellow " and red circle." *Hiſt. Gen. des Voy. tom.* xix, *p.* 45.

XX. The *tornoviarſuk* of the ſame frozen ſeas of Greenland, which is a maritime bird of the ſize of a pigeon, and approaching the genus of the duck. It is difficult to determine the family of this bird, of which Egede ſays nothing more. *Dict. Groënl. Hafniæ*, 1750.

XXI. Beſides the birds of Poland known to naturaliſts, and enumerated by Rzaczynſki, there are ſome " which he knows only by the vulgar

* Petit de la Croix remarks in the ſame place, that the *chon-kui* is a bird of prey, which is preſented to the king of the country, decked with many precious ſtones, as a mark of homage; and that the Ruſſians, as well as the Tartars of the Crimea, are bound by their treaties with the Ottomans to ſend one every year to the Porte, decorated with a certain number of diamonds.

" name,

"name, and which he refers to no ſpecies." Three of theſe particularly ſeem, from their natural habits, to belong to the tribe of cloven-footed water fowl.

The *derkacz* "ſo called from its cry, *der*, *der*, "frequently repeated. It inhabits the low and "wet meadows; it approaches the ſize of the "partridge; its legs are tall, and its bill long." (This may be a rail).

The *hayſtra*, which is pretty large, of a dark brown colour, with a thick and long bill: it fiſhes in rivers, like the heron, and neſtles on trees.

The third is the *krzyczka*, which lays ſpotted eggs in the ruſhes among bogs.

XXII. The *arau* or *kara* of the northern ſeas; "it is a bird larger than a duck; its eggs are "very good to eat, and its ſkin ſerves for furs: "its head, neck, and back, are black; its belly "blue; its bill long, *ſtraight*, black, and *pointed*." *Hiſt. Gen. des Voy. tom.* xix. *p.* 270. From theſe characters the *arau* or *kara* muſt be a ſpecies of diver.

XXIII. The *John-van-Ghent* or *John-de-Gand*, of the Dutch navigators at Spitzbergen (*Recueil des Voyages du Nord, tom.* ii. *p.* 110). which, they ſay, is at leaſt as large as a ſtork, and has the ſame figure; its feathers are white and black; it cuts the air without almoſt ſtirring its wings; and as ſoon as it approaches the ice,

ice, it turns back again: it is a ſort of bird calculated for falconry; it darts ſuddenly, and from a great height, upon the water, which makes us preſume that it has a very quick ſight. The ſame birds are ſeen in the Spaniſh ſea, and almoſt through the whole of the north ſea, but chiefly near the herring fiſheries.

This *John-de-Gand* ſeems to be the great mew or great gull, which we denominated the *black mantle*.

XXIV. The *hav-ſule*, "which the Scots," ſays Pontoppidan, "call the *gentleman*;" which appears to us a ſpecies of mew or gull, perhaps the *ratzher* or *counſellor* of the Dutch. We ſhall tranſcribe what Pontoppidan relates on this ſubject, though we can repoſe little confidence in the Norwegian biſhop, ever near the marvellous in his anecdotes, and far from accuracy in his deſcriptions: "This bird," ſays he, "ſerves as a ſign to the herring-fiſhers; it appears in Norway about the end of January, "when the herrings begin to enter the gulfs, "and it follows them at the diſtance of a league "from the coaſt. It is ſo greedy of this fiſh, "that the people need only lay herrings on the "edge of their boats to catch the *gentlemen*. "This bird reſembles a gooſe; its head and "neck are like thoſe of the ſtork, the bill ſhorter "and thicker; the feathers of the back, and of "the under ſide of the wings, are light white; "it has a red creſt; its head is greeniſh and "black;

"black; its neck and breaſt are white." *Hiſt. Nat. de Norwege, par Pontoppidan*; *Journal Etranger, Fevrier*, 1757.

XXV. The *pipelines*, of which I find the name in Frezier, and *which bear a reſemblance*, he ſays, *to the ſea-bird mauve:* The mauve is the ſame with the mew or maw; but what he adds, *that they are very well taſted*, agrees not with mews, which are very bad meat.

XXVI. The *margaux*, of which the name uſed among ſailors ſeems to denote a booby or cormorant, or perhaps both the one and the other. "The wind not being fair for coming "out of Saldana Bay," ſays Flaccourt, we "ſent twice to the iſlet of *Margaux*, and each "trip the boat was filled with theſe birds "and their eggs. Theſe birds, which are as "large as a gooſe, are there ſo numerous, that, "walking on ſhore, one cannot avoid trampling "on them. When they ſtruggle to take wing, "they entangle one another. They are knocked "down with a ſtick as they riſe in the air." *Voy. a Madagaſcar, par Flaccourt*; *Paris*, 1661, *p.* 250.

"There were at the ſame *iſland*" (that *of birds*, near the Cape of Good Hope) ſays Francis Cauche, "*margots*, bigger than a goſling, with "gray feathers, the bill hooked at the point "like a hawk's; the foot ſmall and flat, with "a pellicle between the toes. They reſt on the "ſea;

" ſea; they have broad wings; they make their " neſts in the middle of the iſland, among herbs, " and never lay more than two eggs." *Voy. à Madagaſcar; Paris*, 1651, *p.* 135.

" In a diſtrict of the *iſle*" (*of birds*, on the tract to Canada), ſays Sagar Theodat, " were birds " living ſeparate from one another, and very " difficult to catch, for they bit like dogs; they " are called *margaux*." *Voy. au Pays des Hurons; Paris*, 1632, *p.* 37.

From theſe circumſtances we are diſpoſed to take the *margau* for the ſhag or little cormorant, which we have deſcribed.

XXVII. Theſe ſame ſhags appear to us to have been mentioned by ſeveral voyagers, under the name of *alcatraz* *, very different from the true and great alcatraz of Mexico, which is the pelican. (*See the article of the pelican*).

XXVIII. The *fauchets*, which we ſhall refer to the family of ſea-ſwallows. " The commo-" tion of the elements (in a great ſtorm)," ſays Forſter, " never drove theſe birds from us; at " times, a black *fauchet* fluttered on the agitated " ſurface

* Hiſtoire des Incas; *Paris*, 1744, *tom.* ii. *p.* 277.—Voyage de Coreal; *Paris*, 1722, *tom.* i. *p.* 345.—Hiſt. Gen. des Voy. *tom.* i. *p.* 448, & *tom.* iv. *p.* 533. In the latter place it is ſaid, that during the night the alcatraz fly as high as poſſible, and then, putting their head under the one wing, they ſupport themſelves ſome time with the other, till their body approaching the water, they reſume their flight to the heavens; thus repeating frequently the ſame action, they may be ſaid to ſleep flying. It is ſcarce neceſſary to add, that the whole of this relation is a fable.

" ſurface of the ſea, and broke the force of the " waves, by expoſing itſelf to their action. The " aſpect was then threatening and terrible." *(Cook's Second Voyage.)*—" We perceived the " high grounds (or the weſt entrance to the " Straits of Magellan) drifted and covered with " ſnow almoſt to the water's edge; but great " flocks of *fauchets* made us hope to find re- " freſhments, if we could meet with a haven." *Idem.*—Fauchets, in 27° 4′ lat. ſouth, and 103° 56′ long. weſt, about the firſt of March. *Idem* *.

XXIX. The *backer* or *pecker*, of the inhabitants of Oëland and Gotnland, which we recognize more certainly to be a ſea-ſwallow, from the particulars we learn of its inſtinct. " If " any perſon goes to the place where theſe " birds neſtle, they fly round his head, and ſeem " diſpoſed to peck or bite him; at the ſame " time they emit a cry, *tirr, tirr*, repeated in- " ceſſantly. The backer comes every year to " Oëland, there paſſes the ſummer, and leaves " that country in autumn: its neſt coſts it leſs " trouble than that of the ordinary ſwallows; " it lays two eggs, and drops them on the flat " ground in the firſt place it meets; yet it never " depoſits them among tall herbs; if it lays on a " ſandy plain, it only excavates a little ſhallow

* The bird here alluded to is the ſhearwater or puffin, *Procellaria Puffinus*, deſcribed in the body of the work. The French tranſlator renders *ſhearwater* by the word *fauchet*. *T.*

" hole; its eggs are of the ſize of pigeons', gray-" iſh, and ſpotted with black: this bird ſits " four weeks; if little hens' eggs be placed un-" der her, ſhe will hatch them in three weeks; " and *the chickens thus hatched are very miſ-" chievous, eſpecially the males.* In the ſtrongeſt " wind, it can hold itſelf motionleſs in the air; " and when it marks its prey, it deſcends ſwifter " than an arrow, and accelerates or retards its " force, according to the depth it ſees the fiſh " to be at in the water; ſometimes it only dips " its bill, and ſometimes it plunges till the " points of its wings only, and a part of its tail, " appear above the ſurface: its plumage is gray; " all the upper half of its head is pitch-black; " its bill and legs are fire-coloured; its tail is " like that of the ſwallow. When plucked it " is hardly ſo large as a thruſh." *Deſcription of a water-fowl of the iſle of Oëland; Journal Etranger, Fevrier*, 1758.

XXX. The *vourouſambé* of Madagaſcar, or *gri-ſet*, of the voyager Flaccourt (*p.* 165), is proba-bly alſo a ſea-ſwallow.

XXXI. The *ferret* of the iſlands Rodrigue and Maurice, which Leguat mentions in two places of his voyages. " Theſe birds," ſays he, " are of the bulk and nearly of the figure of " a pigeon: their general reſort in the evening, " was to a ſmall iſlet entirely naked. We found " their eggs lying on the ſand, and quite near

"each

" each other; yet they have only a ſingle egg at " each hatch . . . We carried off three or four " dozens of young, and as they were very fat, " we had them roaſted: we found they had " nearly the taſte of the ſnipe, but we were " hurt by them, and never afterwards were " tempted to taſte them . . . Having returned " ſome days after to the iſland, we found that " the *ferrets* had forſaken their eggs and their " young in the whole of the diſtrict which we " had viſited . . . The goodneſs of the eggs made " amends for the bad quality of the fleſh of the " young. During our ſtay we ate many thou- " ſands of theſe eggs: they are ſpotted like " thoſe of a pigeon." *Voyage de François Leguat*; *Amſterdam*, 1708, *tom.* i. *p.* 104, *and tom.* ii. *pp.* 43 & 44.

Theſe *ferrets* appear to be ſea-ſwallows; and it would be doubly intereſting to know the ſpecies, on account of the goodneſs of their eggs, and of the bad quality of their fleſh.

XXXII. The *collier (charbonnier)*, ſo called by Bougainville, and which, from the firſt characters, we might take for a ſea-ſwallow, but in the laſt ones, if they be exact, it ſeems to differ. " The *collier*," ſays Bougainville, " is of the " ſize of a pigeon; its plumage is of a deep gray, " and the upper ſide of the head white encircled " with a gray cord, more inclined to black than " the reſt of the body; the bill is ſlender, two " inches long, and a little curved at the end; " the eyes are bright, the toes yellow, reſem-

" bling those of ducks; the tail is abundantly " furnished with feathers, rounded at the end; " the wings are much cut out, and each of about " eight or nine inches extent. The following " days we saw many of these birds (it was in " the month of January, and before his arrival " at the river de la Plata)." *Voyage autour du Monde, tom.* i. *pp.* 22 & 23.

XXXIII. The *velvet sleeves, mangas de velado,* of the Portugueze, which, according to the dimensions and the characters that some give, seem to be pelicans, and, according to other notices, present more analogy to the cormorant. It is in the creek at the Cape of Good Hope, that these birds are found. They owe their name to the resemblance of their plumage to velvet (*Hist. Gen. des Voy. tom.* i. *p.* 248), or to their tips being velvet black (*Tachard, p.* 58.), and that in flying their wings appear to fold like the arm. (*Hist. des Voy. ibid.*) According to some, they are all white, except the end of the wing, which is black; they are as large as the swan, or, more exactly, as the goose (*Merolla, in the Hist. Gen. des Voyages, tom.* iv. *p.* 534); according to others, they are blackish above and white below. (*Tachard.*)

M. de Querhoënt says, that they fly heavily, and scarcely ever leave the deep water; he believes them to be of the same genus with the *margaux d'Ouessan.* (*Remarks made on board his Majesty's ship Victory, by the Viscount de Querhoënt*);

boënt); but these margaux, as we have said, must be cormorants.

XXXIV. The *stariki* and *gloupichi* of Steller, "which," he says, "are reckoned unlucky "birds at sea; their belly is white, and the "rest of their plumage is of a black, sometimes "verging on blue: there are some entirely black, "with a vermilion bill, and a white crest on the "head.

"The last, which derive their name from "their stupidity, are as large as a river-swallow. "The islands, or the rocks, situated in the strait "which separates Kamtschatka from America, "are all covered with them. It is said, that "they are black as painters' umber, with white "spots over their whole body: the Kamtscha-"dales, to catch them, have only to sit near "their retreat, clothed in a *pelisse* with hanging "sleeves: when these birds come in the evening "to their holes, they creep into the pelisse of "the hunter, who takes them without trouble.

"In the species of *starikis* and *gloupichis*," adds Steller, "they reckon the *kaiover* or *kaior* "which is said to be very cunning: it is a black "bird, with red bill and toes: the Cossacks call "it *iswoschiki*, because it whistles like horse-"drivers." *Hist. Gen. des Voy. tom.* xix. *p.* 271.

Neither these characters nor these peculiarities, of which a part savours of fable, are sufficient to discriminate these birds.

XXXV. The *tavon* of the Philippines, of which the name *tavon* signifies, it is said, to *cover with earth*, becauſe this bird lays a great number of eggs, and depoſits them in the ſand, with which it covers them. Its deſcription and hiſtory, of which Gemelli Carreri was the firſt author (*Voyage autour du Monde*; *Paris*, 1719, *tom.* v. *p.* 286) are filled with ſo many incongruities, that we cannot admit it into the text, but throw it into a note *.

XXXVI. The *parginia*, a name which the Portugueze, according to Kœmpfer, give to a kind of bird which the Japaneſe call *kanjemon*: it is found in an iſland on the track from Siam to Manilla. The eggs of theſe birds are almoſt

* Of many ſingular birds on theſe iſlands, the moſt wonderful by its properties is the *tavon*. It is a ſea-fowl, black, and ſmaller than a hen, but its legs and neck pretty long; it lays its eggs on ſandy ground, and theſe are nearly as large as thoſe of a gooſe: what is moſt ſurprizing, after the young are hatched, the yolk is ſtill found without any of the white . . . the young are roaſted before they are covered with feathers, and they are as good as the beſt pigeons. The Spaniards often eat, from the ſame diſh, the young and the yolk of the egg; but what follows merits much more admiration: the female gathers her eggs, to the number of forty or fifty, into a ſmall ditch, which ſhe covers with ſand, and of which the heat of the ſun makes a ſort of furnace: at length, when the brood have ſtrength to ſhake off the ſhell, and open the ſand to come out, ſhe perches on the neighbouring trees; ſhe makes ſeveral circuits round the neſt, ſcreaming with all her might, and the young, rouzed by this ſound, make ſuch motions and efforts, as to burſt through every obſtacle, and find their way to her. The tavons make their neſts in the months of March, April and May, the time when, the ſea being more placid, the waves do not riſe ſo high as to hurt them: the ſailors ſeek eagerly for theſe neſts along the beach; when they find the ſand thrown up, they open the ſpot with a ſtick, and take out the eggs and the young, which are equally prized. *Hiſt. Gen. des Voy. tom.* x. *p.* 411.

as large as hens' eggs. They are found the whole year in that iſland, and they proved a great reſource for the ſubſiſtence of the crew in this traveller's ſhip. *Kœmpfer, Hiſt. Nat. du Japon, tom.* i. *pp.* 9 & 10. It is obvious that this curſory mention will not aſcertain the *parginia* of the Portugueze.

XXXVII. The *miſago* or *biſago*, which the ſame Kœmpfer compares to a hawk (*tom.* i. *p.* 113). It is ſcarce more recognizable than the preceding; however, we think that it ſhould be ranged among the aquatic birds, ſince it feeds on fiſh. "The *miſago*," ſays he, "lives principally upon fiſh; it makes a hole in ſome rock on the coaſts, and there lays its prey or its proviſions, which, it is remarked, preſerve as well as the pickled fiſh, *altiar*; and for this reaſon it is called *biſagonohuſi*, or *altiar* of Biſago: it taſtes extremely ſalt, and ſells very dear. Thoſe who diſcover this kind of larder, may draw great profit from its ſtore, provided they do not rob it completely at once."

XXXVIII. Finally, the *azores*, of which we have only this notice. "The name *Azores* was given to the iſlands, on account of the great number of birds of this kind that were ſeen or diſcovered on them." *Hiſt. Gen. des Voy. tom.* i. *p.* 12.

 Theſe

Theſe Azore birds certainly are not an unknown ſpecies; but it is impoſſible to recognize it under this name, which we can meet with no-where elſe *.

* The Portugueze diſcovered theſe iſlands, and in their language *açor* ſignifies a falcon.—*T.*

APPENDIX,

APPENDIX,

BY THE

TRANSLATOR.

APPENDIX, I.

OF SYSTEMS IN ORNITHOLOGY.

THE moſt valuable work tranſmitted from the ancients on the ſubject of Ornithology, is contained in Ariſtotle's Hiſtory of Animals. That great and univerſal genius, aſſiſted by the liberality of his pupil Alexander the Great, conducted the vaſt undertaking with admirable ſucceſs. He poſſeſſed the rare faculty of acute perception; and the happy flexibility of the Greek language enabled him to mark with preciſion the diſtinguiſhing features of animals. Yet that philoſopher affects a dry and conciſe ſtile, that frequently borders on obſcurity; nor is he always at ſufficient pains to diſcuſs and reject popular notions. The natural Hiſtory of Pliny is a compilation which oftener diſplays the taſte and elegance of its author than his critical diſcernment. Heſiod, Ælian, Columella, Aulus Gellius, and other writers, have left us ſome hints reſpecting the œconomy of animals. The Chriſtian fathers indulged much in turgid figurative language, and occaſionally drew

drew their comparisons from the current opinions in natural history. But the sun of science was now set, and that dismal night succeeded, which overspread the nations of Europe. After the lapse of twelve centuries, a ray of light burst in upon the Christian world; and men of the greatest abilities laboured with enthusiasm to restore the noble remains of antiquity. The commentators on the treatises of natural history were not in general so well qualified for acquitting themselves with credit: yet in that line of criticism, Turner, and the celebrated Joseph Scaliger, deserve applause. At this period, America had been discovered and explored, settlements formed along the coast of Africa, and an extensive intercourse established with India. From these countries were imported birds of singular forms and wonderful beauty, which, while they increased the subjects of Ornithology, incited powerfully to the study of it. Prompted by a love of science, the learned and sagacious Belon travelled into Greece, and Egypt, and Asia Minor. Upon his return to France, he published his observations; but his History of Birds was not given to the world till the year after his death, in 1555. Gesner composed, in 1557, a Treatise on the Birds found in Switzerland. Various other productions appeared; and from all these sources, Aldrovandus, with industry and erudition, but with little taste or judgment, compiled his voluminous History of of Birds, in 1599. Marcgrave's account o the birds diſcovered in Brazil, was publiſhed 1648. Mr. Ray, with the aſſiſtance of his friend, Francis Willughby, Eſquire, wrote a Syſtem of Ornithology in 1667, though it was not printed till 1678; a work of conſiderable merit. Barrere publiſhed his Syſtem in 1745; Klein, in 1750; Moehring, in 1753; and Briſſon, in 1760. Linnæus attempted a claſſification of birds in his *Fauna Suecica*, in 1746, which he improved in his *Syſtema Naturæ*, in 1758; but it has been greatly altered and enlarged in the ſubſequent editions. One of the neateſt ſyſtems of Ornithology was compoſed in our own language, by the ingenious Thomas Pennant, Eſquire, in 1772, and publiſhed in 1781. He contents himſelf, however, with the outlines We proceed to give an abſtract of his method.

Mr. Pennant diſtinguiſhes birds into the Land Birds and the Water Fowl. The firſt Diviſion comprehends ſix Orders. Theſe are:

I. The Rapacious. Including three Genera:—The Vulture, the Falcon, and the Owl.

II. The Pies. Including twenty-ſix Genera:—The Shrike, the Parrot, the Toucan, the Motmot, the Hornbill, the Beef-eater, the Ani, the Wattle, the

the Crow, the Roller, the Oriole, the Grakle, the Paradise, the Curucui, the Barbet, the Cuckoo, the Wry-neck, the Woodpecker, the Jacamar, the Kingsfisher, the Nuthatch, the Tody, the Bee-eater, the Hoopoe, the Creeper, the Honeysucker.

III. The GALLINACEOUS. Including ten Genera:—The Cock, the Turkey, the Pintado, the Curasso, the Peacock, the Pheasant, the Grous, the Partridge, the Trumpeter, and the Bustard.

IV. The COLUMBINE. Containing only one Genus:—The Pigeon.

V. The PASSERINE. Including sixteen Genera:—The Stare, the Thrush, the Chatterer, the Coly, the Grosbeak, the Bunting, the Tanager, the Finch, the Flycatcher, the Lark, the Wagtail, the Warblers, the Manakin, the Titmouse, the Swallow, and the Goatsucker.

VI. The STRUTHIOUS. Containing only two Genera:—The Dodo, and the Ostrich.

THE Second Divifion comprehends three Orders. Thefe are :—

VII. The CLOVEN-FOOTED. Including feventeen Genera : — The Spoonbill, the Screamer, the Jabiru, the Boatbill, the Heron, the Umbre, the Ibis, the Curlew, the Snipe, the Sandpiper, the Plover, the Oyfter-catcher, the Jacana, the Pratincole, the Rail, the Sheath-bill, and the Gallinule.

VIII. The PINNATED-FEET. Containing three Genera :—The Phalarope, the Coot, and the Grebe.

IX. The WEB-FOOTED. Including feventeen Genera :—The Avofet, the Courier, the Flammant, the Albatrofs, the Auk, the Guillemot, the Diver, the Skimmer, the Tern, the Gull, the Petrel, the Merganfer, the Duck, the Pinguin, the Pelican, the Tropic, and the Darter.

In this diftribution, Mr. Pennant attends fometimes to the Ornithology of Briffon ; but in general he adheres to that of Linnæus. Of this work we fhall now give a full view, with occafional occasional hints respecting the œconomy and habits that obtain in certain genera.

Linnæus divides the birds into six orders, which he thus defines:

I. The ACCIPITRES.

BILL, somewhat curved: *upper mandible* dilated on both sides behind the tip, and armed with a half-tooth: *nostrils* wide.

FEET, close-seated, short, robust: *toes*, warty under the joints, with nails bent, and very sharp.

BODY, with muscular head and neck; *skin* adhesive. Impure.

FOOD, the rapine and carnage of carcases.

NEST, placed in lofty situations; *eggs* about four: *female* the larger.—Monogamous.

II. The PICÆ.

EILL, knife-shaped, with a convex back.

FEET, furnished with thiee toes before and one behind, short and stout.

BODY, stringy and impure.

FOOD, gathered from dirt and rubbish.

NFST, built on trees; the *male* feeding the female during incubation.—Monogamous.

III. The

III. The ANSERES.

BILL, ſmooth, covered with an epidermis, enlarged at the tip.

FEET, adapted for ſwimming, the toes palmated by a membrane: *legs* ſhort and compreſſed.

BODY, plump; ſkin adheſive; plumage valuable. Rankiſh.

FOOD, procured in the water from plants, fiſh, &c.

NEST, uſually on land. The *mother* ſeldom nurſes her young. For the moſt part polygamous.

IV. The GRALLÆ.

BILL, inclined to cylindrical.

FEET, adapted for walking, with thighs half naked.

BODY, compreſſed with a very thin ſkin: *tail* ſhort. Sapid.

FOOD, gathered in marſhes from inſects.

NEST, uſually on land: nuptials various.

V. The GALLINÆ.

BILL, convex: *upper mandible* arched above the lower: *noſtrils* arched with a cartilaginous membrane.

FEET, adapted for running: the *toes* rough beneath.

BODY, fat, muſcular. Pure.

FOOD,

FOOD, collected on land from seeds, and macerated in a craw. Pulverent.

NEST, on the ground, inartificial: *eggs* numerous. Food pointed out to the *young*. Polygamous.

VI. The PASSERES.

BILL, sharpened conically.

FEET, adapted for hopping, tender, cleft.

BODY, slender. Pure in the *granivorous* kinds: impure in the *carnivorous*.

NEST, artificial. Food crammed into the *young*. Monogamous. Song.

THE First Order, that of the ACCIPITRES, comprehends four Genera. These are:—

I. VULTUR.

Characters. BILL straight, blunt at the tip.
HEAD featherless, covered behind with naked skin.
TONGUE bifid.
NECK retractile.

This genus contains thirteen species, besides varieties. Their natural habits are these:— They are very voracious; prefer dead carcases, even though putrid, and will not attack living animals, unless urged by famine; fly slowly,

ſlowly, except when riſen to a certain height, and in flocks; are endowed with a moſt acute ſmell.

II. FALCO.

Characters. Bill hooked, furniſhed at its baſe with a cere.
Head cloſely beſet with feathers.
Tongue bifid.

This is a very extenſive genus, containing one hundred and twenty ſpecies, excluſive of a multitude of varieties. It admits of four ſub-diviſions, and includes ſeveral of the vultures, the eagles, the kites, the hobbies, the falcons, and the hawks.

III. STRIX.

Characters. Bill hooked, and without a cere.
Nostrils oblong, concealed by reclining briſtly feathers.
Head large, with great ears and eyes.
Tongue bifid.

This genus contains the owls, which form forty-three ſpecies, beſides many varieties, and ranged in two ſub-diviſions; *the eared* and the *earleſs*. Theſe birds are nocturnal, and prey on ſmall birds, mice and bats; moſt of them have woolly feet; their outer toe can be turned back; their ears are broad; their eyes large and glaring.

IV. LANIUS.

Characters. Bill pretty ſtraight, with a tooth on each ſide near the tip, and naked at the baſe.
Tongue jagged.

This genus contains the butcher-birds or ſhrikes, forming fifty-three ſpecies, beſides a few varieties. Their middle toe is connected to the firſt joint.

The Second Order, that of the PICÆ, comprehends twenty-three Genera:—Of theſe eleven have *ambulatory* feet; that is, have three diſtinct toes before and one behind; eight have *ſcanſory* feet, that is, have two toes before and two behind; and four have *greſſory* feet, that is, have two fore toes connected, but without a membrane.

I. PSITTACUS.

Characters. Bill hooked, the upper mandible moveable, and furniſhed in many with a cere.
Nostrils at the baſe of the bill, and round.
Tongue fleſhy, obtuſe, entire.
Feet ſcanſory.

This genus contains the parrots, parrakeets, macaos, and lories, amounting to one hundred and forty-one ſpecies, beſides numerous varieties.

§ Theſe

These birds are sub-divided into those with *short* and those with *long* tails. Their head is large, the summit flat, their feet short: they are garrulous, docile, long-lived: subsist chiefly on nuts, acorns, the seeds of pompions, &c.: they climb by means of the bill, and when angry they erect their feathers: they are not found in high latitudes; they occur however in the thirty-fourth degree, but are most frequent in the zone extending twenty-five degrees on each side of the equator. In their natal regions they are often eaten.

II. RAMPHASTOS.

Characters. Bill exceeding large, hollow, convex, serrated outwards; both mandibles curved at the tip.

Nostrils behind the base of the bill, long and narrow.

Tongue feathery.

Feet in most of the species scansory.

This singular genus contains the toucans and motmots, distributed into sixteen species. These birds occur in South America between the tropics: they cannot bear cold; live chiefly on dates, and are easily tamed; in their native climate they fly in little companies of eight or ten; nestle in holes made in trees by the woodpeckers, and lay two eggs; the individuals are numerous.

 III. BU-

III. BUCEROS.

Characters. BILL convex, curved, knife-ſhaped, large, and ſerrated outwards: the front bare, and ſwelling with bone.
NOSTRILS behind the baſe of the bill.
TONGUE ſhort and ſharp.
FEET greſſory.

This genus contains the hornbills, which form twelve ſpecies: they correſpond in their habits, and even in their ſtructure, to thoſe of the preceding genus, and inhabit the ſame parallels in the old world.

IV. BUPHAGA.

Characters. BILL ſtrait and ſubquadrangular; the mandibles ſwelling and entire, ſwelling ſtill more outwards.
FEET ambulatory.

Only one ſpecies has yet been found; and this the African beef-eater.

V. CROTOPHAGA.

Characters. BILL compreſſed, ſemi-oval, arched, and keel-ſhaped on the ridge; the upper mandible angled at both margins.
NOSTRILS pervious.

This genus includes the anis, of which there are only three ſpecies.

VI. GLAU-

VI. GLAUCOPIS.

Characters. Bill curved, vaulted; the under mandible ſhorter, and carunculated at the baſe.

Nostrils flat, half covered with a ſemi-cartilaginous membrane.

Tongue ſub-cartilaginous, notched and ciliated at the tip.

This genus contains only a ſingle ſpecies, the cinereous wattle-bird, a native of New Zealand. It walks on the ground, and ſeldom perches on trees. It has a piping or murmuring voice. Its fleſh is well taſted. Length fifteen inches.

VII. CORVUS.

Characters. Bill convex, knife-ſhaped.
Nostrils hid beneath reclining briſtly feathers.
Tongue cartilaginous and bifid.
Feet ambulatory.

This genus contains the ravens, the crows, the rooks, and the jays: the number of ſpecies is forty-ſix, and there are ſeveral varieties. Moſt of theſe birds occur in every climate; are exceedingly noiſy; neſtle upon trees, and lay ſix eggs; and take both animal and vegetable food.

VIII. CORACIAS.

Characters. Bill knife-ſhaped, curved at the tip, bare of feathers at the baſe.
Tongue cartilaginous and bifid.
Feet ambulatory.

This genus contains the rollers, of which there are ſeventeen ſpecies. They are diſperſed over the whole globe, and are remarkable for their ſhort legs.

IX. ORIOLUS.

Characters. BILL conical, convex, very ſharp and ſtrait; the upper mandible ſomewhat longer, and ſlightly notched.
TONGUE bifid and ſharp.
FEET ambulatory.

This genus contains the orioles, which are ranged in fifty-two ſpecies, excluſive of ſeveral varieties. Theſe birds are found chiefly in America, and have pendulous neſts: they are numerous and gregarious; noiſy and voracious, ſubſiſting on grain.

X. GRACULA.

Characters. BILL convex, knife-ſhaped, ſomewhat naked at the baſe.
TONGUE entire, ſomewhat enlarged and fleſhy.
FEET ambulatory.

This genus contains the grakles, which amount to twelve ſpecies. None of theſe inhabit Europe: they are remarkable for their thick bill compreſſed at the ſides; their minute noſtrils placed at its baſe; their hooked ſharp nails; and the middle of their fore toes is connected with the exterior one.

XI. PA-

XI. PARADISEA.

Characters. BILL covered with the downy feathers of the bridle.

FLANK-FEATHERS longer.

TAIL-QUILLS, the two upper detached and unwebbed.

This ſingular and beautiful genus contains the paradiſe birds, which amount to nine ſpecies. They chiefly inhabit New Guinea, from which they remove in the dry ſeaſon to the adjacent iſlands: their noſtrils are ſmall, and covered with feathers; their tail conſiſts of ten quills, of which the two mid-ones are webbed only at the root and the tip; their feet are large and ſtout; the middle of the fore toes is connected to the outer at the firſt joint.

XII. TROGON.

Characters. BILL ſhorter than the head, knife-ſhaped, hooked, ſerrated at the margin of the mandibles.

FEET ſcanſory.

This genus contains the curucuis, of which there are ſeven ſpecies, beſides ſome varieties. They are natives of the hotter parts of America, where they live ſolitary in the cloſe, ſwampy foreſts, and ſit on the lower boughs: they take very ſhort flights; ſubſiſt upon inſects: their body is long ſhaped; their feet ſhort; their tail very long, and containing twelve quills.

XIII. BUCCO.

Character. BILL knife-ſhaped, compreſſed laterally, notched on each ſide at the tip, bent, with a chap ſtretching forward below the eyes.
NOSTRILS hid under reclining feathers.
FEET ſcanſory.

This genus contains the barbets, which form ſeventeen ſpecies. They occur in Africa, but chiefly inhabit Aſia and the hotter parts of America. They are reckoned ſtupid birds: their bill is ſtout and ſomewhat ſtrait, covered almoſt completely with briſtles.

XIV. CUCULUS.

Characters. BILL ſomewhat taper.
NOSTRILS ſlightly protuberant at the margin.
TONGUE arrow-ſhaped, flat, entire.
FEET ſcanſory.

This genus contains the cuckoos, which amount to forty-eight ſpecies beſides varieties. They occur in both continents.

XV. YUNX.

Characters. BILL ſomewhat taper and ſharpened, faintly bent for a ſhort ſpace.
NOSTRILS concave and naked.
TONGUE taper, worm-ſhaped, very long, and pointed at the tip.
TAIL-QUILLS are ten in number, flexible.
FEET ſcanſory.

This

This genus contains only two ſpecies, the wryneck and the minute woodpecker; the former a native of Europe and Aſia, the latter of America.

XVI. PICUS.

Characters. BILL many-ſided, ſtrait, wedged at the tip.
NOSTRILS hid under reclining briſtly feathers.
TONGUE taper, worm-ſhaped, very long, bony, miſſile, pointed, beſet at the tip with reflected briſtles.
TAIL-QUILLS amount to ten, ſtiff and pointed.
FEET ſcanſory.

This genus contains the woodpeckers, of which there are fifty-three ſpecies. They are common to both continents: they ſettle on decayed rotten trees, and ſometimes bore into ſuch as are freſh in ſearch of inſects and larvæ; they cut with their bill, and make a hideous, grating noiſe; they are guided to their prey by the ear, and extract it from the cavities by injecting the bill.

XVII. SITTA.

Characters. BILL awl-ſhaped and ſomewhat taper, ſtrait, extended, and very entire; the upper mandible a little broader, compreſſed at the tip.
TONGUE notched and jagged, ſhort, with a horny tip.
NOSTRILS ſmall, covered with whiſkers.
FEET ambulatory.

This genus contains the nuthatches and loggerheads, which are ranged in eight ſpecies, excluſive of varieties: they are found in both continents.

XVIII. TO-

XVIII. TODUS.

Characters. BILL awl-ſhaped, flattiſh, obtuſe, ſtrait, with broad briſtles at the baſe.
FEET greſſory.

This genus contains the todies, of which there are ſixteen ſpecies: they inhabit the warmer parts of America; are much analogous to the fly-catchers, only in the latter the mid fore-toe is detached from its origin.

XIX. ALCEDO.

Characters. BILL three-ſided, thick, ſtrait, long, pointed.
TONGUE fleſhy, very ſhort, flat, and ſharp.
FEET for the moſt part greſſory.

This genus contains the kingfiſhers, which, excluſive of varieties, amount to forty ſpecies. They are diſperſed over the whole globe; inhabit chiefly the water, and live upon fiſh, which they catch with ſurprizing alertneſs, ſwallowing them entire, and afterwards rejecting the undigeſted parts: though their wings are ſhort, they fly ſwiftly: their prevailing colour is ſky-blue: their noſtrils are ſmall, and generally covered.

XX. MEROPS.

Characters. BILL curved, four-ſided, flattened, keel-ſhaped, ſharp.
NOSTRILS ſmall, ſituated at the baſe of the bill.
TONGUE ſlender, for the moſt part fringed at the tip.
FEET greſſory.

This

This genus contains the bee-eaters, which make twenty-one ſpecies, beſides ſeveral varieties. Theſe birds inhabit America, and are unfrequent: they live upon inſects, eſpecially bees and waſps; imitate the kingfiſhers in the conſtruction of their neſts: moſt of them have a harſh voice.

XXI. UPUPA.

Characters. BILL arched, long, ſlender, convex, ſomewhat compreſſed, and rather blunt.
NOSTRILS minute, ſituated at the baſe of the bill.
TONGUE obtuſe, very entire, triangular, and very ſhort.
FEET ambulatory.

This genus contains the hoopoes and the promeropſes, ranged in eight ſpecies.

XXII. CERTHIA.

Characters. BILL arched, thin, ſomewhat triangular, ſharp.
TONGUE ſharp.
FEET ambulatory.

This genus contains the creepers, which amount to fifty-four ſpecies. They are ſpread over the whole globe; live chiefly on inſects; have minute noſtrils, and are conſpicuous by their twelve tail-quills, their tall legs, their large hind-toe, and their long hooked nails: in many ſpecies the tongue is ſharp, in others it is flat at the tip, in others ciliated, and in a few tubulated.

XXIII. TRO-

XXIII. TROCHILUS.

Characters. BILL awl-ſhaped, thread-like, the tip tubulated, longer than the head: the upper mandible ſheaths the under.

TONGUE thread-like, tubulated with two coaleſcing threads.

FEET ambulatory.

This exquiſite miniature genus contains the various humming-birds, which form no leſs than ſixty-five ſpecies. They admit of a ſub-diviſion into thoſe with *curved* bills and thoſe with *ſtrait* bills. They inhabit the new world, and, except two ſpecies that migrate to the north, they are all confined to South America. Their bill and feet are feeble, their noſtrils minute; their tongue darts out: they have ten tail-feathers, which are beſpangled with the moſt glowing colours: they are forward and quarrelſome; fly very ſwift; feed hovering upon their wings, and ſuck the nectar from the flowers. The whirring of their wings is louder than the notes of their voice: they are gregarious; build an elegant hemiſpherical neſt of the woolly ſubſtance of plants, and lay two white eggs, about the ſize of peas, upon which the male and female ſit by turns: the young ones are attacked by ſpiders.

THE

The Third Order, that of the ANSERES, comprehends thirteen Genera: — Of theſe four have the bill furniſhed with a tooth; in the other nine it is plain.

I. ANAS.

Characters. Bill lamellar and toothed, convex, obtuſe.
Tongue ciliated, obtuſe.

This very extenſive genus includes the ſwans, the geeſe, the ducks, the ſheldrakes, the ſhovelers, the gadwalls, the wigeons, the garganeys, and the teals, forming in all one hundred and twenty-four ſpecies.

II. MERGUS.

Characters. Bill denticulated, of a cylindrical awl-ſhape, hooked at the tip.

This genus contains the merganſers, the dundivers, and the ſmews, which amount to ſeven ſpecies, with ſeveral varieties.

III. ALCA.

Characters. Bill plain, ſhort, compreſſed, convex, often furrowed tranſverſely; the lower mandible ſwelled before the baſe.
Nostrils behind the bill.
Feet, in moſt of the ſpecies, three-toed.

This genus contains the auks, which are ranged in twelve ſpecies. They inhabit the northern ſeas; they are ſilly birds; remain concealed

cealed during the night; neſtle in burrows, or in the holes and clefts of rocks, and lay only a ſingle egg, which is very large in proportion to their ſize: they are pretty uniform in their colours, black above and white below; they are ſhaped like a gooſe, their feet being placed behind the point of equilibrium; the bill is large and conical, ſtretching, in curved lines and ſurfaces, to a ſharp tip.

IV. APTENODYTES.

Characters. BILL ſtrait, ſmooth, flattiſh, and ſomewhat knife-ſhaped; the upper mandible marked longitudinally with oblique furrows, the lower truncated at the tip.

FEET fettered and palmated.

WINGS conſiſt of pinions, without ſhafts.

This genus contains the penguins, of which there are eleven ſpecies. They are analogous to the *alcæ* or auks in their colour, their food, their habits, their ſtupidity, the neſts and eggs, and the remote poſition of their feet: but they are found only in the South Seas; they are utterly incapable of flying, the feathers of their wings reſembling ſcales; their feet conſiſt of four toes; their plumage is ſofter, of a different texture, and reſiſts the water better: their fatneſs enables them to ſupport cold: they ſwim very faſt and alertly; ſometimes they are diſcerned walking in companies on land: they hatch ſtanding; make a clangorous noiſe like geeſe, but hoarſer: their noſtrils are ſlits concealed

cealed in the furrow of the bill; the palate and bill are planted with ſeveral rows of reflected briſtles; their body is fleſhy; the wings are covered with a dilated ſtrong membrane; the tail is wedge-ſhaped and ſhort, its feathers very ſtiff.—The name of this genus is formed from α, *privat*. and πτημι, to fly.

V. PROCELLARIA.

Characters. Bill plain, flattiſh: the mandibles equal; the upper with a hooked tip, the lower with a flat channelled tip.

Nostrils in a truncated cylinder, leaning above the baſe of the bill.

Feet palmated; the hind-nail cloſe ſet, and without any toe.

This genus contains the petrels, which amount to twenty-three ſpecies. Theſe birds keep on the ſea in the moſt tempeſtuous weather, and ſeldom repair to the ſhores: their legs are naked a little above the knees.—The name of the genus formed from *Procella*, a ſtorm.

VI. DIOMEDEA.

Characters. Bill ſtrait; the upper mandible hooked at the tip, the under truncated.

Nostrils oval, broad, prominent, and lateral.

Tongue extremely ſmall.

Feet furniſhed with three toes.

This genus contains the albatroſſes, of which there are only four ſpecies.

VII. PE-

VII. PELECANUS.

Characters. BILL ſtrait; the tip conſiſting of a hooked nail.
NOSTRILS obſcure chinks.
FACE ſomewhat naked.
FEET balanced; all the four toes palmated.

This genus contains the pelicans, the man-of-war birds, the cormorants, the ſhags, the boobies, and the gannets, amounting to thirty-two ſpecies, which are ſub-divided into thoſe with *plain* bills and thoſe with *ſerrated* bills. Theſe birds are ſo dextrous at fiſhing, that they have ſometimes been trained for that purpoſe. Moſt of them inhabit the ſeas, though ſome occur on land; they have a long bill, in the lateral furrow of which the noſtrils are ſeated: they are gregarious and very voracious; the nail of their mid-toe is generally ſerrated.

VIII. PLOTUS.

Characters. BILL ſtraight, ſharpened, toothed.
FACE AND CHIN naked.
FEET ſhort, palmated, all the toes connected.

This genus contains the darters, which form three ſpecies, and as many varieties. Their head is ſmall, their neck ſlender, and extremely long; and they are eaſily diſtinguiſhed by their noſtrils, which are placed like long chinks at the baſe of the bill: they occur in the warm countries of the ſouth; and live upon fiſh alone, which they catch by wreathing their neck like a ſerpent, and then darting their bill.

IX. PHAE-

IX. PHAETON.

Characters. Bill knife-ſhaped, ſtraight, ſharpened, with chaps gaping behind the bill.
Nostrils oblong.
Hind-toe turned forwards.

This genus contains the tropic-birds, which form three ſpecies. They are diſtinguiſhed by their flat bill, bent a little downwards, by the lower mandible being angled, by their four-toed and palmated feet, by their wedge-ſhaped tail, by the two middle quills of the tail being exceeding long.

X. COLYMBUS.

Characters. Bill plain, awl-ſhaped, ſtraight, ſharpened.
Chaps toothed.
Nostrils ſlits at the baſe of the bill.
Feet fettered.

This genus conſiſts of twenty-eight ſpecies, which are ſubdivided into thoſe with *three toes*, correſponding to the guillemots; thoſe with *four toes* and *palmated*, correſponding to the divers; thoſe with *four toes* and *lobed*, correſponding to the grebes. The birds of this genus cannot walk, but they run very ſwiftly on the water, and ſwim and dive with the utmoſt agility: their ſkin is adheſive, and their tail ſhort. The *guillemots* live generally at ſea; have a ſlender tongue, of the ſize of their bill, which is flat, and covered at its baſe with ſhort feathers; their upper mandible ſomewhat bent at the tip: their fleſh is commonly ſtringy, and their eggs nauſeating; they keep together in flocks, and lay on the bare rocks. The *divers* in the northern climates inhabit alſo the lakes: their bill is ſtrong, not ſo ſharp, cylindrical; the margin of the mandibles bent inwards, the upper mandible exceeding the under; the noſtrils parted by little membranes; the tongue long, ſharp, ſerrated on both ſides at the root; the legs ſmall and flattened; they have black ſtripes on their thighs, and twenty tail-quills. They are monogamous; lay their eggs on the turf; fly difficultly, and paſs the time of incubation in freſh water. The *grebes* have no tail; their bill is ſtrong; their ſtraps bald; their tongue ſlightly cleft at the tip; their body ſquat, and thickly clothed with ſoft ſhining feathers: their wings are ſhort, their legs compreſſed. They inhabit chiefly the lakes of the ſouth of Europe, and are ſubject to much variety of colour.

XI. LARUS.

Characters. BILL plain, ſtraight, knife-ſhaped, and ſomewhat hooked at the tip; the under mandible ſwelled below the tip.

NOSTRILS ſlits, broader before, and ſeated in the middle of the bill.

This genus contains the mews and gulls, which amount to twenty ſpecies, beſides ſome varieties. They are natives of the northern climates; their body light, their wings long, their tongue

tongue ſomewhat cleft, their bill ſtrong, their legs ſhort, and naked above the knees: they live chiefly on fiſh, even on ſuch as are dead, and reject the undigeſted portions; they are reſtleſs and unquiet; their eggs may be eaten, but their fleſh is loathſome. The young continue ſometimes ſpotted till the third year, which occaſions a confuſion in the claſſification.

XII. STERNA.

Characters. BILL plain, awl-ſhaped, ſomewhat ſtraight, ſharp, flattiſh.
NOSTRILS ſlits placed at the baſe of the bill.

This genus contains the terns and noddies, ranged in twenty-ſix ſpecies. They live for the moſt part on the ſea, ſubſiſting chiefly on fiſh; are not ſhy; their tail is forked, their wings very long, their hind-toe ſmall, their tongue ſlender and ſharp: the young are ſpotted.

XIII. RYNCHOPS.

Characters. BILL ſtraight: upper mandible much the ſhorter; the lower truncated at the tip.

This genus contains only a ſingle ſpecies, together with a variety; both natives of North America. In their habits and figure they reſemble much the gulls: their legs are weak, and their noſtrils pervious.

The Fourth Order, that of the GRALLÆ, comprehends twenty Genera: Two of these have three toes on each foot, and the rest four toes.

I. PHŒNICOPTERUS.

Characters. BILL bare, with a broken curvature, and toothed.
NOSTRILS slits.
FEET palmated, three-toed.

This genus contains the flamingos, of which there are only two species: they rarely occur in the warmer parts of Europe, but are found chiefly in Africa and in South America. They seem to occupy the gradation between the order of ANSERES and that of GRALLÆ: their bill is large and thick; the upper mandible keel-shaped, toothed at the margin; the under mandible compressed, furrowed transversely; their nostrils covered with a thin membrane, and pervious; their hind toe very small, the membrane connecting the fore toes being extended to the nails.

II. PLATALEA.

Characters. BILL flattish, long, thin; the tip dilated, orbiculated, and plain.
NOSTRILS minute, placed at the base of the bill.
TONGUE small, sharpened.
FEET four-toed, semi-palmated.

This genus contains the spoonbills, which form only three species.

III. PA-

III. PALAMEDEA.

Characters. BILL conical; the upper mandible hooked.
NOSTRILS oval.
FEET four-toed, cleft, only a very ſhort membrane connecting the toes at their origin.

This genus contains the ſcreamers, which are only two in number, and found in South America.

IV. MYCTERIA.

Characters. BILL ſomewhat riſing, ſharp: upper mandible three-ſided, and very ſtraight; the under triangular, ſharpened, riſing.
FRONT bald.
NOSTRILS ſlits.
TONGUE wanting.
FEET three-toed.

Of this genus a ſingle ſpecies only has been diſcovered; the jabiru, a native of South America.

V. CANCROMA.

Characters. BILL ſwelled; upper mandible ſhaped like an inverted boat.
NOSTRILS minute, placed in the furrow of the bill.
TONGUE ſmall.
FEET cleft.

This genus contains the boatbills, which form only two ſpecies, both natives of America.

VI. SCOPUS.

Characters. BILL thick, compreſſed, long, ſtraight.
NOSTRILS linear, oblique.
FEET three-toed, cleft.

 This

This genus contains only a ſingle ſpecies, the umbre, a native of Africa, and of the ſize of a rook.

VII. ARDEA.

Characters. Bill ſtraight, ſharp, long, flattiſh, with a furrow extending from the noſtrils to the tip.
Nostrils linear.
Tongue ſharpened.
Feet four-toed.

This extenſive genus contains the herons, the ſtorks, the cranes, the egrets, and the bitterns, amounting in all to eighty-eight ſpecies. They are ranged in five ſubdiviſions. 1. The *crowned*, whoſe bill is ſcarcely longer than the head. 2. The *cranes*, whoſe head is bald. 3. The *ſtorks*, whoſe orbits are naked. 4. The *herons*, whoſe mid-toe is ſerrated inwards. 5. Thoſe which have *the bill gaping in the middle*. The firſt ſubdiviſion includes two ſpecies; the ſecond, five; the third, three; the fourth, ſeventy-five; and the fifth, three.

VIII. TANTALUS.

Characters. Bill long, awl-ſhaped, ſomewhat taper, ſomewhat arched.
Face naked beyond the eyes.
Tongue ſhort, and broad.
Jugular Pouch naked.
Nostrils oval.
Feet four-toed, palmated at the baſe.

This genus contains the ibiſes, and ſome of the curlews: the number of ſpecies is twenty-one.

IX. COR-

IX. CORRIRA.

Characters. BILL ſhort, ſtraight, plain.
FEET long, four-toed, and palmated; the toes very ſhort.

This genus contains only a ſingle ſpecies, the *trochilus* of Aldrovandus, which is a native of Italy, and remarkable for its ſwift running.

X. SCOLOPAX.

Characters. BILL ſomewhat taper, obtuſe, longer than the head.
NOSTRILS linear.
FACE clothed.
FEET four-toed; the hind toe reſting upon many joints.

This genus contains ſeveral curlews, the whimbrels, the ſnipes, the woodcocks, the godwits, the red-ſhanks, the green-ſhanks, and the yellow-ſhanks; which form in all forty-ſix ſpecies.

XI. TRINGA.

Characters. BILL ſomewhat taper, of the length of the head
NOSTRILS linear.
TONGUE ſlender.
FEET four-toed; the hind one conſiſting of a ſingle joint, and raiſed from the ground.

This genus contains the lapwings, the ſandpipers, the gambets, the purres, the dotterels, the knots, and the phalaropes; amounting in all to forty-one ſpecies. Theſe birds run on the plains and the ſhores, ſcarce reſting on their hind-toe; whereas thoſe of the preceding genus reſt on all their four toes, and wade in the marſhes.

XII. CHARADRIUS.

Characters. BILL ſomewhat taper, obtuſe.
NOSTRILS linear.
FEET curſory, three-toed.

This genus contains the plovers, ranged in thirty ſpecies.

XIII. RECURVIROSTRA.

Characters. BILL flat and depreſſed, awl-ſhaped, curved back, ſharpened, flexible at the tip.
FEET palmated, four-toed; the hind toe very ſhort, and placed very high.
NOSTRILS narrow, pervious.
TONGUE ſhort.

This genus contains the avoſets, which form only three ſpecies.

XIV. HEMATOPUS.

Characters. BILL compreſſed; the tip of an equal wedge-ſhape.
NOSTRILS linear.
TONGUE a third ſhorter than the bill.
FEET curſory, three-toed, cloven.

This genus contains only a ſingle ſpecies, the oyſter-catcher.—The name derived from αἱμα blood, and πȣς the foot.

XV. GLAREOLA.

Characters. BILL ſtrong, ſhort, ſtraight, hooked at the tip.
NOSTRILS at the baſe of the bill, linear, oblique.
GAP wide.
FEET four-toed; the toes long, ſlender, connected to each other at the baſe by a membrane.
TAIL forked, with twelve quills.

This

This genus contains the pratincoles, which form three ſpecies and as many varieties.—The name formed from *glarea*, gravel.

XVI. FULICA.

Characters. BILL convex; the upper mandible vaulted at its margin, over the under, which ſwells behind its tip.
NOSTRILS oblong.
FRONT bald.
FEET four-toed, ſomewhat pinnated.

This genus contains twenty-five ſpecies, ranged in two ſubdiviſions; thoſe with *cloven feet*, correſponding to the gallinules, and thoſe with *pinnated* feet, correſponding to the coots: the former amount to eighteen ſpecies, the latter to ſeven. Theſe birds inhabit the water, and live upon worms, inſects, and ſmall fiſh: in the compreſſed form of their body, they reſemble the rails; their bill is thick, their tail and wings ſhort.

XVII. VAGINALIS.

Characters. BILL ſtrong, thick, conically-convex, compreſſed; the upper mandible covered by a horny ſheath notched and jagged.
NOSTRILS ſmall, conſpicuous before the ſheath.
TONGUE taper above, flattened below, ſharpened at the tip.
FACE naked, covered with papillæ.
WINGS ſtrengthened under the flexure, by an obtuſe knot.
FEET ſtrong, curſory, naked a little way above the knees; the toes rough below; the nails furrowed.

Only

Only one ſpecies has hitherto been diſcovered, the ſheathbill, a native of New Zealand.—The name derived from *vagina*, a ſheath.

XVIII. PARRA.

Characters. BILL ſomewhat taper, ſomewhat obtuſe.
NOSTRILS oval, placed in the middle of the bill.
FRONT carunculated; the caruncles parted into lobes.
BASTARD WINGS ſpinous.

This genus contains the jacanas, and ſeveral of the ſandpipers; the number of ſpecies is fifteen.

XIX. RALLUS.

Characters. BILL thicker at the baſe, compreſſed, attenuated on the back near the tip, equal and ſharp.
NOSTRILS oval.
FEET four-toed, cloven.
BODY compreſſed.

This genus contains the rails, which are comprized in thirty-one ſpecies. They are remarkable for the ſlight inflection of their bill, their ſmall noſtrils, their rough tongue, and their very ſhort tail.

XX. PSOPHIA.

Characters. BILL of a form between the cone and cylinder, convex, ſomewhat acute; the upper mandible the longer.
NOSTRILS oval and broad.
TONGUE cartilaginous, flattened, fringed at the tip.
FEET four-toed and cloven.

This

This genus contains only two ſpecies, the agamis or trumpeters; the one a native of South America and the Weſt Indies, the other a native of Africa.

The Third Order, that of the GALLINÆ, comprehends ten Genera: They are—

I. OTIS.

Characters. BILL ſomewhat convex.
NOSTRILS oval, pervious.
TONGUE bifid, ſharp.
FEET curſory, three-toed, legs tall, naked above the thighs.

This genus contains the buſtards; of which there are nine ſpecies.

II. STRUTHIO.

Characters. BILL ſomewhat conical.
NOSTRILS oval.
WINGS uſeleſs for flying.
FEET curſory.

This genus contains only three ſpecies, viz. the oſtrich, the caſſowary, and the *nandaguaca*.

III. DIDUS.

Characters. BILL ſtraitened in the middle by two tranſverſe wrinkles; the tip of each mandible inflected.
NOSTRILS oblique, near the margin of the middle of the bill.
FACE naked beyond the eyes.
FEET ſhort, thick, cloven.
WINGS uſeleſs for flying.
TAIL wanting.

This genus contains the dodos, which form three ſpecies.

IV. PAVO.

Characters. HEAD crested.
BILL convex, ſtrong.
NOSTRILS broad.
QUILLS of the rump elongated, broad, expanſible, ſpangled with eyes.

This genus contains the peacocks, which form four ſpecies.

V. MELEAGRIS.

Characters. BILL ſhort and ſtrong.
HEAD covered with ſpongy caruncles.
THROAT, at its upper part, furniſhed with a longitudinal membranaceous caruncle.
TAIL broad and expanſible.

This genus contains the turkey, of which only one ſpecies has yet been diſcovered.

VI. PENELOPE.

Characters. BILL naked at the baſe.
HEAD covered with feathers.
THROAT naked at its upper part.
TAIL conſiſting of twelve quills.

This genus contains ſix ſpecies of curaſſos.

VII. CRAX.

Characters. BILL ſtrong and thick, covered at the baſe with a cere in each mandible, or ſwelled.
NOSTRILS ſmall, placed in the cere.
FEATHERS that cover the head.
TAIL large and ſtraight.

This genus contains five other ſpecies of curaſſos.

VIII. PHA-

VIII. PHASIANUS.

Characters. Bill ſhort and ſtrong.
Cheeks ſmoothed, with naked ſkin.
Feet, for the moſt part, ſpurred.

This genus contains not only the pheaſants, which form nine ſpecies and five varieties, but alſo the cock, which includes fourteen varieties.

IX. NUMIDA.

Characters. Bill ſtrong and ſhort, furniſhed at the baſe with a caruncúlated cere receiving the noſtrils.
Head horned, the neck compreſſed and coloured.
Tail ſhort, bending down.
Body ſpeckled.

This genus contains the Guinea-hens, or pintadoes, of which there are three ſpecies.

X. TETRAO.

Characters. Spot near the eyes naked, or papillous, or ſometimes covered with feathers:

This genus includes three ſubdiviſions: thoſe with a naked ſpot above the eyes, and their feet ſhaggy; comprehending the grous and ptarmigans: thoſe with a papillous ſkin about the eyes, and with naked feet; comprehending the partridges and quails: and thoſe with the ſpace about the eyes covered thinly with feathers, and their feet imperfect; comprehending the tinamous. There are ſixty-ſix ſpecies in all. In this genus, the young, for the moſt part, follow their mother the inſtant after they are hatched: the

the fleſh, and even the eggs, are well-taſted. The grous and ptarmigans, the partridges, and the quails, have a convex bill; the two firſt are deſtined to inhabit the coldeſt countries, their noſtrils are ſmall and concealed among feathers; their tongue is ſharp at the tip; their legs are ſtrong, and their tail is long: the partridges and quails are fitted for mild or warm climates; they are ſmaller-ſized; their tail is ſhorter, and their noſtrils are covered with an excreſcence: the quails have a longer bill than the partridges. The tinamous are peculiar to Guiana, and reſemble the pheaſant in their habits; their bill is long, and blunt at the tip; their noſtrils are placed in the middle with a very wide gap; their throat is ſprinkled with feathers; their tail is very ſhort; their hind-toe curtailed, and uſeleſs for running.

The Sixth Order, that of the PASSERES, comprehends the ſeventeen remaining Genera. Theſe are ranged in four nearly equal diviſions: the *thick-bills*, the *curved-bills*, the *notched-bills*, and the *ſimple-bills*.

I. COLUMBA.

Characters. BILL ſtraight, ſinking at the tip.
NOSTRILS oblong, half-covered with a ſoft ſwelling membrane.
TONGUE entire.

This

This extenſive genus includes the turtles and pigeons, which amount to ſeventy-one ſpecies, beſides numerous varieties. Theſe birds are remarkable for their delicate bill, and ſhort legs; their toes are generally red, and divided to the origin. They inhabit only the temperate and hot regions; they are monogamous, and diſplay tenderneſs and ſenſibility in their courtſhips, and in the education of their young.

II. ALAUDA.

Characters. BILL, cylindrical-awl-ſhaped, ſtraight, and ſtretching right forwards: the mandibles equal, and parted aſunder at the baſe.

TONGUE bifid.

HIND NAIL rather ſtraight, longer than the toe

This genus includes the larks, of which there are thirty-three ſpecies.

III. STURNUS.

Characters. BILL awl-ſhaped, depreſſed at the corners, ſomewhat blunt: upper mandible very entire, the margins rather open.

NOSTRILS marginated above.

TONGUE notched, ſharp.

This genus contains the ſtares, which form ſeventeen ſpecies.

IV. TURDUS.

Characters. BILL ſlender, knife-ſhaped: upper mandible deflected at the tip, and notched.

NOSTRILS

NOSTRILS naked, hal covered above with a little membrane.
CHAPS ciliated.
TONGUE jagged and notched.

This genus includes the thrushes and blackbirds, which amount to one hundred and twenty-six. species.

V. AMPELIS.

Characters. BILL straight, convex: upper mandible longer, somewhat bent inwards, and notched on both sides.
NOSTRILS beset with bristles.
TONGUE sharp, cartilaginous, bifid.

This genus includes the chatterers, of which one species inhabits Europe, and the remaining ten, the hotter parts of America.

VI. COLIUS.

Characters. BILL short, thick, convex above, plane below: upper mandible curved apart.
NOSTRILS small, generally covered with feathers at the base of the bill.
TONGUE fringed at the tip.
TAIL wedge-shaped, and long.

This genus includes the colies, ranged in five species, all natives of Africa.

VII. LOXIA.

Characters. BILL conically-bunched, at the base of the front rounded towards the head: under mandible inflected at its lateral margin.
NOSTRILS placed in the base of the bill, minute, and rounded.
TONGUE entire.

This

This genus includes the grosbeaks, the cross-bills, the wax-bills, and the bull-finches, in all ninety-two species.

VIII. EMBERIZA.

Characters. BILL conical.
MANDIBLES parting asunder at the base: the under hemmed by the inflected sides; the upper narrower.

This genus includes the buntings, which amount to seventy-five species.

IX. TANAGRA.

Characters. BILL, conical, sharpened, notched, somewhat triangular at the base, the tip sloping down.

This genus contains the tanagres, of which there are forty-six species, almost all of them natives of America.

X. FRINGILLA.

Characters. BILL conical, straight, sharpened.

This extensive and multifarious genus includes the finches, the canaries, the siskins, the linnets, and the sparrows, which amount in all to one hundred and eight species, exclusive of many varieties.

XI. PHYTOTOMA.

Characters. BILL conical, straight, serrated.
NOSTRILS oval.
TONGUE short, blunt.

 Only

Only one ſpecies, the *rara*, is known. It inhabits Chili, and is nearly of the ſize of a quail. It ſcreams with a raucous interrupted voice, crops and tears up the tender plants, and makes deſtructive viſits in gardens. It neſtles in ſhady places on leafy trees, and lays white eggs, ſpotted with black.—The name formed from φυλον, a plant, and τεμνω, to cut.

XII. MUSCICAPA.

Characters. Bill ſomewhat triangular, notched on both ſides, curved inwards at the tip; *whiſkers* expanding towards the chaps.

Nostrils roundiſh.

This genus includes the fly-catchers, which amount to ninety-two ſpecies, and are generally natives of the hot climates.

XIII. MOTACILLA.

Characters. Bill awl-ſhaped, ſtraight; the mandibles ſomewhat equal.

Nostrils ovaliſh.

Tongue jagged and notched.

This genus includes a prodigious variety of birds; the warblers, the petty-chaps, the nightingale, the wag-tails, the white-ears, the whinchats, the ſtone-chats, the black-cap, the redſtart, the gray-ſtart, the red-breaſt, the wrens; the number of ſpecies being no leſs than one hundred and ſeventy-four.

XIV. PIPRA.

Characters. BILL ſhorter than the head, ſtout, hard, ſomewhat triangular at the baſe, very entire, curved inwards at the tip.
NOSTRILS, in moſt of the ſpecies, bare.
FEET greſſory.
TAIL ſhort.

This genus includes the manakins, of which there are twenty-ſix ſpecies, all natives of the hotter parts of America.

XV. PARUS.

Characters. BILL very entire, narrow, ſomewhat compreſſed, ſtout, hard, ſharpened, beſet at the baſe with briſtles.
TONGUE truncated, terminated with briſtles.
TOES disjoined to their origin, the hind one large and ſtrong.

This genus includes the titmice, which form thirty-two ſpecies. Theſe birds are remarkably prolific, laying eighteen or twenty eggs at each hatch. They feed on ſeeds, fruits, inſects, and even fleſh, particularly the brains of other birds: they are petulant, reſtleſs, quarrelſome, and their voice is generally diſagreeable.

XVI. HIRUNDO.

Characters. BILL very ſmall, curved inwards, awl-ſhaped, depreſſed at the baſe, broad.
GAPE wider than the head.
TONGUE ſhort, broad, cleft.
WINGS long.
TAIL, in moſt of the ſpecies, forked.

This genus contains the ſwallows and martins. ranged in thirty-ſeven ſpecies.

XVII. CAPRIMULGUS.

Characters. BILL moderately curved inwards, very ſmall, awl-ſhaped, depreſſed at the baſe.
WHISKERS, in a row at the mouth.
GAPE very wide.
EARS very wide.
TONGUE ſharp, very entire.
TAIL not forked; its quills ten in number.
FEET ſhort; the margin of the mid-toe broad and ſerrated.

This genus includes the goatſuckers, forming fifteen ſpecies, all of them, except one, natives of America. Theſe birds appear only in the duſk, and make a loud dull noiſe. They drop two eggs on the naked ground.

IN his late work, the *Index Ornithologicus*, Mr. Latham has, upon the whole, cloſely followed Linnæus and Gmelin: I ſhall only mark the inſtances where he has ventured to differ from them.

In the land-birds he has added two new Orders, the *Columbæ*, and the *Struthiones*; in the water-fowl, he has rejected the Order of the *Anſeres*, and revived the old diviſion into the *Pinnatipedes* and the *Palmipedes*. So that he follows Mr. Pennant in admitting nine Orders: He thus delineates theſe:—

COLUMBÆ.

COLUMBÆ.

BILL ſomewhat ſtraight, ſwelling at the baſe.
FEET ambulatory, ſhort, the nails ſimple.
FOOD from grain, ſeeds, and fruits, by ſwallowing.
NEST artleſs, in trees and holes; two eggs; the young feed from the mother's craw. Monogamous.

STRUTHIONES.

BODY vaſt, ponderous, hardly eatable.
WINGS ſmall, uſeleſs for flying, or wanting.
FEET curſory, ſtrong, with various toes.
FOOD grain and vegetables.
NEST on the ground. Monogamous.

The Order of the *Columbæ* contains only the pigeons: that of the *Struthiones* comprehends the oſtrich, the caſſowary, the dodo, and the touyou.

PINNATIPEDES.

BILL, BODY, and FOOD, as in the GRALLÆ of Linnæus.
FEET wading, thighs half naked, toes cleft, pinnated their whole length.
NEST large, formed of leaves and graſs, in marſhes. Monogamous.

This Order contains the phalaropes, the coots, and the grebes.

PALMIPEDES, admit of a ſub-diviſion:

1. Thoſe with long feet.

BODY ſomewhat depreſſed, conical, the fleſh of the young birds well-taſted.

FEET very long, wading, greateſt part of the thighs naked, toes ſemi-palmated by a membrane.
FOOD in the water, from ſmall fiſh and various inſects.
NEST on land. Monogamous.

2. Thoſe with ſhort feet.

BILL ſmooth, covered with an epidermis, enlarged at the tip.
FEET fitted for ſwimming, the toes palmated with a membrane, the legs compreſſed and ſhort.
BODY fat; the ſkin adheſive, the feathers valuable: rankiſh.
NEST ofteneſt on land. The mother ſeldom feeds the young. Generally polygamous.

The firſt ſub-diviſion includes the avoſets, the courier, and the flamingos. The ſecond ſub-diviſion comprehends the albatroſſes, the auks, the guillemots, the divers, the ſkimmer, the terns, the gulls, the petrels, the merganſer, the ſwans and geeſe, ducks and teals, &c. the penguins, the pelicans, including the cormorant, the ſhags, the boobies, and the gannets, the tropic-birds, and the darter.

Mr. Latham has alſo made ſeveral alterations in the Genera. He has removed the genus *Lanius* from the order of the ACCIPITRES to that of the PICÆ: and in this order he has alſo erected the motmot, *Galbula*, into a genus under the name *Momotus*, and the jacamar under that; and he has added the *Scythrops*, a bird lately brought from Botany Bay: it is thus characterized: *Bill* large, convex, knife-ſhaped, hooked at at the tip; *noſtrils* rounded, naked, placed at the baſe of the bill; *tongue* cartilaginous, bifid at the tip; *feet* ſcanſory. It is about the ſize of a raven, being twenty-ſeven inches long.—He has changed the name *Glaucopis* of a new genus into *Callæas*.

In the order of the PASSERES, he has ſplit the numerous genus *Motacilla* into two, the *Moticilla* and the *Sylvia*; the former containing only the wagtails, and the latter comprehending the nightingale, the warblers, the chats, and the wrens. The genus *Sylvia* is thus characterized: *Bill* awl-ſhaped, ſtraight, ſlender, the mandibles ſomewhat equal; *noſtrils* ovaliſh, rather depreſſed; *outer toe* connected to the mid one at the baſe; *tail* middling-ſized.

APPENDIX, II.

BY THE TRANSLATOR.

BIRDS omitted by the COMTE DE BUFFON, or ſince diſcovered.

THE late voyages round the world, the expeditions to New South Wales, and the journies performed by Gmelin, Pallas, and Jacquin, have introduced ſeveral new ſpecies of birds, which require to be particularly noticed. Cabinets of Natural Hiſtory furniſh ſingle ſpecimens of many others, which, though reckoned diſtinct ſpecies by ſyſtematic writers, are probably varieties only. The bare mention of theſe will ſuffice; or at moſt, ſome general hint of their peculiarities.—In forming this catalogue, I have followed the arrangement of Mr. Latham.

In the Genus VULTUR.

1. The Bearded Vulture, *Vultur Barbatus*; from Barbary.
2. The Arabian Vulture, *Vultur Monachus*.
3. The Black Vulture, *Vultur Niger*. It is frequent

frequent in Egypt: its quills are brown, and its feet feathered.

4. The Pondicherry Vulture, *Vultur Ponticerianus*. It is black, its head and neck rather naked and carnation, a red fleſhy caruncle on the ſides of the neck. It is of the ſize of a gooſe.

5. The Angola Vulture, *Vultur Angolenſis*. It is white, and of the ſame ſize with the preceding.

6. The Indian Vulture, *Vultur Indicus*. It is brown, its upper ſide marked with pale ſtripes; its head and neck naked and rufous; its quills black. It inhabits the coaſts of India, and is exceedingly voracious.

7. The Gingi Vulture, *Vultur Gingianus*. It is white, its wing quills black, its legs and bill gray. It is known in India by the name of *wild turkey*.

8. The Tawny Vulture, *Vultur Ambuſtus*; from the Falkland Iſlands.

9. The Plaintive Vulture, *Vultur Plancus*; from Tierra del Fuego.

10. The Cheriway Vulture, *Vultur Cheriway*; from the iſland Aruba.

In the Genus FALCO.

1. The Oronooko Eagle, *Falco Harpyia*. Its head is creſted by the production of the feathers; the body variegated, and white below.

below. It is ſaid to be as large as a ram, and to cleave a man's ſkull at one ſtroke. It inhabits the hotter parts of America.

2. The White-bellied Eagle, *Falco Leucogaſter*.
3. The Japoneſe Hawk, *Falco Japonicus*.
4. The Fierce Eagle, *Falco Ferox*; from Aſtracan.
5. The Black-cheeked Eagle, *Falco Americanus*; from North America. It is of the bulk of the Ring-tail Eagle.
6. The Cheela Falcon, *Falco Cheela*; from India.
7. The Aſiatic Falcon, *Falco Aſiaticus*; from China.
8. The Spotted Eagle, *Falco Maculatus*; a native of Europe.
9. The Statenland Eagle, *Falco Auſtralis*.
10. The Chilian Falcon, *Falco Tharus*. Its cere and legs are yellow, its body blackiſh white, its top creſted. It breeds on lofty trees, laying five eggs.
11. The Black-backed Eagle, *Falco Melanonotus*.
12. The White-crowned Eagle, *Falco Leucoryphos*.
13. The Ruſſian Eagle, *Falco Mogilnik*; from the deſerts on the Tanais.
14. The Caracca Falcon, *Falco Criſtatus*.
15. The Leverian Falcon, *Falco Leverianus*.
16. The Rough legged Falcon, *Falco Lagopus*; from the North of Europe and of America.

17. The

17. The Booted Falcon, *Falco Pennatus*.
18. The Javan Falcon, *Falco Maritimus*.
19. The Arabian Kite, *Falco Forſkalii*. Inhabits Egypt in winter.
20. The Auſtrian Kite, *Falco Auſtriacus*.
21. The Equinoctial Eagle, *Falco Æquinoctialis*; from Cayenne.
22. The Oriental Hawk, *Falco Orientalis*; from Japan.
23. The Speckled Buzzard, *Falco Variegatus*; from North America.
24. The Sclavonian Falcon, *Falco Marginatus*.
25. The Barred-breaſted Buzzard, *Falco Lineatus*; from North America.
26. The Collared Falcon, *Falco Ruſticulus*; from Sweden and Siberia.
27. The Long-tailed Falcon, *Falco Macrourus*; from Ruſſia.
28. The Northern Falcon, *Falco Hyemalis*; from New York.
29. The Rhomboidal Falcon, *Falco Rhombeus*; from the Ganges.
30. The Black-necked Falcon, *Falco Nigricollis*; from Cayenne.
31. The White-necked Falcon, *Falco Albicollis*; from Cayenne.
32. The Rufous-headed Falcon, *Falco Meridionalis*; from Cayenne.
33. The Black and White Falcon, *Falco Melanoleucos*. Its legs are yellow, its body white; its head, neck, back, axillæ, and wing-

wing-quills black. It inhabits Ceylon, and is called *Kaloe Koeroelgoya*. Its length is sixteen inches, its weight ten ounces.

34. The Surinam Falcon, *Falco Sufflator*.

35. The Laughing Falcon, *Falco Cachinnans*; from South America.

36. The Streaked Falcon, *Falco Melanops*; from Cayenne.

37. The Notched Falcon, *Falco Bidentatus*; from Cayenne.

38. The Marsh Hawk, *Falco Uliginosus*; from America.

39. The Behree Falcon, *Falco Calidus*; from India.

40. The Plumbeous Falcon, *Falco Plumbeus*; from Cayenne.

41. The Bohemian Falcon, *Falco Bohemicus*.

42. The Brown Hawk, *Falco Badius*; from Ceylon.

43. The Dusky Falcon, *Falco Obscurus*; from North America.

44. The Guiana Falcon, *Falco Superciliosus*.

45. The Ingrian Falcon, *Falco Vespertinus*. Its cere, its legs, and its eye-lids, are yellow; its vent and its thighs ferruginous. It flies in the dusk of the evening, and preys chiefly on quails. It nestles on the summits of trees. It is found Russia and Siberia.

46. The Criard Falcon, *Falco Vociferus*; from India.

47. The Siberian Falcon, *Falco Regulus*.

In

In the Genus of the STRIX.

1. The Virginian Eared Owl, *Strix Virginianus.*
2. The Ceylonese Eared Owl, *Strix Ceylonensis.*
3. The Chinese Eared Owl, *Strix Sinensis.*
4. The Coromandel Eared Owl, *Strix Coromanda.*
5. The Scandinavian Eared Owl, *Strix Scandiaca.* It lives in the mountains of Lapland, and is as large as a peacock.
6. The Mottled Owl, *Strix Nævia*; from New York.
7. The Indian Eared Owl, *Strix Bakkamuna*; from Ceylon.
8. The Siberian Eared Owl, *Strix Pulchella.*
9. The Wapachtu Owl, *Strix Wapachtu*; from Hudson's Bay.
10. The Cinereous Owl, *Strix Cinerea*; from Hudson's Bay. It flies in pairs.
11. The Swedish Owl, *Strix Tengmalmi.*
12. The Acadian Owl, *Strix Acadica*; from North America.
13. The New Zealand Owl, *Strix Fulva.*

In the Genus LANIUS.

1. The Chesnut-backed Shrike, *Lanius Castaneus.*
2. The Rufous-tailed Shrike, *Lanius Phœnicurus.*
3. The Surinam Shrike, *Lanius Atricapillus.*

4. The

4. The Magpie Shrike, *Lanius Leverianus*; from South America.
5. The Black Shrike, *Lanius Niger*; from Jamaica.
6. The Chinese Shrike, *Lanius Schach*.
7. The Pacific Shrike, *Lanius Pacificus*; from the islands in the Pacific Ocean.
8. The Black-headed Shrike, *Lanius Melanocephalus*; from the Sandwich islands.
9. The Northern Shrike, *Lanius Septentrionalis*; from North America.
10. The Black-capped Shrike, *Lanius Pileatus*; from Cayenne.
11. The Short-tailed Shrike, *Lanius Brachyurus*; from Hungary.
12. The Ferruginous-bellied Shrike, *Lanius Ferrugineus*; from the Cape of Good Hope.
13. The Tabuan Shrike, *Lanius Tabuensis*; from the island of Tongataboo.
14. The White-shouldered Shrike, *Lanius Varius*; from Brazil.
15. The Panayan Shrike, *Lanius Panayensis*.
16. The Red Shrike, *Lanius Ruber*; from Surinam.
17. The Orange Shrike, *Lanius Aurantius*; from Cayenne.
18. The Nootka Shrike, *Lanius Natka*.
19. The Boulboul Shrike, *Lanius Boulboul*; from India.
20. The Dusky Shrike, *Lanius Obscurus*.

In

In the Genus PSITTACUS.

1. The Obſcure Parrot, *Pſittacus Obſcurus*; from Africa.

2. The Noble Parrot, *Pſittacus Nobilis.* It is long-tailed and green; its cheeks naked; its ſhoulders ſcarlet. It inhabits Surinam, and is of the ſize of the turtle.

3. The Javan Parrakeet, *Pſittacus Javanicus.* It is long-tailed and green; its head variegated with blue and bright yellow; its temples black; its throat and breaſt red; a bright yellow ſpot on its coverts.

4. The Tabuan Parrot, *Pſittacus Tabuenſis*; from Tongataboo.

5. The Beautiful Lory, *Pſittacus Elegans.* It is long-tailed and brown; its feathers edged with red and green; its head, its neck, and the under ſide of its body, ſcarlet; its ſhoulders, and the margins of its quills, edged with blue. It inhabits the Moluccas.

6. The Variegated Lory, *Pſittacus Variegatus*; from India.

7. The Pennantian Parrot, *Pſittacus Pennantii.* It is long-tailed and ſcarlet; the fore part of its back black, waved with ſcarlet; the ſides of the body, and the throat, blue; a white ſpot on the inſide of the wing-quills. It is found in New South Wales.

8. The

8. The Black Lory, *Psittacus Novæ Guineæ.*
9. The Crimson-vented Parrot, *Psittacus Erythropygius*; from Asia.
10. The Chilian Parrot, *Psittacus Jaguilma.* It is long-tailed and green; its wing-quills tipt with brown, its orbits fulvous. It lives in numerous flocks during summer on the Cordilleras, and crops herbs and the buds of trees.
11. The Varied-winged Parrot, *Psittacus Marginatus*; from the isle of Luçon.
12. The Scaly-breasted Parrakeet, *Psittacus Squamosus*; from Cayenne.
13. The Horned Parrot, *Psittacus Bisetis*; from New Caledonia.
14. The Caledonian Parrot, *Psittacus Caledonicus.*
15. The Red-rumped Parrot, *Psittacus Zealandicus*; from New Zealand.
16. The Crested Parrakeet, *Psittacus Novæ Hollandiæ*; from New Holland.
17. The Society Parrot, *Psittacus Ulietanus*; from Ulietea.
18. The White-collared Parrot, *Psittacus Multicolor*; from India.
19. The Lineated Parrot, *Psittacus Lineatus.*
20. The Pacific Parrot, *Psittacus Pacificus*; from Otaheite.
21. The Peregrine Parrakeet, *Psittacus Peregrinus*; from the islands in the South Sea.

22. The

22. The Palm Parrot, *Psittacus Palmarum*; from the island of Tanna.

23. The Blue-crested Parrakeet, *Psittacus Pipilans*; from the Sandwich islands. It is of the size of a lark, and beautiful: it has a piping voice.

24. The New South Wales Parrakeet, *Psittacus Pusillus.*

25. The Pygmy Parrakeet, *Psittacus Pygmæus*; from the islands in the South Sea.

26. The Crowned Cockatoo, *Psittacus Coronatus*; from Guiana.

27. The Bankian Cockatoo, *Psittacus Banksii*; from New Holland.

28. The New South Wales Cockatoo, *Psittacus Galeritus.*

29. The Southern Brown Parrot, *Psittacus Meridionalis*; from New Holland.

30. The South American Parrot, *Psittacus Fringillaceus.*

31. The Robust Parrot, *Psittacus Robustus.*

32. The Cochin-China Parrot, *Psittacus Cochinsinensis.*

33. The Yellow-breasted Lory, *Psittacus Guineensis*; from Guinea.

34. The Grisled Parrot, *Psittacus Nasutus*; from China.

35. The White-crowned Parrot, *Psittacus Albifrons.*

36. The New Guinea Green Parrot, *Psittacus Viridis.*

37. The Eastern Parrot, *Psittacus Orientalis*; from India.
38. The Blue-cheeked Parrot, *Psittacus Adscitus*.
39. The Amber Parrot, *Psittacus Batavensis*; from Batavia.
40. The Crimson-winged Parrot, *Psittacus Erythropterus*; from New South Wales.
41. The Purple-tailed Parrakeet, *Psittacus Purpuratus*; from Cayenne.

In the Genus BUCEROS.

1. The White Hornbill, *Buceros Albus*. Caught near the island of Tinian.
2. The New Holland Hornbill, *Buceros Orientalis*.
3. The Gray Hornbill, *Buceros Griseus*; from New Holland.

In the Genus CORVUS.

1. The South Sea Raven, *Corvus Australis*.
2. The New Caledonian Crow, *Corvus Caledonicus*.
3. The Pacific Crow, *Corvus Pacificus*; from the islands in the Pacific Ocean.
4. The Tropic Raven, *Corvus Tropicus*; from the island Owhyhee.
5. Steller's Crow, *Corvus Stelleri*; from Nootka Sound. It bears a sort of crest.
6. The White-eared Jay, *Corvus Auritus*; from China.

7. The

7. The Purple-headed Jay, *Corvus Purpurascens*; from China.
8. The Macao Crow, *Corvus Sinensis*.
9. The Rufous Crow, *Corvus Rufus*; from China.
10. The African Crow, *Corvus Africanus*. It is brown and somewhat crested.

In the Genus CORACIAS.

1. The Indian Roller, *Coracias Indica*; from Ceylon.
2. The Cape Roller, *Coracia Caffra*.
3. The Ultramarine Roller, *Coracias Cyanea*.
4. The Fairy Roller, *Coracias Puella*; from India.
5. The Blue-striped Roller, *Coracias Striata*; from New Caledonia.
6. The Gray-tailed Roller, *Coracias Vagabunda*; from India.
7. The Docile Roller, *Coracias Docilis*; from the South of Asia.
8. The Black Roller, *Coracias Nigra*.
9. The African Roller, *Coracias Afra*.
10. The Black-headed Roller, *Coracias Melanocephala*; from China.
11. The Obstreperous Roller, *Coracias Strepera*. It is black; the spot on its wings, its vent, and the base and tip of its tail, white. It is very numerous in Norfolk island: is a silly bird, noisy and restless during the night.

In the Genus ORIOLUS.

1. The Rice Oriole, *Oriolus Orizyvorus.* It is black; its head, neck, and breaſt, of a gloſſy purple. It inhabits Cayenne.
2. The Ruſty Oriole, *Oriolus Ferrugineus*; from New York.
3. The Red Oriole, *Oriolus Ruber*; from the iſland Panay.
4. The Antiguan Yellow Oriole, *Oriolus Flavus*; from Panay and South America.
5. The Oonalaſkan Oriole, *Oriolus Aoonalaſchkenſis.*
6. The Sharp-tailed Oriole, *Oriolus Caudacutus*; from North America.

In the Genus GRACULA.

1. The Fetid Grakle, *Gracula Fœtida.* It is black; the outſide of its wing-quills blueiſh; a naked bar on its neck: from North America.
2. The Boat-tailed Grakle, *Gracula Banta.* It is grayiſh; its ſhoulders blue; the outſide of its wing-quills green. It inhabits the warmer parts of America and the Weſt Indies.
3. The Egyptian Grakle, *Gracula Atthis.* It is blue-green; its belly ferruginous; its legs blood-coloured. It is of the ſize of a lark; it feeds on inſects.

4. The

4. The Long-billed Grakle, *Gracula Longiroſtra*; from Surinam.
5. The Daurian Grakle, *Gracula Sturnina.*
6. The Yellow-faced Grakle, *Oriolus Icterops*; from New Holland.

In the Genus PARADISEA.

1. The Gorget Bird of Paradiſe, *Paradiſea Gularis.* Its length twenty-eight inches.
2. The White-winged Paradiſe Bird, *Paradiſea Leucoptera.* Length twenty-five inches.
3. The White Paradiſe Bird, *Paradiſea Alba*; from the Papuan iſlands.

In the Genus TROGON.

1. The Faſciated Curucui, *Trogon Faſciatus*; from Ceylon.
2. The Spotted Curucui, *Trogon Maculatus*; from Ceylon.
3. The Blue-cheeked Curucui, *Trogon Aſiaticus*; from India.
4. The Blackiſh-ſpotted Curucui, *Trogon Indicus*; from India.

In the Genus BUCCO.

1. The Buff-faced Barbet, *Bucco Lathami.*
2. The Red-crowned Barbet, *Bucco Rubricapillus*; from Ceylon.
3. The Yellow-cheeked Barbet, *Bucco Zeylonicus.* It is green; its head and neck pale brown; the coverts of its wings

 ſpotted

ſpotted with white. It inhabits Ceylon: it ſits on trees murmuring like the turtle; and is thence named by the natives *Kottorea*.

4. The White-breaſted Barbet, *Bucco Fuſcus*; from Cayenne: ſize of a lark.
5. The Blue Barbet, *Bucco Gerini*; from India.

In the Genus CUCULUS.

1. The Panayan Spotted Cuckoo, *Cuculus Panayus*.
2. The Eaſtern Black Cuckoo, *Cuculus Indicus*. Its tail is rounded; its body black; its wings and its tail-quills marked at the tip with three black croſs lines. It inhabits India, and goes in flocks: it is ſaid to ſing delightfully: its fleſh is delicate. Held in great veneration by the Mahometans. Its length ſixteen inches.
3. The Creſted Black Cuckoo, *Cuculus Serratus*; from the Cape of Good Hope.
4. The Shining Cuckoo, *Cuculus Lucidus*; from New Zealand.
5. The Punctated Cuckoo, *Cuculus Punctulatus*; from Cayenne.
6. The Red-headed Cuckoo, *Cuculus Pyrrhocephalus*. It inhabits the woods of Ceylon, and lives on fruits. The natives call it *Malkoha*. It is ſixteen inches long, and weighs four ounces. Its body is black.

In

In the Genus PICUS.

1. The Buff-crested Woodpecker, *Picus Melanoleucos*; from Surinam.
2. The Red-breasted Woodpecker, *Picus Ruber*; from Surinam.
3. The White-rumped Woodpecker, *Picus Obscurus*; from North America.
4. The Striped-bellied Woodpecker, *Picus Fasciatus*; from Otaheite.
5. The Red-winged Woodpecker, *Picus Miniatus*; from India.
6. The Malacca Woodpecker, *Picus Malaccensis*. Its tuft and shoulders are scarlet; its throat reddish-yellow; its tail black.
7. The Gold-winged Woodpecker, *Picus Cafer*; from the Cape of Good Hope.
8. The Crimson-breasted Woodpecker, *Picus Olivaceus*; from the Cape of Good Hope.
9. The Chilian Woodpecker, *Picus Pitius*. It is brown, with drops of white; its tail short. It has the appearance of a pigeon. It is said not to nestle in hollow trees, but on the banks of rivers and the sides of hills, and to lay four eggs. Its flesh is esteemed by the natives.

In the Genus GALBULA.

The White-billed Jacamar, *Galbula Albirostris*; from South America.

In the Genus ALCEDO.

1. The Egyptian Kingfisher, *Alcedo Ægyptia.* It is long-tailed and brown, with ferruginous spots; its throat lighter ferruginous; its belly and thighs whitish, with ash-spots; its tail ashy. It is of the size of a crow.
2. The New Guinea Kingfisher, *Alcedo Novæ Guineæ.* It is black, spotted with white.
3. The Yellowish Kingfisher, *Alcedo Flavicans.*
4. The Sacred Kingfisher, *Alcedo Sacra.* It is blue green; below white; its eye-brows and a streak below its eyes ferruginous; its wing-quills and its tail blackish. It inhabits the Society Islands.
5. The Venerated Kingfisher, *Alcedo Venerata.* It is brown variegated with green, below pale; a stripe above the eyes whitish green. It inhabits the Friendly Islands.
6. The Respected Kingfisher, *Alcedo Tuta.* It is long-tailed, green-olive, below white; a green-black collar; the eye-brows white. Found in Otaheite.
7. The Violet Kingfisher, *Alcedo Coromanda*; from Coromandel.
8. The Spotted Kingfisher, *Alcedo Inda*; from Guiana.
9. The Surinam Kingfisher, *Alcedo Surinamensis.* It is short-tailed and blue, and below rufous-white.

10. The

10. The Three-toed Kingfisher, *Alcedo Tridactyla*. It is short-tailed and small. Found in India.

In the Genus SITTA.

1. The Surinam Nuthatch, *Sitta Surinamensis*. It is chesnut-rufous, below rusty white; its wings black; its coverts spotted with white; its tail black tipt with white. It is the smallest in the genus, being only three inches and a half long.
2. The Cape Nuthatch, *Sitta Caffra*. Above varied with yellow and black; below bright yellow; its legs black. It is eight inches and a half long.
3. The Long-billed Nuthatch, *Sitta Longirostris*. It is blueish, below pale rusty; its primary wing-quills brown at the tip; its straps black. From Batavia. Length eight inches.
4. The Green Nuthatch, *Sitta Chloris*. Its body is green above, bright white below; tail black, the extreme tip yellowish. It inhabits the country about the Cape of Good Hope, and is there called *Akter Brunties*. Its length hardly exceeds three inches and a half.

In the Genus TODUS.

1. The Short-tailed Tody, *Todus Brachyurus*; from North America. It is black above, and white below.

2. The

2. The Plumbeous Tody, *Todus Plumbeus*; from Surinam.

3. The Duſky Tody, *Todus Obſcurus*. It is olive-brown, below light yellowiſh; its throat pale. It inhabits dead trees in the foreſts of North America, and ſings pleaſantly.

4. The Ferruginous-bellied Tody, *Todus Ferrugineus*; from North America.

5. The Broad-billed Tody, *Todus Roſtratus*.

6. The Yellow-bellied Tody, *Todus Flavigaſter*; from New Holland.

In the Genus MEROPS.

1. The Coromandel Bee-eater, *Merops Coromandus*.

2. The Surinam Bee-eater, *Merops Surinamenſis*.

3. The Poe Bee-eater, *Merops Cincinnatus*. It is of a dark gloſſy green; a tuft on either ſide of the throat, and a ſtripe on the wings, white. It inhabits New Zealand, where it is held in veneration by the natives. It has an agreeable ſong, and its fleſh is well taſted.

4. The Yellow-tufted Bee-eater, *Merops Faſciculatus*. The people of the Sandwich Iſlands, where it is found, weave its yellow feathers into various ſorts of dreſſes.

5. The New Holland Bee-eater, *Merops Carunculatus*. It has fleſhy wattles.

6. The Horned Bee-eater, *Merops Corniculatus*; from New Holland.

In the Genus UPUPA.

1. The Red-billed Promerops, *Upupa Erythrorynchos*.
2. The Blue Promerops, *Upupa Indica*; from India.

In the Genus CERTHIA.

1. The Green Creeper, *Certhia Viridis*; from Carniola.
2. The Great Hook-billed Creeper, or Hoohoo, *Certhia Pacifica*; from the Sandwich Iſlands.
3. The Hook-billed Green Creeper, *Certhia Obſcura*. It is very frequent in the Sandwich Iſlands.
4. The Hook-billed Red Creeper, *Certhia Veſtiaria*. Common in the Sandwich Iſlands: its red feathers, with the olive ones of the preceding ſpecies, are preſerved by the natives for making their robes of ceremony.
5. The Sickle-billed Creeper, *Certhia Falcata*.
6. The Fulvous Creeper, *Certhia Fulva*; from South America.
7. The Cinereous Creeper, *Certhia Cinerea*; from the Cape of Good Hope.
8. The Crimſon Creeper, *Certhia Sanguinea*; from the Sandwich Iſlands.

9. The

9. The Brown Creeper, *Certhia Fuſca*; from the Southern Archipelago.
10. The Waved Creeper, *Certhia Undulata.*
11. The Wattled Creeper, *Certhia Caranculata*; from the iſland Tongataboo. It ſings ſweetly.
12. The Yellow-cheeked Creeper, *Certhia Ocrochlora*; from Surinam.
13. The Blue-throated Creeper, *Certhia Cyanogaſtra*; from Cayenne.
14. The Orange-breaſted Creeper, *Certhia Aurantia*; from Africa.
15. Mocking Creeper, *Certhia Sannio*; from New Zealand. It feeds on the honey of flowers.
16. The New Holland Creeper, *Certhia Novæ Hollandiæ*. It is black, and ſtriped below with white.
17. The Browniſh Creeper, *Certhia Incana*; from New Caledonia.
18. The Olive Creeper, *Certhia Peregrina.*
19. The Bracelet Creeper, *Certhia Armillata*; from Surinam.
20. The Cinnamon Creeper, *Certhia Cinnamomea.*
21. The Aſh-bellied Creeper, *Certhia Verticalis*; from Africa.
22. The Indigo Creeper, *Certhia Parietum*; from India.
23. The Yellow-bellied Creeper, *Certhia Lepida*; from India.

24. The

24. The Orange-backed Creeper, *Certhia Cantillans*; from China. Only three inches long. Its ſong agreeable.

25. The Tufted Creeper, *Certhia Erythrorynchos*; from India.

26. The Yellow-winged Creeper, *Certhia Chryſoptera*; from Bengal.

27. The Long-billed Creeper, *Certhia Longiroſtra*; from Bengal.

28. The Barred-tail Creeper, *Certhia Griſea*; from China.

In the Genus TROCHILUS.

1. The Aſh-bellied Colibri, *Trochilus Cinereus*. Length ſix inches.

2. The Harlequin Colibri, *Trochilus Multicolor*. Length four inches and a half.

3. The Yellow-fronted Colibri, *Trochilus Flavifrons*.

4. The Purple-crowned Colibri, *Trochilus Torquatus*.

5. The Orange-headed Colibri, *Trochilus Aurantius*.

6. The Little Colibri, *Trochilus Exilis*. Length an inch and a half: weight ſcarce fifty grains.

7. The Duſky-crowned Fly-bird, *Trochilus Obſcurus*. Length four inches and a half.

8. The Black and Blue Fly-bird, *Trochilus Bancrofti*. Length four inches. From the Weſt Indies.

9. The

9. The Ruff-necked Fly-bird, *Trochilus Collaris*. Length three inches and three quarters. From Nootka Sound.
10. The Blue-headed Fly-bird, *Trochilus Cyanocephalus*. Size of a walnut. From Chili.
11. The Patch-necked Fly-bird, *Trochilus Maculatus*.

In the Genus STURNUS.

1. The Wattled Stare, *Sturnus Carunculatus*; from New Zealand. Length ten inches. Has a weak piping voice.
2. The Cock's-comb Stare, *Sturnus Gallinaceus*; from the Cape of Good Hope. Length ſix inches.
3. The Silk Stare, *Sturnus Sericeus*; from China. Length eight inches.
4. The Green Stare, *Sturnus Viridis*; from China.
5. The Brown Stare, *Sturnus Olivaceus*; from China.
6. The Alpine Stare, *Sturnus Moritanicus*; from Perſia. It is cinereous and ſpotted. It breeds in the holes of rocks.
7. The Chilian Stare, *Sturnus Loyca*. It is ſpotted with brown and white; its breaſt ſcarlet. It makes its neſt careleſsly in holes in the ground; is eaſily tamed, and is venerated by the natives.
8. The Daurian Stare, *Sturnus Dauricus*.

In

In the Genus TURDUS.

1. The Jamaica Thrush, *Turdus Jamaicensis*. It is ash-brown, and white below.
2. The Oonalaschka Thrush, *Turdus Aoonalashkæ*.
3. The Ruby-throat, *Turdus Calliope*. It is brown ferruginous, below yellowish-white; its throat cinnabar, edged with black and white; its straps black; its eye-brows white. Inhabits the wilds of Siberia, and pours its sweet note from the highest sprays.
4. The Tawny Thrush, *Turdus Mustelinus*; from North America.
5. The Yellow-backed Thrush, *Turdus Striatus*; from Surinam.
6. The Variegated Thrush, *Turdus Variegatus*; from Surinam.
7. The Pagoda Thrush, *Turdus Pagodarum*; from India. It is crested and gray.
8. The Rufous-tailed Thrush, *Turdus Ruficaudus*; from the Cape of Good Hope.
9. The Dark Thrush, *Turdus Obscurus*. It inhabits the forests in the southern parts of Siberia, and has a ringing voice.
10. The Red-necked Thrush, *Turdus Ruficollis*. Inhabits the summits of Dauria.
11. The White-browed Thrush, *Turdus Sibiricus*; from the north of Russia and Siberia.

12. The

12. The Pale Thrush, *Turdus Pallidus*; from Siberia.
13. The Thick-billed Thrush, *Turdus Crassirostris*; from New Zealand.
14. The Bay Thrush, *Turdus Ulietensis*; from the island of Ulietea.
15. The Crescent Thrush, *Turdus Arcuatus*; from China.
16. The New Holland Thrush, *Turdus Novæ Hollandiæ*.
17. The Black-faced Thrush, *Turdus Shanhu*. It inhabits the woods of China.
18. The Surat Thrush, *Turdus Suratensis*.
19. The Pacific Thrush, *Turdus Pacificus*.
20. The Sandwich Thrush, *Turdus Sandwichensis*.
21. The Yellow-bellied Thrush, *Turdus Brasiliensis*; from Brazil.
22. The White-chinned Thrush, *Turdus Americanus*; from America.
23. The Chilian Thrush, *Turdus Curæus*. It is glossy black; its bill somewhat streaked; its tail wedge-shaped. It is of the size of the blackbird; commonly breeds in holes: it is noisy and imitative, and has a fine song.
24. The Labrador Thrush, *Turdus Labradorus*.
25. The Persian Thrush, *Turdus Persicus*.
26. The White-tailed Thrush, *Turdus Leucurus*. Inhabits the south of Europe.

27. The

27. The Violet Thrufh, *Turdus Violaceus*; from China.
28. The White-headed Thrufh, *Turdus Leucocephalus*; from China.
29. The Songfter Thrufh, *Turdus Cantor*; from the Philippine Iflands.
30. The Black-necked Thrufh, *Turdus Nigricollis*; from China.
31. The Yellow-fronted Thrufh, *Turdus Malabaricus*; from Malabar.
32. The Chanting Thrufh, *Turdus Boubil.* It is brown, with a black ftripe behind the ears. Inhabits China.
33. The Yellow Thrufh, *Turdus Flavus*; from China.
34. The Orange-headed Thrufh, *Turdus Citrinus*; from India.
35. The Green Thrufh, *Turdus Virefcens*; from China.
36. The Gray Thrufh, *Turdus Grifeus*; from Coromandel.
37. The White-fronted Thrufh, *Turdus Albifrons*; from New Zealand.
38. The Long-tailed Thrufh, *Turdus Macrourus*; from Malabar.
39. The Yellow-crowned Thrufh, *Turdus Ochrocephalus*; from Ceylon and Java.
40. The Margined Thrufh, *Turdus Africanus*; from Africa.
41. The Hudfonian Thrufh, *Turdus Hudfonicus*.

42. The New York Thrush, *Turdus Noveboracensis.*
43. The Gingi Thrush, *Turdus Gingianus*; from India.
44. The Dauma Thrush, *Turdus Dauma*; from India.
45. The Black and Scarlet Thrush, *Turdus Speciosus*; from India.

In the Genus AMPELIS.

1. The Coppery Chatterer, *Ampelis Cuprea*; from Surinam.
2. The Red-winged Chatterer, *Ampelis Phœnicea*; from Africa.
3. The Crested Chatterer, *Ampelis Cristata*; from America.

In the Genus COLIUS.

1. The White-backed Coly, *Colius Leuconotus*; from the Cape of Good Hope.
2. The Green Coly, *Colius Viridis*; from New Holland.
3. The Indian Coly, *Colius Indicus.* It is cinereous; below rufous.

In the Genus LOXIA.

1. The White-winged Cross-bill, *Loxia Falcirostra*; from North America.
2. The Parrot-billed Grosbeak, *Loxia Psittacea*; from the Sandwich islands.

3. The

3. The Caucaſian Groſbeak, *Loxia Rubicilla.* It is ſcarlet ſpotted with white.

4. The Siberian Groſbeak, *Loxia Siberica.* It is ſcarlet ſpotted with brown; below pale ſcarlet; the wings ſtriped with black and white. It frequents orchards near water. It is of the ſize of a linnet. Its voice is hoarſe and grating. It is perpetually fluttering.

5. The Creſted Groſbeak, *Loxia Criſtata*; from Æthiopia. It is very large.

6. The Spotted Groſbeak, *Loxia Maculata*; from North America.

7. The Duſky Groſbeak, *Loxia Obſcura*; from New York.

8. The Hudſonian Groſbeak, *Loxia Hudſonica.* It is brown; its belly white. Called by the natives *Atick-oom-aſhiſh.*

9. The Social Groſbeak, *Loxia Socia.* It is rufous brown, below yellowiſh, its bridle black, its tail ſhort. Its length five inches and a half. Inhabits the country back from the Cape of Good Hope. It breeds on the large boughs of the *Mimoſa*; and ſometimes a flock of eight hundred or a thouſand ſit together in the ſame neſt, which they occaſionally weave to a great extent.

10. The Yellow Groſbeak, *Loxia Flavicans*; from Aſia. Size of a canary.

11. The Yellow-rumped Groſbeak, *Loxia*

Hordeacea; from India. Size of a wagtail.

12. The Eaſtern Groſbeak, *Loxia Undulata.* It is duſky red, and waved below with brown.

13. The Northern Groſbeak, *Loxia Septentrionalis.* It is deep black, with a white ſpot on the wings. Found in the North of Europe.

14. The Brown-headed Groſbeak, *Loxia Ferruginea.*

15. The Gray-necked Groſbeak, *Loxia Melanura*; from China.

16. The Brown Groſbeak, *Loxia Fuſca*; from Aſia.

17. The Thick-billed Groſbeak, *Loxia Craſſiroſtris.*

18. The Black-breaſted Groſbeak, *Loxia Pectoralis.*

19. The Black-headed Groſbeak, *Loxia Erythromelas*; from Cayenne.

20. The Blue-ſhouldered Groſbeak, *Loxia Virens*; from Surinam.

21. The White-tailed Groſbeak, *Loxia Leucura*; from Brazil. Length three inches.

22. The Totty Groſbeak, *Loxia Totta.* Of a brick brown, below whitiſh. Found among the Hottentots.

23. The Aſh-headed Groſbeak, *Loxia Indica*; from India. Very ſmall.

24. The Malabar Groſbeak, *Loxia Malabarica.*

It

It is cinereous; its quills black; its throat and vent white.

25. The Black-bellied Groſbeak, *Loxia Afra*; from Africa.

26. The Aſiatic Groſbeak, *Loxia Aſiatica*; from China. It is cinereous-reddiſh; below cinereous; the belly pale red.

27. The Brown-cheeked Groſbeak, *Loxia Canora*; from Mexico.

28. The Radiated Groſbeak, *Loxia Lineata.*

29. The Faſciated Groſbeak, *Loxia Faſciata*; from Africa.

30. The Warbling Groſbeak, *Loxia Cantans*; from Africa. It is marked with croſs lines of brown and blackiſh; below white; the tail wedge-ſhaped.

31. The Javan Groſbeak, *Loxia Praſina.* It is olive, the rump red, the legs yellow.

32. The Dwarf Groſbeak, *Loxia Minima*; from India and China. It is brown; below brick-coloured.

In the Genus EMBERIZA.

1. The Chineſe Bunting, *Emberiza Sinenſis.* It is reddiſh; below yellow; its quills brown.

2. The Yellow-winged Bunting, *Emberiza Chryſoptera*; from the Falkland Iſlands.

3. The Paſſerine Bunting, *Emberiza Paſſerina*; from Ruſſia.

4. The Angola Bunting, *Emberiza Angolenſis*. It is black; its breaſt fire-coloured.

5. The Barred-tail Bunting, *Emberiza Fuſca*; from China.

6. The Weaver Bunting, *Emberiza Textrix*.

7. The Scarlet Bunting, *Emberiza Coccinea*. Found in the foreſts of Germany.

8. The Flame-coloured Bunting, *Emberiza Rutila*; from Siberia.

9. The Ruſty Bunting, *Emberiza Ferruginea*; from North America.

10. The Black-throated Bunting, *Emberiza Americana*; from Hudſon's Bay.

11. The Military Bunting, *Emberiza Militaris*, found near Malta. It is yellowiſh brown; below white.

12. The Black-headed Bunting, *Emberiza Melanocephala*.

13. The Brumal Bunting, *Emberiza Brumalis*; from Tyrol. It is yellow-brown; the under ſide of its body yellow; its wing-quills brown.

14. The White-crowned Bunting, *Emberiza Leucophrys*. Found in Canada, where it is migratory. Its ſong pleaſant.

15. The Pine Bunting, *Emberiza Pithyornus*; from Siberia. It is rufous; its belly hoary; with a white ſpot on its cheeks, its temples, and its breaſt.

16. The Daurian Bunting, *Emberiza Ruſtica*. Its head is black, with three longitudinal white

white bars. Appears ſo early as the month of March.

17. The Wreathed Bunting, *Emberiza Luctuoſa.*

18. The Yellow-breaſted Bunting, *Emberiza Aureola.* Frequent in all the pine and poplar foreſts of Siberia.

19. The Dwarf Bunting, *Emberiza Puſilla.* Haunts the rills on the Daurian Alps.

20. The Sandwich Bunting, *Emberiza Arctica.*

21. The Black-crowned Bunting, *Emberiza Atricapilla*; from the Sandwich Iſlands.

22. The Surinam Bunting, *Emberiza Surinamenſis.* Above cloudy brown; below yellowiſh; breaſt ſpotted with black.

23. The Gaur Bunting, *Emberiza Aſiatica*; from India. It is cinereous; its wings and tail brown.

24. The Stained Bunting, *Emberiza Fucata.* Frequent in the humid parts of Siberia.

25. The Aſh-headed Bunting, *Emberiza Spodocephala.* Found in the ſpring near brooks on the Daurian Alps.

26. The Gold-browed Bunting, *Emberiza Chryſophrys.* Inhabits the ſame tracts with the preceding.

In the Genus TANAGRA.

1. The Variable Tanagre, *Tanagra Variabilis.*

2. The Black Tanagre, *Tanagra Atrata*; from India.

3. The Capital Tanagre, *Tanagra Capitalis.* Above green; below yellow; the head and under part of the neck, black.

In the Genus FRINGILLA.

1. The Scarlet Finch, *Fringilla Coccinea*; from the Sandwich Islands.
2. The Red-breasted Finch, *Fringilla Punicea*; from North America.
3. The Ferruginous Finch, *Fringilla Ferruginea*; from Pensylvania.
4. The White-throated Finch, *Fringilla Pensylvanica.*
5. The Fasciated Finch, *Fringilla Fasciata*; from New York.
6. The Grass Finch, *Fringilla Graminea*; from New York.
7. The Norton Finch, *Fringilla Nortoniensis.* It is black; below white; its throat spotted with ferruginous.
8. The Striped-headed Finch, *Fringilla Striata*; from New York.
9. The Surinam Finch, *Fringilla Surinama.* It is gray; its wing-quills white on both sides.
10. The Black-headed Finch, *Fringilla Melanocephala*; from China.
11. The Brown Finch, *Fringilla Fusca*; from China.
12. The Red-faced Finch, *Fringilla Afra*; from Angola.

13. The

13. The Parrot Finch, *Fringilla Psittacea*; from New Caledonia.

14. The Red-headed Finch, *Fringilla Erythrocephala*; from the isle of France.

15. The Saffron-fronted Finch, *Fringilla Flaveola*.

16. The Autumnal Finch, *Fringilla Autumnalis*; from Surinam. It is greenish, with a rusty cap; its vent brick-coloured.

17. The Lepid Finch, *Fringilla Lepida*. It is dun-green; the stripe above and below the eyes, and its throat, fulvous; its breast black. It inhabits the woods of Havannah, and sings perpetually with an exceedingly slender voice. It is easily tamed.

18. The Bearded Finch, *Fringilla Barbata*. It inhabits the mountains of Chili near the ocean.

19. The Chilian Finch, *Fringilla Diuca*. It is blue; its throat white. It haunts the neighbourhood of dwellings, and sings delightfully to the rising sun.

20. The Sharp-tailed Finch, *Fringilla Caudacuta*. Found in the back parts of Georgia.

21. The Long-tailed Finch, *Fringilla Macroura*; from Cayenne.

22. The White-eared Finch, *Fringilla Leucotis*; from China.

23. The Ceylon Finch, *Fringilla Zeylonica*. It is yellow; its back greenish; its head black.

24. The

24. The Brown-throated Finch, *Fringilla Fuscicollis*; from China.
25. The Blue-faced Finch, *Fringilla Tricolor*; from Surinam.
26. The Fire Finch, *Fringilla Ignita*; from Africa.
27. The Lunar Finch, *Fringilla Torquata*; from India. It is reddiſh; its rump blue, with a black creſcent on its throat.
28. The Green-rumped Finch, *Fringilla Multicolor*; from Ceylon.
29. The Yellow-throated Finch, *Fringilla Flavicollis*; from North America.
30. The Carthagena Finch, *Fringilla Carthaginienſis*. It is entirely cinereous, ſpotted with brown and yellow.
31. The Ochre Finch, *Fringilla Ochracea*; found in Auſtria.
32. The Teſtaceous Finch, *Fringilla Teſtacea*.
33. The Imperial Finch, *Fringilla Imperialis*; from China. It is roſe-coloured; its top and under ſide bright yellow.
34. The Ruſty-collared Finch, *Fringilla Auſtralis*; from Tierra del Fuego.

In the Genus MUSCICAPA.

1. The White-fronted Flycatcher, *Muſcicapa Albifrons*; from the Cape of Good Hope.
2. The Black and White Flycatcher, *Muſcicapa Melanoleuca*. Found in the plains of Georgia, in the Ruſſian dominions.

3. The

3. The Leucomele Flycatcher, *Muſcicapa Leucomela*. Found near the Volga. Neſtles in crags. Has a motion with its tail.

4. The Black-fronted Flycatcher, *Muſcicapa Nigrifrons*.

5. The White-tailed Flycatcher, *Muſcicapa Leucura*; from the Cape of Good Hope.

6. The Spotted Yellow Flycatcher, *Muſcicapa Afra*; from the Cape of Good Hope.

7. The Flammeous Flycatcher, *Muſcicapa Flammea*; from India.

8. The Society Flycatcher, *Muſcicapa Nigra*; from Otaheite. It is deep black.

9. The Tufted Flycatcher, *Muſcicapa Comata*; from Ceylon.

10. The Red-vented Flycatcher, *Muſcicapa Hæmorrhouſa*; from Ceylon.

11. The Yellow-breaſted Flycatcher, *Muſcicapa Melanietera*; from Ceylon.

12. The Green Flycatcher, *Muſcicapa Nitens*; from India.

13. The Gray-necked Flycatcher, *Muſcicapa Griſea*; from China.

14. The Yellow-necked Flycatcher, *Muſcicapa Flavicollis*; from China.

15. The Orange-vented Flycatcher, *Motacilla Fuſceſcens*; from China.

16. The Blue-headed Flycatcher, *Muſcicapa Cyanocephala*; from Manilla.

17. The Yellow-throated Flycatcher, *Muſcicapa Manillenſis*.

18. The

18. The Fan-tailed Flycatcher, *Muscicapa Flabellifera*; from New Zealand and Tanna. Spreads its tail like a fan when it flies.

19. The Supercilious Flycatcher, *Muscicapa Superciliosa*. It is cinereous, and below carnation.

20. The Ferruginous Flycatcher, *Muscicapa Ferruginea*; from Carolina.

21. The Long-tailed Flycatcher, *Muscicapa Aëdon*. It is frequent among the rocks and warm situations in Dauria, and sings delightfully even in the night.

22. The New Holland Flycatcher, *Muscicapa Novæ Hollandæ*. It is brown, and below whitish.

23. The Sooty Flycatcher, *Muscicapa Deserti*. Found in the deserts of Africa.

24. The Olive Flycatcher, *Muscicapa Caledonica*; from New Caledonia.

25. The Luteous Flycatcher, *Muscicapa Lutea*; from Otaheite.

26. The Yellow-headed Flycatcher, *Muscicapa Ochrocephala*; from New Zealand.

27. The Yellow-fronted Flycatcher, *Muscicapa Flavifrons*; from the island of Tanna.

28. The Clouded Flycatcher, *Muscicapa Nævia*; from New Caledonia.

29. The Red-bellied Flycatcher, *Muscicapa Erythrogastra*; from Norfolk Island.

30. The Sandwich Flycatcher, *Muscicapa Sandwichensis*.

31. The Dusky Flycatcher, *Muscicapa Obscura*; from the Sandwich Islands.

32. The Spotted-winged Flycatcher, *Muscicapa Maculata*; from the Sandwich Islands.

33. The Striped Flycatcher, *Muscicapa Striata*; from North America.

34. The Dun Flycatcher, *Muscicapa Sibirica*; from Kamtschatka.

35. The Red-faced Flycatcher, *Muscicapa Erythropis*. Found near the river Jenesei.

36. The Cinnamon Flycatcher, *Muscicapa Cinnamomea*; from Cayenne.

37. The Yellow-rumped Flycatcher, *Muscicapa Spadicea*; from Cayenne.

38. The Surinam Flycatcher, *Muscicapa Surinama*.

39. The Phœbe Flycatcher, *Muscicapa Phœbe*; from New York. It is ash-olive; below yellowish.

40. The Golden-throat Flycatcher, *Muscicapa Ochroleuca*; from North America.

41. The Nitid Flycatcher, *Muscicapa Nitida*; from Chma.

42. The Lesser Crested Flycatcher, *Muscicapa Acadica*; from Nova Scotia.

43. The Hanging Flycatcher, *Muscicapa Noveboracensis*; from New York.

44. The Passerine Flycatcher, *Muscicapa Passerina*; from the island of Tanna.

45. The Double-coloured Flycatcher, *Muscicapa Dichroa*; from the south of Africa.

46. The

46. The Javan Flycatcher, *Muſcicapa Javanica.* Its tail is very long and round.

In the Genus ALAUDA.

1. The Malabar Lark, *Alauda Malabarica.* It is brown; its feathers edged with rufous, and ſpotted with white at the tip. A beautiful ſpecies.
2. The Gingi Lark, *Alauda Gingica*; from Coromandel. Its head is cinereous; its under ſide black.
3. The Black Lark, *Alauda Tartarica.* Found in Tartary.
4. The Yelton Lark, *Alauda Yeltonienſis.* It is black, variegated with rufous and white. Found at the lake Yelton, beyond the Volga. Is gregarious; and in the month of Auguſt is fat and delicious.
5. The New Zealand Lark, *Alauda Novæ Zealandiæ.* It is dun; its feathers edged with aſhy; its belly white; its eye-brows white; a black bar on its eye.
6. The Teſtaceous Lark, *Alauda Teſtacea*; from Gibraltar.
7. The Portugal Lark, *Alauda Luſitana*; from Portugal.

In the Genus MOTACILLA.

1. The Hudſonian Wagtail, *Motacilla Hudſonica.* It is ruſty brown; below whitiſh; duſky

dusky ſtreaks on the neck and the under ſide.

2. The Indian Wagtail, *Motacilla Indica.* It is greeniſh gray, below yellowiſh; two black creſcents on the breaſt.

3. The Yellow-headed Wagtail, *Motacilla Citreola*; found in Siberia.

4. The Tſchutki Wagtail, *Motacilla Tſchutſchenſis.* It is olive-brown, below white.

5. The Green Wagtail, *Motacilla Viridis*; from Ceylon.

In the Genus SYLVIA.

1. The Sardinian Warbler, *Sylvia Moſchita.* It is lead-coloured, with a tawny cap.

2. The Aquatic Warbler, *Sylvia Aquatica*; found in Italy, where it is migratory. It is ruſty, ſpotted with brown, and a white bar on the wings.

3. The Cheſnut-bellied Warbler, *Sylvia Erythrogaſtra.* Haunts the gullies in the Caucaſian mountains.

4. The Guiana Red-tail, *Sylvia Guianenſis.*

5. The Black Red-tail, *Sylvia Atrata.*

6. The Leſſer White-throat, *Sylvia Sylviella.* It is aſh-brown, below dirty white; the two middle tail-quills ſhorter and awl-ſhaped. This ſpecies is pretty frequent in England among the hedges, though ſeldom obſerved, being exceedingly ſmall. It neſtles in orchards near the ground.

7. The

7. The Patagonian Warbler, *Sylvia Patagonica*. It is cinereous, ſpotted below with white.
8. The White-breaſted Warbler, *Sylvia Dumetorum*. Inhabits the buſhes in Germany and Ruſſia.
9. The Black-jawed Warbler, *Sylvia Nigriroſtris*.
10. The Ruſty-headed Warbler, *Sylvia Borealis*; from Kamtſchatka.
11. The Buff-faced Warbler, *Sylvia Luteſcens*.
12. The Siberian Warbler, *Sylvia Montanella*. It is brick-coloured, ſpotted with brown, below yellowiſh.
13. The Moor Warbler, *Sylvia Maura*; from Ruſſia. It is black, edged with gray, below white.
14. The Yellow-browed Warbler, *Sylvia Supercilioſa*; from Ruſſia.
15. The Gilt-throat Warbler, *Sylvia Ferruginea*. Found about the river Tunguſka.
16. The Blue-tailed Warbler; *Sylvia Cyanura*. Inhabits the ſhady humid places near the river Jeneſei.
17. The Daurian Warbler, *Sylvia Aurorea*. It is black; its top gray-white.
18. The Black-poll Warbler, *Sylvia Striata*; from New York.
19. The Gray-poll Warbler, *Sylvia Incana*, from New York.
20. The Yellow-fronted Warbler, *Sylvia Flavifrons*; from Penſylvania.

21. The

21. The Blackburnian Warbler, *Sylvia Blackburniæ*; from New York. It has a black cap, with a black bar acroſs the eyes.

22. The Murine Warbler, *Sylvia Murina*.

23. The Thorn-tailed Warbler, *Sylvia Spinicauda*; from Tierra del Fuego.

24. The Citrine Warbler, *Sylvia Citrina*; from New Zealand.

25. The Long-legged Warbler, *Sylvia Longipes*; from New Zealand.

26. The Black-hooded White-ear, *Sylvia Pileata*. Found at the Cape of Good Hope, and in China.

27. The White-crowned Warbler, *Sylvia Albicapilla*; from China.

28. The Pink Warbler, *Sylvia Caryophyllacea*; from Ceylon.

29. The Cingaleſe Warbler, *Sylvia Cingalenſis*; from Ceylon. It is variegated green, below bright yellow.

30. The China Warbler, *Sylvia Sinenſis*. It is green; a pale ſpot behind the eyes.

31. The Tailor Warbler, *Sylvia Sutoria*. Its colour is light yellow; its length three inches; its weight ninety grains. It ſews with delicate fibres a dead leaf to the ſide of a living one, and lines the cavity with feathers, goſſamer, and down. Its eggs are white, and not larger than thoſe of ants. It is found in India.

32. The Black-throated Warbler, *Sylvia Gularis*; from South America.
33. The Long-billed Warbler, *Sylvia Kamtschatkensis*.
34. The Ochry-tailed Warbler, *Sylvia Ochrura*; from Persia.
35. The Awatcha Warbler, *Sylvia Awatcha*. It is brown, below white; its breast spotted with black.
36. Van Diemen's Warbler, *Sylvia Canescens*. It is hoary, below white; head black; front streaked with white.
37. The Black-necked Warbler, *Sylvia Nigricollis*; from India.
38. The Plumbeous Warbler, *Sylvia Plumbea*. Very small.

In the Genus PIPRA.

1. The Superb Manakin, *Pipra Superba*.
2. The White-headed Manakin, *Pipra Leucocephala*; from Surinam.
3. The Little Manakin, *Pipra Minuta*; from India.
4. The Crimson-vented Manakin, *Pipra Hæmorrhoa*.
5. The Black-throated Manakin, *Pipra Nigricollis*.
6. The Orange-bellied Manakin, *Pipra Capensis*; from the Cape of Good Hope.
7. The Cinereous Manakin, *Pipra Cinerea*.

In

In the Genus PARUS.

1. The Norway Titmoufe, *Parus Stroemei.* It is very like the ox-eye, only its head is yellowifh-green inftead of black.
2. The black-breafted Titmoufe, *Parus Afer*; from the Cape of Good Hope and India.
3. The Hudfon's Bay Titmoufe, *Parus Hudfonicus.* It is reddifh-brown, its back cinereous, its throat jet black, its flanks rufous.
4. The Chinefe Titmoufe, *Parus Sinenfis.* It is rufty-brown, its wings and tail brown, edged with black.
5. The Great-headed Titmoufe, *Parus Macrocephalus*; from New Zealand. It is black; its belly and front white.
6. The New Zealand Titmoufe, *Parus Novæ Zealandiæ.* It is afh-red, below rufous-gray; its eye-brows white.
7. The White Titmoufe, *Parus Kujaefeik*; found in the oak-woods of Siberia.

In the Genus HIRUNDO.

1. The Otaheite Swallow, *Hirundo Tahitica.* It is blackifh-brown; its front, its neck, and its under fide, purple-fulvous; its tail fomewhat forked and black.
2. The Daurian Swallow, *Hirundo Daurica.* It is blue, below white; its temples and

 rump

rump ferruginous; its outermost tail-quill very long, and marked on the inside with a white spot. It inhabits the lofty rocks and the mountain-caves of Siberia. Its nest is large and hemispherical, constructed elegantly with pellets of pure mud, and having an entrance of some inches length.

3. The Red-headed Swallow, *Hirundo Erythrocephala*; from India. A small species.

4. The Oonalaschkan Swallow, *Hirundo Aoonalaschkensis*. It is blackish, below ashy; its rump whitish.

5. The Chinese Swift, *Hirundo Sinensis*. It is brown, below tawny-gray; its cap rufous; its throat and orbits white. Length eleven inches and a half.

In the Genus CAPRIMULGUS.

1. The Bombay Goatsucker, *Caprimulgus Asiaticus*. It is ashy, clouded with black and ferruginous; cinereous bars on the breast.

2. The Crested Goatsucker, *Caprimulgus Novæ Hollandiæ*; from New Holland. Rather smaller than the European.

In the Genus COLUMBA.

1. The White-crowned Pigeon, *Columba Leucocephala*; found in North America and in Jamaica.

2. The

2. The White-winged Pigeon, *Columba Leucoptera*; from India.
3. The Leſſer Crowned Pigeon, *Columba Criſtata*; from India. It neſtles among graſs and reeds.
4. The Gray-headed Pigeon, *Columba Albicapilla*; from the iſland Panay.
5. The Purple-ſhouldered Pigeon, *Columba Phœnicoptera*; from India.
6. The Garnet-winged Pigeon, *Columba Erythroptera*; from the iſland Eimeo.
7. The Green-winged Pigeon, *Columba Indica*; from Amboyna.
8. The Jamboo Pigeon, *Columba Jamboo*; from Sumatra and Java. It is green; its front red; its breaſt white.
9. The Purple Pigeon, *Columba Purpurea*; from Java.
10. The Purple-breaſted Pigeon, *Columba Eimenſis*; from the iſland Eimeo.
11. The Hook-billed Pigeon, *Columba Curviroſtra*; from the iſland of Tanna.
12. The Ferrugineous-vented Pigeon, *Columba Specifica*; from the Friendly Iſlands.
13. The White Nutmeg Pigeon, *Columba Alba*; from New Guinea.
14. The New Zealand Pigeon, *Columba Zealandica*. It is red; its belly white; its rump blue; its tail black.
15. The Brown Pigeon, *Columba Brunnea*; from New Zealand.

16. The Bronze-winged Pigeon, *Columba Chalcoptera*; from Norfolk Iſland.
17. The Hackled Pigeon, *Columba Franciæ*; from the Iſle of France.
18. The Spotted Green Pigeon, *Columba Maculata.*
19. The Gray Pigeon, *Columba Corenſis*; from Coro, in South America.
20. The Egyptian Turtle, *Columba Ægyptiaca.* It is reddiſh; its throat ſpotted with black feathers.
21. The Surinam Turtle, *Columba Surinamenſis.* It is cinereous, below white; its bill blue.
22. The Surat Turtle, *Columba Suratenſis.* It is gray; the upper ſide of its neck black; its nape white.
23. The Blue-crowned Turtle, *Columba Cyanocephala*; from India and China.
24. The Red-breaſted Turtle, *Columba Cruenta*; from Manilla.
25. The Sanguine Turtle, *Columba Sanguinea*; from Manilla.
26. The Malacca Pigeon, *Columba Malaccenſis.* The ſides of its neck are white. It is of the ſize of a ſparrow, and very beautiful.
27. The Melancholy Turtle, *Columba Bantamenſis*; from Java. Its tail is wedge-ſhaped; its orbits naked and fleſhy.
28. The Black-winged Turtle, *Columba Melanoptera*; from Chili.

In

In the Genus PENELOPE.

The Piping Curaſſow, *Penelope Pipile*; from Brazil. Has a blue caruncle on its throat; its belly white; its back brown, ſpotted with deep black.

In the Genus NUMIDA.

1. The Mitred Pintado, *Numida Mitrata*; from Madagaſcar and Guinea.
2. The Creſted Pintado, *Numida Criſtata*; from Africa.

In the Genus CRAX.

1. The Globoſe Curaſſow, *Crax Globicera*; from Guiana.
2. The Galeated Curaſſow, *Crax Galeata*; from Curaçoa.

In the Genus PHASIANUS.

1. The Superb Pheaſant, *Phaſianus Superbus*; from China.
2. The African Pheaſant, *Phaſianus Africanus*. It is aſh-blue, below white; its head creſted.
3. The Impeyan Pheaſant, *Phaſianus Impeyanus*; from India. It is creſted, and purple with gloſſy green; below black.
4. The Coloured Pheaſant, *Phaſianus Leucomelanos*; from India. It is creſted and black; the feathers on the body edged with white.

In the Genus TETRAO.

1. The Rock Grous, *Tetrao Rupeſtris*; from Hudſon's Bay. It is orange, variegated with black ſtripes and white blotches; its toes feathered; its tail-quills black tipt with white; its ſtraps black. It is much ſmaller than the white grous. It frequents not the woods, but ſitting on the rocks with its neck extended, it utters a noiſe like a perſon ſneezing.
2. The Rehuſak Grous, *Tetrao Lapponicus*. Its back is black variegated with ferruginous; its neck ferruginous ſpotted with black; its breaſt and vent white. Size of a hen. Found in the Lapland Alps.
3. The Helſingian Grous, *Tetrao Canus*. Its body is hoary waved with brown; its bill and legs black. Reſembles ſomething the Hazel Grous.
4. The Sand Grous, *Tetrao Arenarius*; from the deſerts about the Caſpian Sea. Its collar, belly, and vent, are deep black; its tail-quills ſtriped with brown and gray, and tipt with white; the two middle ones tawnyiſh.
5. The Namaqua Grous, *Tetrao Namaqua*; from the Cape of Good Hope. Its feet ſhaggy; its back cheſnut; its belly blackiſh; its two middle tail-quills projecting and awl-ſhaped.

6. The

6. The Heteroclite Grous, *Tetrao Paradoxus.* Its feet three-toed and ſhaggy; its back waved with gray and black; its belly black, with pale ſpots; the ſides of its neck marked with a fulvous ſpot.

In the Genus PERDIX.

1. The Cape Partridge, *Perdix Capenſis.* It is almoſt double-ſpurred; its breaſt ſtreaked with white; its legs red.
2. The Ceylon Partridge, *Perdix Ceylonenſis.* Size of a hen: double-ſpurred; its head and neck variegated with black and white.
3. The Brown African Partridge, *Perdix Spadiceus*; from Madagaſcar.
4. The Arragon Partridge, *Perdix Aragonica.* It is ſpurred; its wings, belly, and thighs black.
5. The Pintado Partridge, *Perdix Madagaſcarienſis.*
6. The Pearled Partridge, *Perdix Afra*; from the Cape of Good Hope.
7. The Gingi Partridge, *Perdix Gingica.* It is rufous-gray; its rump ſpotted with black.
8. The Green Partridge, *Perdix Viridis.*
9. The Javan Partridge, *Perdix Javanicus.* It is cinereous, with duſky creſcents; its cheeks black.
10. The Madagaſcar Quail, *Perdix Striata.* It is twice as large as the common quail.

11. The Gray-throated Quail, *Perdix Griseus*; from Madagascar. Size of the common quail.

12. The New Guinea Quail, *Perdix Novæ Guineæ*. Its body brown; the coverts of its wings edged with yellow. One half smaller than the common quail.

13. The Manilla Quail, *Perdix Manillensis*. Its body blackish above, yellowish below, with blackish stripes; its throat white. Size of a sparrow.

14. The Hudsonian Quail, *Perdix Hudsonica*. Its body pale-rusty; its neck spotted with white; its wings, its back, and its tail, marked with cross white lines widely parted. A small species.

15. The Kakerlik Quail (so called from its cry) *Perdix Kakerlik*; from Bucharia. Its bill, eye-brows, and legs, are scarlet; its breast cinereous.

16. The Caspian Quail, *Perdix Caspius*. It is cinereous spotted with scarlet; its nostrils, orbits, and temples yellow.

17. The Gibraltar Quail, *Perdix Gibraltarica*. Its body above brown striped with black; below yellowish-white; black crescents on the breast.

18. The Luzonian Quail, *Perdix Luzoniensis*. Brown above; yellow below; head variegated with black and white.

19. The Andalusian Quail, *Perdix Andalusicus*.

Its

Its body rufous waved with black; below yellowiſh.

In the Genus PSOPHIA.

The Undulated Trumpeter, *Pſophia Undulata*; from Africa. Size of a gooſe.

In the Genus OTIS.

1. The Chilian Buſtard, *Otis Chilenſis*. Its head and throat ſmooth; its body white; its top and tail cinereous.
2. The White-chinned Buſtard, *Otis Indica*.

In the Genus PLATALEA.

The Dwarf Spoonbill, *Platalea Pygmea*; from Guiana and Surinam. Its body brown above, and white below. Size of a ſparrow.

In the Genus MYCTERIA.

The Indian Jabiru, *Myсteria Aſiatica*. It is white; a ſtripe acroſs its eyes; the lower part of its back, and its quills, black.

In the Genus ARDEA.

1. The Gigantic Crane, *Ardea Argala*. It is cinereous; its head, its neck, and jugular pouch naked; its belly and ſhoulders bright white. It is five or ſeven feet long,

long, and exceſſively voracious. Found in Aſia and Africa, and particularly near the mouths of rivers in the province of Bengal.

2. The Duſky Crane, *Ardea Obſcura.* Size of a bittern. Found in Sclavonia.

3. The Dwarf Heron, *Ardea Pumila.* Found in the Caſpian Sea. It is cheſnut; the middle-quills of its wings are variegated with white and yellow. It is nineteen inches long.

4. The Minute Bittern, *Ardea Exilis.* Its neck rufous; a creſcent on its breaſt, and its quills black. Found in Jamaica and in North America. Hardly larger than a thruſh.

5. The Ferruginous Heron, *Ardea Ferruginea.* Frequent on the Tanais; neſtles on trees. Length twenty-one inches.

6. The Red-headed Heron, *Ardea Erythrocephala.* Found in Chili. Its creſt reaches to its back.

7. The Blue-headed Heron, *Ardea Cyanocephala.* Found in Chili. Its wings are black, edged with white.

8. The Striated Heron, *Ardea Striata*; from Guiana.

9. The Wattled Heron, *Ardea Carunculata*; from the Cape of Good Hope. Length five feet and a half.

10. The

10. The Rufous Heron, *Ardea Rufa.* Found ſometimes near the pools in Auſtria.

11. The Ruſty-crowned Heron, *Ardea Rubiginoſa*; from North America. Size of a bittern.

12. The Aſh-coloured Heron, *Ardea Cana*; from North America.

13. The Streaked Heron, *Ardea Virgata*; from North America.

14. The Snow Heron, *Ardea Nivea.* Length two feet. Found in moſt parts of the world. Neſtles on lofty trees.

15. The Galeated Heron, *Ardea Galeata.* Its body milky; its bill yellow; its legs ſcarlet. Found in Chili.

16. The Sacred Heron, *Ardea Sacra.* It is white; its head ſmooth, the feathers on its back jagged and white. Found at Otaheite, where it is held ſacred.

17. The Chineſe Heron, *Ardea Sinenſis.* It is brown with paler ſtreaks; its quills black. Small ſpecies.

18. The Johanna Heron, *Ardea Johannæ.* A black creſt; the body gray above and white below; the wings black.

19. The Lohaujung Heron, *Ardea Indica.* It is brown variegated with green; its tail black.

20. The Yellow-necked Heron, *Ardea Flavicollis*; from India. Length two feet.

21. The White-fronted Heron, *Ardea Novæ Hollandiæ*. Length twenty-eight inches.

In the Genus TANTALUS.

1. The Black-faced Ibis, *Tantalus Melanopis*. Found in New Year's Iſland, where it breeds on the rocks. Length twenty-eight inches.
2. The White-headed Ibis, *Tantalus Leucocephalus*; from Ceylon. A broad band of black croſſes the breaſt; the wings are black; the coverts of the tail long, and of a fine pink. Its roſy feathers loſe their colour during the rainy ſeaſon. It makes a ſnapping noiſe with its bill. A very large ſpecies.
3. The Ethiopian Ibis, *Tantalus Æthiopicus*, the *Abou Hannes* of Mr. Bruce. It is white; the head and upper ſide of the neck brown; the hind part of its back and its wing-quills black.
4. The Green Ibis, *Tan.alus Viridis*. Found in Ruſſia; flies in flocks, and neſtles in trees.
5. The Gloſſy Ibis, *Tantalus Igneus*. Reſembles the preceding, and found likewiſe in Ruſſia. One was killed in Cornwall.
6. The Leſſer Ibis, *Tantalus Minutus*. Its face, bill, and legs greeniſh; its body ferruginous, and white below. Found in Surinam.

7. The

7. The Black-headed Ibis, *Tantalus Melanocephalus*; from India. Length twenty-one inches.

8 The Pillan Ibis, *Tantalus Ibis*. bill, and legs, are brown; its body white; its quills black. Inhabits the lakes and rivers of Chili, and frequently ſits upon the trees. Size of a gooſe.

9. The Hagedaſh Ibis, *Tantalus Hagedaſh*. It is cinereous; its back variegated with green and yellow; its wings blue-black; its leſſer coverts violet. Found at the Cape of Good Hope. Feeds on roots; paſſes the night on trees. Larger than a hen.

In the Genus NUMENIUS.

1. The Otaheitan Curlew, *Numenius Tahitenſis*. It is tawny-white; its neck ſtreaked with black; its back and the coverts of its wings waved with blackiſh and whitiſh. Length twenty inches.

2. The Eſkimaux Curlew, *Numenius Borealis*. Its bill and legs black; its body brown ſpotted with gray. Inhabits the wet meadows in the country of Hudſon's Bay.

3. The Cape Curlew, *Numenius Africanus*. It is cinereous; its neck, its belly, and its rump, white.

4. The

4. The Pygmy Curlew, *Numenius Pigmeus.* *Size of a lark.* Inhabits Europe, and occurs ſometimes in England.

In the Genus SCOLOPAX.

1. The Little Woodcock, *Scolopax Minor*; from North America. Length eleven inches and a half.
2. The Great Snipe, *Scolopax Major.* Its back and coverts are brick-coloured ſpotted with black, and edged with white; its neck and breaſt yellowiſh white, with creſcents of black; its ſides waved with black. Inhabits Siberia, and fôund likewiſe in England and Germany. Length ſixteen inches: weight eight ounces.
3. The Cayenne Snipe, *Scolopax Cayannenſis.* It is cinereous brown, variegated with brick-colour; the under ſide of its body and its rump white. Length thirteen inches.
4. The Straight-billed Snipe, *Scolopax Belgica.* Found in the Netherlands.
5. The Marbled Goodwit, *Scolopax Marmorata*; from Hudſon's Bay. Size of the American goodwit.
6. The Semipalmated Snipe, *Scolopax Semipalmata*; from North America. Length fourteen inches.
7. The Stone Snipe, *Scolopax Melanoleuca.* Its tail and rump ſtriped with black and white;

white; its legs yellow. Found in North America. Twice as large as the common ſnipe.

8. The Yellow-ſhank Snipe, *Scolopax Flavipes.* It is whitiſh, ſpotted with black; its wings brown; its belly, and the coverts of its tail, white. Appears in autumn in the ſtate of New York. Length eleven inches.

9. The Nodding Snipe, *Scolopax Nutans.* It is cinereous, variegated with ferruginous; its belly, its rump, and its tail, white. Found on the ſhores of Labrador.

10. The Black Snipe, *Scolopax Nigra*; from the Northern Archipelago.

11. The Red-breaſted Snipe, *Scolopax Noveboracenſis.* Inhabits the coaſts of New York.

12. The Brown Snipe, *Scolopax Griſea.* Found on the coaſts of New York. Length eleven inches.

13. The Aſh-coloured Snipe, *Scolopax Incana*; from the Eimeo and Palmerſton Iſlands. Length eleven inches.

14. The Terek Snipe, *Scolopax Terek.* It is cinereous, ſpotted with brown, and white below. Found near the Caſpian Sea: flies in flocks. Length nine inches.

In the Genus TRINGA.

1. The Red-legged Sandpiper, *Tringa Erythropus*. Larger than the ruff.
2. The Wood Sandpiper, *Tringa Glareola*. Its bill is smooth; its legs greenish; its body brown dotted with white; its breast whitish. Found in the swamps of Sweden. Size of a stare.
3. The White-winged Sandpiper, *Tringa Leucoptera*. Found in the islands of the South Sea.
4. The Selninger Sandpiper, *Tringa Maritima*. It is variegated with gray and black; below white; its throat and tail duskish. Inhabits the shores of Norway and Iceland.
5. The Waved Sandpiper, *Tringa Undata*. Found in Denmark and Norway.
6. The Uniform Sandpiper, *Tringa Uniformis*; from Iceland.
7. The Brown Sandpiper, *Tringa Fusca*. Found in England. Size of a jack snipe.
8. The Black Sandpiper, *Tringa Lincolniensis*; from Lincolnshire.
9. The New York Sandpiper, *Tringa Noveboracensis*. It is blackish; its feathers edged with whitish; below white; its tail cinereous.
10. The Streaked Sandpiper, *Tringa Virgata*; from Sandwich Bay.

11. The

11. The Boreal Sandpiper, *Tringa Borealis*. It is cinereous; its neck, its sides, and its breast, waved with a paler hue. Found in King George's Sound.

12. The Newfoundland Sandpiper, *Tringa Novæ Terræ*. It is blackish, marginated with brown; below cinereous white.

13. The Variegated Sandpiper, *Tringa Variegata*; from Nootka Sound.

14. The Little Sandpiper, *Tringa Pusilla*. Found in the northern parts of Europe. Size of a sparrow.

15. The Red Sandpiper, *Tringa Islandica*. Size of a turtle. Its bill and legs brown. Found in the northern parts of Europe, Asia, and America, and sometimes in Great Britain.

16. The Southern Sandpiper, *Tringa Australis*. Its bill and legs black; its belly and rump whitish. Found at Cayenne. Length eleven inches.

17. The Banded Sandpiper, *Tringa Fasciata*; from Astracan.

18. The Black-topped Sandpiper, *Tringa Keptuscha*. Inhabits the pools of Siberia.

In the Genus CHARADRIUS.

1. The Ruddy Plover, *Charadrius Rubidus*; from Hudson's Bay.

2. The Black-crowned Plover, *Charadrius Atricapillus*;

Atricapillus; from New York. Length ten inches.

3. The New Zealand Plover, *Charadrius Novæ Zealandiæ*. It is ash-green; its face and collar black. Larger than the ringed plover, being eight inches long.

4. The Gregarious Plover, *Charadrius Gregarius*. It is cinereous; below white; the quills of its tail white, with a black bar. Abounds on the meadows near the Volga and the Jaik.

5. The Asiatic Plover, *Charadrius Asiaticus*. It is gray-brown; its front, its eye-brows, its throat, and its belly, are white. Found sometimes in the salt marshes in South Tartary. Larger than the ringed plover.

6. The Rusty-crowned Plover, *Charadrius Falklandicus*; from the Falkland Islands. Length seven inches and a half.

7. The Dusky Plover, *Charadrius Obscurus*. Its legs blueish. Found in New Zealand.

8. The Fulvous Plover, *Charadrius Fulvus*. Found in the marshes of Otaheite. Length twelve inches.

9. The White-bellied Plover. *Charadrius Leucogaster*. Length six inches.

10. The Red-necked Plover, *Charadrius Rubricollis*; from Van Diemen's Land.

11. The Indian Plover, *Charadrius Indicus.* It is brown; below white; two brown ſtripes on the breaſt. Size of a lark.

In the Genus RALLUS.

1. The Clapper Rail, *Rallus Crepitans*; from North America. It is olive-brown; its throat white. Length fourteen or ſixteen inches.
2. The Troglodyte Rail, *Rallus Auſtralis*; from New Zealand. Its wings and tail deep brown; its feathers ſtriped with black. Length fifteen or ſeventeen inches.
3. The Cape Rail, *Rallus Capenſis.* It is ferruginous; below ſtriped with black and white. Size of the land rail.
4. The Blue-necked Rail, *Rallus Cæruleſcens*; from the Cape of Good Hope. Length ſeven inches.
5. The Ceylon Rail, *Rallus Zeylanicus.* Its head is blackiſh; its bill and legs red. Larger than the water rail.
6. The Pacific Rail, *Rallus Pacificus.* It is black, dotted with white; its wings ſtriped; its breaſt blueiſh-aſh. Found in Otaheite.
7. The Tabuan Rail, *Rallus Tabuenſis.* Entirely black, red about the eyes, the tail extremely ſhort. Found in the Society Iſlands.

8. The Otaheite Rail, *Rallus Taitienſis*. It is cinereous; its tail black; its throat white. Length ſix inches.

9. The Dwarf Rail, *Rallus Puſillus*. In ſize, colour, and form, it reſembles a lark. Frequents the ſalt marſhes of Dauria.

In the Genus PARRA.

1. The Luzonian Jacana, *Parra Luzonienſis*. It is brown. Smaller than the lapwing.

2. The Chineſe Jacana, *Parra Sinenſis*. It is wine-cheſnut. Size of the painted pheaſant.

3. The African Jacana, *Parra Africana*. It is cinnamon-coloured; its neck white below. Length nine inches and a half.

4. The Faithful Jacana, *Parra Chavaria*. Its creſt hangs from the back of its head; its body is brown above. Found near Carthagena. Feeds on herbs, and is eaſily tamed.

5. The Indian Jacana, *Parra Indica*. It is blackiſh-blue; its back and wings brown. Builds a floating neſt with herbs near the brinks of pools.

6. The Chilian Jacana, *Parra Chilenſis*. Its legs brown; its head ſomewhat creſted behind. Feeds on inſects and worms: is vociferous: builds its neſt among the graſs,

grafs, and lays four fulvous eggs, dotted with black.

In the Genus GALLINULA.

1. The Carthagena Gallinule, *Gallinula Carthagena.* Its front blue; its body rufous.
2. The Black-bellied Gallinule, *Gallinula Ruficollis.* Its body is black below; its back dusky-green; its breast rufous. Length seventeen inches.
3. The White Gallinule, *Gallinula Alba.* Its front, bill, and legs, red. Found in Norfolk Island. Length two feet.
4. The Yellow-breasted Gallinule, *Gallinula Noveboracensis*; from New York. Smaller than a quail.
5. The Crested Gallinule, *Gallinula Cristata.* Found in China and India. Length eighteen inches.

In the Genus PHALAROPUS.

1. The Plain Phalarope, *Phalaropus Glacialis.* Inhabits the Icy Sea.
2. The Brown Phalarope, *Phalaropus Fuscus*; from North America.
3. The Barred Phalarope, *Phalaropus Cancellatus*; from Christmas Sound. Length seven inches and a half.

In the Genus FULICA.

The Cinereous Coot, *Fulica Americana*; from North America.

In the Genus PHŒNICOPTERUS.

The White-winged Red Flamingo, *Phœnicopterus Chilensis*. Frequents the lakes in Chili.

In the Genus DIOMEDEA.

1. The Chocolate Albatross, *Diomedea Spadicea*; from the South Sea. Its bill is whitish; its body deep chesnut-brown; its belly pale; its face and upper side of its wings white.
2. The Yellow-nosed Albatross, *Diomedea Chlororhynchos*. Its bill is black above, and yellow at its base; its body above dark blue; its under side and the rump white. Found in the South Sea and at the Cape of Good Hope. Size of a goose.
3. The Sooty Albatross, *Diomedea Fuliginosa*. A white crescent behind the eyes. Found within the Antarctic Circle. Nearly three feet long.

In the Genus ALCA.

1. The Labrador Auk, *Alca Labradora*. Size of the Puffin. Its bill keel-shaped; its lower

lower mandible ſwelling; a black ſpot at the tip; its orbits and temples whitiſh; its belly white.

2. The Creſted Auk, *Alca Criſtatella*. Size of the redwing. Its bill is compreſſed, and ſomewhat furrowed; its body blackiſh; ferruginous ſpots on its back; a creſt on its front leaning backwards. Found in the iſlands near Japan.

3. The Ancient Auk, *Alca Antiqua*. Its bill is black, whitiſh at the baſe; its body blackiſh, its belly white. Found near Kamtſchatka and the Kurile Iſlands.

4. The Flat-billed Auk, *Alca Pygmea*. Its body is deep black, below cinereous. Found in the Iſle of *Aves*, between Aſia and America. Is gregarious. Length ſeven inches.

In the Genus URIA.

1. The White Guillemot, *Uria Lacteola*. Size of the black guillemot. Found on the weſt coaſt of Holland.

2. The Marble Guillemot, *Uria Marmorata*; from Kamtſchatka. Length ten inches.

In the Genus COLYMBUS.

1. The Striped Diver, *Colymbus Striatus*. Found in the lakes of North America. Weighs between two and three pounds.

2. The Chineſe Diver, *Colymbus Sinenſis.* It is greeniſh-brown with darker ſpots; its breaſt and belly rufous white, with rufous ſpots.

In the Genus STERNA.

1. The Surinam Tern, *Sterna Surinamenſis.* It is cinereous; below white, its legs red.
2. The African Tern, *Sterna Africana.* It is white; its body blueiſh above; its top black; its wings ſpotted with brown.
3. The Philippine Tern, *Sterna Philippina.* It is wine-gray, a white cap; the fillet acroſs the eyes, the wing-quills, the tail, and the bill, black. Twice as large as the greater tern.
4. The Simple Tern, *Sterna Simplex.* It is inclined to lead-colour, white below, its top whitiſh. Found in Cayenne. Size of the noddy.
5. The Egyptian Tern, *Sterna Nilotica.* It is cinereous, below white; its orbits black, ſpotted with white. Size of a pigeon.
6. The Striated Tern, *Sterna Striata*; from New Zealand.
7. The Wreathed Tern, *Sterna Vittata*; from Chriſtmas Sound. Length fifteen inches.
8. The Brown Tern, *Sterna Spadicea*; from Cayenne. Its vent white. Length fifteen inches.

9. The

9. The White Tern, *Sterna Alba.* Its bill and legs are black. Found in the Eaſt Indies, at the Cape of Good Hope, and in the South Sea.

10. The Chineſe Tern, *Sterna Sinenſis.* It is white, its back cinereous; a black bar on its top. Length eight inches.

11. The Southern Tern, *Sterna Auſtralis*; from Chriſtmas Sound. It is cinereous, below gray; its wing-quills white. Length ſeven inches and a half or nine inches.

12. The Hooded Tern, *Sterna Metopoleucos.* Found in Ruſſia and the South of Siberia. Goes in pairs. Length eight inches and a half.

In the Genus LARUS.

1. The Great Gull, *Larus Icthyætus.* Its head and the top of its neck black; its back and wings grayiſh; its eye-lids and tail white. Size of the barnacle. In flying, utters a deep croak. Found on the Caſpian Sea.

2. The Little Gull, *Larus Minutus.* It is ſnowy; its head black; its wings dirty white; its legs ſcarlet. Size of the miſſel. Frequents the large rivers in Siberia.

3. The Eſquimaux Keeaſk, *Larus Keeaſk.* It is brown; the coverts of its wings variegated

riegated with white; its tail black, ſpotted and tipt with white. It arrives in Hudſon's Bay in April: makes its neſt with graſs, and lays two pale-ruſty eggs with black ſpots. Length twenty-two inches.

In the Genus PROCELLARIA.

1. The Dark-gray Petrel, *Procellaria Griſea.* The inferior coverts of its wings white; its bill brown; its legs blueiſh before. Length fourteen or fifteen inches. Found in the ſouthern hemiſphere.

2. The Glacial Petrel, *Procellaria Gelida.* It is blueiſh-aſh; its back blackiſh; its throat and breaſt white; its bill yellow; its legs blue. Length nineteen inches. Found on the utmoſt verge of the Antarctic Ocean.

3. The White-breaſted Petrel, *Procellaria Alba.* It is duſky-blackiſh; its belly and vent white. Length ſixteen inches. Found in the iſlands of the Pacific Ocean.

4. The Cinereous Petrel, *Procellaria Cinerea.* White below; tail blackiſh; bill yellow; legs aſhy. Length twenty inches and a half. Found within the Antarctic Circle.

5. The Black-toed Petrel, *Procellaria Melanopus.* It is dark cinereous; its bridle and

and throat gray, with minute blackiſh ſpots. Length thirteen inches. From North America.

6. The Brown-banded Petrel, *Procellaria Deſolata.* It is blueiſh-aſh; below white; the tips of its tail-quills blackiſh. Length eleven inches. Found at Deſolation Iſland.

7. The Sooty Petrel, *Procellaria Fuliginoſa.* Its tail is notched. Length eleven inches. From Otaheite.

8. The Fork-tail Petrel, *Procellaria Furcata.* It is ſilver-gray; its throat pale; its vent white. Length ten inches. Inhabits the Northern Archipelago.

9. The Diving Petrel, *Procellaria Urinatrix.* Length eight inches and a half. Found at New Zealand.

10. The Pacific Petrel, *Procellaria Pacifica.* Deep black; below duſky; legs pale. Length twenty-two inches. Found near the iſlands of the Pacific Ocean.

11. The Duſky Petrel, *Procellaria Obſcura.* Length thirteen inches. From Chriſtmas Sound.

In the Genus MERGUS.

1. The Imperial Merganſer, *Mergus Imperialis.* Size and form of a gooſe. Its tongue ciliated.

2. The Brown Merganser, *Mergus Fuscus*; from Hudson's Bay. Length seventeen inches and a half.

3. The Blue Merganser, *Mergus Cæruleus*; from Hudson's Bay. Length fourteen inches.

In the Genus ANAS.

1. The Black-necked Swan, *Anas Nigricollis*; from the Falkland Islands. Size of the common swan.

2. The Black Swan, *Anas Atrata*; from New Holland. Larger than the common swan.

3. The Hybrid Goose, *Anas Hybrida*. Its bill semi-cylindrical; its cere red; its tail somewhat sharp. Size of the common goose. Appears in pairs in the sea about Chiloë. Lays eight eggs in the sand.

4. The Coscoroba Goose, *Anas Coscoroba*. Its bill enlarged and rounded at the end; its body white. Found in Chili. Large, and easily tamed.

5. The Antarctic Goose, *Anas Antarctica*. Length twenty-four or twenty-six inches.

6. The Variegated Goose, *Anas Variegata*; from New Zealand. Size of a large duck.

7. The Snow Goose, *Anas Hyperborea*. Its body snowy; its front yellowish; the ten

first quills of its wings black; its bill and legs red. Inhabits the Arctic regions. Length thirty-two inches.

8. The Great Goose, *Anas Grandis.* Its body blackish, below white; its bill black; its legs scarlet. Found in Siberia. Size of the swan.

9. The Barred-headed Goose, *Anas Indica.* In winter these arrive in India, perhaps from Thibet.

10. The Red-breasted Goose, *Anas Ruficollis* Frequent in Russia, and on the northern parts of Siberia.

11. The Ruddy Goose, *Anas Casarca.* Inhabits Astracan. Goes in pairs, and has a pleasant cackle.

12. The Bean Goose, *Anas Segetum.* It is cinereous-brown, below whitish; its wings gray; its greater coverts and its secondary wing-quills tipt with white. Inhabits the northern parts of Europe and of America; found in winter in the fens of Lincolnshire. Length thirty or thirty-six inches.

13. The Bering Goose, *Anas Beringii.* Its bill swelled; its body white; its wings black. Size of a common goose.

14. The Gulaund Duck, *Anas Borealis.* Its bill narrow; its head glossy-green; its breast and belly white. Inhabits the marshes of Iceland.

15. The White-headed Brent, *Anas Torrida.* Size of the tufted duck.

16. The White-fronted Brent, *Anas Albifrons.* Size of a cock.

17. The King Duck, *Anas Spectabilis.* Its bill bunched at the base, and compressed; its head hoary; its body black; its shoulders whitish. Inhabits the northern parts of Europe and of America. Length two feet.

18. The Royal Duck, *Anas Regia:* from Chili. A compressed caruncle on its front; its body blue, and below brown; its collar white.

19. The Georgia Duck, *Anas Georgica.* It is cloudy-ash; a green spangle on its wings edged with white; its quills blackish: from the South Sea. Length twenty inches.

20. The Brown Duck, *Anas Fuscescens*; from Newfoundland. Length sixteen inches.

21. The Spotted-billed Duck, *Anas Poecilorhyncha.* Common in Ceylon.

22. The Curve-billed Duck, *Anas Curvirostra.* Taken in Holland.

23. The Supercilious Duck, *Anas Superciliosa.* The spangle on its wings blueish-green, edged with black. Length twenty-one inches. From New Zealand.

24. The Crimson-billed Duck, *Anas Erythrorhyncha.* It is brown, below white; its

tail

tail black. Length fifteen inches. From the Cape of Good Hope.

25. The Red-breaſted Shoveler, *Anas Rubens*. Its tail is ſhort and white. Sometimes taken in Lincolnſhire.

26. The Jamaica Shoveler, *Anas Jamaicenſis*. Variegated with brown, ſaffron, and ruſty; the under ſide and throat white, with black ſpots. Length ſixteen inches.

27. The Ural Duck, *Anas Leucocephala*. It is cloudy-yellowiſh, powdered with brown; its head and neck white. Larger than a teal. Found in Barbary, and alſo on the Uralian lakes, and on the rivers Irtis and Oby. Cannot walk, but ſwims very faſt. Builds a floating neſt among the reeds.

28. The Pied Duck, *Anas Labradora*; from Labrador. Length nineteen inches.

29. The Lapmark Duck, *Anas Scandiaca*. Its body black above; its breaſt and belly white.

30. The Cape Wigeon, *Anas Capenſis*. It is aſhy; its back reddiſh-brown; its feathers edged with yellow.

31. The Bimaculated Duck, *Anas Glocitans*. Its head green; a round ruſty ſpot between the bill and the eye, and another oblong one behind the ears. It has a clucking voice. Length twenty inches.

 Found

Found on the Lena and the lake Baikal; and ſometimes in England.

32. The Soft-billed Duck, *Anas Malacorhynchos*; from New Zealand. Has a piping voice. Length eighteen inches.

33. Jacquin's Duck, *Anas Jacquini.* Crimſon; its back blackiſh; its bill and feet black. Its voice very ſharp. From St. Domingo.

34. The Weſtern Duck, *Anas Diſpar.* White, below ferruginous; ſpot on the back of the head, and the front, greeniſh. Length ſeventeen inches. From Sweden and Kamtſchatka.

35. The Pink-headed Duck, *Anas Caryophyllacea*; from India. It goes in pairs, and is eaſily tamed. Length twenty-one inches.

36. The New Zealand Duck, *Anas Novæ Zealandiæ.* Reſembles the Tufted Duck.

37. The Creſted Duck, *Anas Criſtata*; from Statenland. Length twenty-eight inches.

38. The Iceland Duck, *Anas Iſlandica.* It is black creſted; its throat, its breaſt, and its belly white.

39. The Duſky Duck, *Anas Obſcura*; from New York. Length two feet.

40. The Baikal Teal, *Anas Formoſa.* It is brown; its top black edged with white; its throat tawnyiſh, ſpotted with black; a black

a black ſpangle on the wings, edged with brick-colour. Length fifteen inches.

41. The Ilina Teal, *Anas Ilina*; from China. It is greeniſh about the eyes.

42. The Black Teal, *Anas Gmelini*. Its breaſt is croſſed with red lines. Found at the Caſpian, and through the whole of the ſouth of Ruſſia.

43. The Alexandrian Teal, *Anas Alexandrina*. Its bill and vent are black; its belly white; its neck cinereous, with black ſemicircles.

44. The Sirſæir Teal, *Anas Sirſæir*. Its bill yellow below; the ſpangle on its wings divided obliquely. Found in Arabia.

In the Genus APTENODYTES.

1. The Papian Penguin, *Aptenodytes Papua*. Its bill and feet reddiſh; a white ſpot on the back of the head. Length two feet and a half.

2. The Antarctic Penguin, *Aptenodytes Antarctica*. Its bill deep black; its feet reddiſh; a black line on its throat.

3. The Collared Penguin, *Aptenodytes Torquata*. Its bill and feet black; a naked bloody ſpace about the eyes. Length eighteen inches. Found in New Guinea, Kerguelen's Land, and New Georgia.

4. The Little Penguin, *Aptenodytes Minor*; from New Zealand. Its bill black;

its feet whitiſh. Length thirteen or fifteen inches.

5. The Woolly-cinereous Penguin, *Aptenodytes Chiloenſis*. Common in the Archipelago of Chiloe. Size of a Gooſe.

6. The Three-toed Penguin, *Aptenodytes Chilenſis*. Found in Chili. Size of the preceding, but longer necked. Lays in the ſand ſix or ſeven eggs whtie, dotted with black.

In the Genus PELECANUS.

1. The Red-backed Pelican, *Pelecanus Ruſeſcens*; from Africa. Length five feet.

2. The Charleſtown Pelican, *Pelecanus Carolinenſis*. Above duſky, below white. Length three feet and a half.

3. The Rough-billed Pelican, *Pelecanus Erythrorhynchos*; from North America. Length four feet and a half.

4. The Saw-billed Pelican, *Pelecanus Thagus*. Inhabits Chili and Mexico. Size of a turkey. Breeds on cliffs.

5. The Palmerſton Frigate Pelican, *Pelecanus Palmerſtoni*. Its tail is forked; its body brown, gloſſed with green, below white; its throat variegated with black and white; its belly white; its vent black. Length thirty-eight inches.

6. The Violet Cormorant, *Pelecanus Violaceus*; from Kamtſchatka.

7. The Red-faced Shag, *Pelecanus Urile*; from Kamtſchatka. Length thirty-one or thirty-four inches.

8. The Spotted Shag, *Pelecanus Punctatus*; ſrom New Zealand. Breed among the rocks or trees. Length twenty-one or twenty-four inches.

9. The Caruncŭlated Shag, *Pelecanus Carunculatus*. Numerous in New Zealand and Statenland. Breeds among the tufts of tall graſs.

10. The Magellanic Shag, *Pelecanus Magellanicus*. A ſpot behind its eyes, and its belly white; its temples and chin reddiſh; its flanks ſtriped with white. Inhabits Terra del Fuego and Statenland. Breeds in holes of the rocks. Length thirty inches.

11. The Pied Shag, *Pelecanus Varius*; from New Zealand. Breeds on trees. Length thirty inches.

12. The Tufted Shag, *Pelecanus Cirrhatus*; from New Zealand. Length thirty-four inches.

13. The African Shag, *Pelecanus Africanus*. Its throat white; the coverts of its wings blue-gray, and black at the edge and tip. Length twenty inches.

14. The Dwarf Shag, *Pelecanus Pygmæus*, Lives among the flocks of Shags on the Caſpian Sea. Hardly ſo large as a teal.

In

In the Genus PHAETON.

The Black-billed Tropic Bird, *Phaeton Melanorynchus*. It is ſtriped with black and white; its under ſide and front white; a bar behind its eyes; its bill and feet is black. Found in Turtle and Palmerſton Iſlands. Length nineteen inches and a half.

ADDENDA.

ADDENDA.

THE manners of the Wood Ibis, *Tantalus Loculator*, are well defcribed by Mr. Bartram:—" This folitary bird does not affociate " in flocks, but is generally feen alone; com- " monly near the banks of great rivers, in vaft " marfhes or meadows, efpecially fuch as are " caufed by inundations; and alfo in the vaft " deferted rice plantations: he ftands alone on " the topmoft limb of tall dead cyprefs trees; " his neck contracted or drawn in upon his " fhoulders, and beak refting like a long fcythe " upon his breaft: in this penfive pofture and " folitary fituation, it looks extremely grave, " forrowful, and melancholy, as if in the deep- " eft thought." *Travels in North and South Carolina, Georgia, &c. p.* 148.

THE King of the Vultures, *Vultur Papa*, is found alfo in the fouthern ftates of America. The Creek Indians, who inhabit the back country adjoining to Georgia and South Carolina, employ

employ the tail-feathers for conſtructing their royal ſtandard. Theſe birds ſeldom appear, except when the deſerts are on fire, in which caſe they gather from all quarters to feaſt on the ſerpents, frogs, and lizards that are roaſted in the hot embers.

END OF THE NINTH VOLUME.

INDEX,

Of the Names *of* BUFFON, LINNÆUS, *and* LATHAM's Synopsis.

ACALOT — VIII. 43
Acatechill — IV. 200
Achbobba — I. 124
Acintli — VIII. 198
Acolchi — III. 181
Agami — IV. 390
Alapi — — 388
Alatli — VII. 203
Alauda Africana — V. 63
Alpestris — 53
Arborea — 23
Arvensis — 7
Calandre — 47
Campestris — 41
Capensis — 51
Cinerea — 62
Cristata — 65
Flava — — 59
Italica — — 45
Ludoviciana — 34
Magna — III. 328
Motellana — V. 57
Nemorosa — 72
Pratensis — 28
Rubra — — 55
Rufa — — 21,61
Senegalensis — 76
Trivialis — 36
Undata — — 74
Albatros, Wandering — 289
Alca, Arctica — IX. 304
Cirrhata — 312
Impennis — 333
Pica — — 335
Torda — — 330
Alcedo Alcyon — VII. 205
Americana — 210
Atricapilla — 183
Bengalensis — 197
Bicolor — — 209
Brasiliensis — 211
Alcedo Cæruleocephala VII. 193
Cancrophaga — 176
Capensis — 178
Cayanensis — 201
Chlorocephala — 184
Collaris — 186
Cristata — 199
Dea — — 192
Erithaca — 191
Fusca — — 174
Galbula — 214
Ispida — 158,188
Leucocephala — 185
Leucoryncha — 195
Maculata — 207
Madagascariensis — 194
Maxima — 182
Paradisea — 216
Purpurea — 194
Rudis — 179
Senegalensis — 188,189
Smyrnensis — 175
Superciliosa — 212
Torquata — 203
Tridactyla — 198
Amazon, Red-headed, or Tarabe — VI. 184
White-headed — 185
Yellow — 186
Yellow-headed — 182
Ampelis, Carnifex — IV. 361
Carunculata — 362
Cayana — 355
Cotinga — 353
Garrulus — III. 389
Maynana — IV. 357
Pompadora — 358
Tersa — 356
Variegata — 364
Anaca — VI. 224
Anas Acuta — IX. 166

I N D E X.

Anas Ægyptiaca — IX. 67
Africana — 229
Albeola — 240
Albifrons — 70
Anſer — 25
Arborea — 156
Bahamenſis — 215
Bernicla — 76
Boſchas — 100
Braſilienſis — 215
Bucephala — 209
Cæruleſcens — 69
Canadenſis — 71
Circia — 225
Clangula — 186
Clypeata — 160
Coromandeliana — 231
Crecca — 222
Cygnoides — 61
Cygnus — 1
Diſcors — 236,237
Dominica — 239
Erythropus — 81
Falcaria — 232
Ferina — 181
Fuſca — 204
Galericulata — 233
Gambenſis — 64
Glacialis — 169
Glaucion — 191
Hiſtrionica — 210
Leucoptera — 58
Madagaſcarienſis — 230
Magellanica — 57
Melanotos — 66
Minuta — 212, 243
Molliſſima — 90
Moſchata — 138
Nigra — 196
Novæ Hiſpaniæ — 241
Penelope — 143
Perſpicillata — 205
Querquedula — 218
Rufina — 153
Ruſtica — 242
Spectabilis — 213
Spinoſa — 238
Sponſa — 206
Strepera — 157
Anas Tadorno — IX. 171
Angala, Dian — V. 502
Angoli — VIII. 195
Anhinga — 406
Melanogaſter — 410
Rufous — ib.
Ani — — VI. 363
Great — 366
Leſſer — 364
Mangrove — 366
Savanna — 364
Anter — IV. 370
Creſted — 381
King of — 374
Nightingale — 387
White-eared — 382
Aourou-couraou — VI. 187
Aptenodytes, Patachonica — } IX. 338
Chryſocome — 346
Ara — VI. 156
Black — 175
Blue — 168
Green — 169
Red — 158
Aracaris — VII. 120
Black-billed — 124
Blue — ib.
Arada — IV. 385
Ardea Æquinoctialis VII. 355,384
Agami — 366
Alba — 350
Americana — 296
Antigone — 295
Atra — 353
Badia — 373
Botaurus — 405
Braſilienſis — 417
Cærulea — 381
Cæruleſcens — 382
Canadenſis — 299
Cayanenſis — 422
Ciconia — 243
Cocoi — 364
Comata — 375,376
Cracra — 386
Cyanopus — 385
Danubialis — 407
Egretta — 361

Ardea

I N D E X.

Ardea Erythropus — VII. 374
  Flava — 412
  Gardeni — 410
  Garzette — 357
  Grus — 277
  Helias — VIII. 161
  Herodias — VII. 369
  Hoactli — 367
  Houhou — 368
  Hudſonias — 370
  Leucocephala — 355
  Leucogaſter — 363
  Lineata — 416
  Ludoviciana — 390
  Maguari — 265
  Major — 329
  Malaccenſis — 377
  Marſigli — 406
  Minuta — 379
  Novæ Guineæ — 377
  Nycticorax — 419
  Pavonina — 306
  Philippenſis — 378
  Pondiceriana — 392
  Purpurea — 354
  Rufeſcens — 362
  Scolopacea — 425
  Senegalenſis — 409
  Solonienſis — 408
  Spadicea — 385
  Squaiotta — 372
  Stellaris — 394, 414
  Tigrina — 415
  Undulata — 413
  Vireſcens 388,389,391
  Violacea — 383
  Virgo — 301
Arimanon — VI. 154
Attagas — II. 221
  White — 230
Auk — IX. 330
  Black-billed — 335
  Great — 333
  Tufted — 312
Avoſet, Scooping VIII. 422
  White — VII. 486
Azurin III. 371—IV. 376
Baglafecht — III. 426
Baker — VI. 407
Balbuzard — I. 70
Balicaſe, Phillipine III. 73
Baltimore — 203
  Baſtard — 205
Bambla — IV. 384
Baniahbou — III. 337
Bannaniſte — V. 336
Barbet, Beautiful VII. 92
  Black-breaſted — 98
  Black-throated — 97
  Cayenne — 90
  Collared — 91
  Doubtful — 126
  Great — 100
  Green — 101
  Little — 99
  Spotted-bellied — 88
  Yellow-throated 96
Barbican — 126
Barge — 476
  Barking — 480
  Brown — 485
  Common — 479
  Rufous — 482
    Great — 483
    of Hudſon's Bay 484
  Variegated — 481
  White — 486
Barnacle — IX. 81
Bartavelle — II. 369
Bee-eater — VI. 411
  Angóla — 428
  Azure-tailed Green 429
  Blue-headed — 430
    Red — ib.
  Braſilian — 409
  Cayenne — 433
  Cheſnut — 420
  Cheſnut and Blue — ib.
  Cinereous — 419
  Gray of Ethiopia — 420
  Gray-headed 419
  Green and Blue Yellow-throated — 427
  Green Blue-throated 424
  Indian — 424
  Little Green and Blue Taper-tailed 428

I N D E X.

Bee-eater, Molucca — VI. 409
 Philippine — 429
 Red and Green Senegal 431
  Green, with Rufous wings and tail — 433
  headed — 433
  winged — 431
  Yellow-headed 434
 Rufous — 407
 Supercilious — 422
 Yellow — 418
  and White — 418
 Yellow-throated — 427
Beef-eater African VII. 154
Belfrey, Great — IV. 376
 Small — 378
Bengal — 81
 Brown — 84
 Punctured — 85
Bentaveo or Cuiriri — 471
Bergeronette, Gray V. 252
 Madras — 267
 of the Island of Timor — 266
 Spring — 256
 Yellow — 259
Bihoreau — VII. 419
 of Cayenne — 422
Bird Saint Martin — I. 164
Biset — II. 439
Bittern, Brasilian VII. 417
 Brown — 411
 Greater — 405
 Hudson's Bay — 414
 Lineated — 416
 Little — 378, 406
 Little, of Cayenne 413
 Rayed Brown — 407
 Rufous — 408
 Senegal — 409
  Little — ib.
 Spotted — 410
 Starred — 411
 Swabian — 406
 Tiger — 415
 Yellow of Brasil — 412
 Zigzag — 413
Blackbird — III. 292
Blackbird, Amboyna III. 354
 Black-headed — 348
 Black and White — 367
 of the Isle of Bourbon 355
 Brown — 336, 344
 Brown of Abyssinia 368
 Brown Jamaica — 351
 Canada — 342
 of China — 326
 Cinereous — 338, 343
 Cravated — 352
 Crescent — 328
 Crested — 324
 Crested of the Cape of Good Hope — 353
 Dominican — 356
 Golden — 359
 Green of Angola — 330
 Green of the Isle of France — 347
 Indian, or Terat Boulan 357
  Olive — ib.
 Madagascar — 345
 Mindanao — 346
 Olive of St. Domingo 364
  of Barbary — 365
 Orange Green — 335
 Rock — 309
 Rufous of Cavenne 363
 Rufous-throated Brown 364
 Surinam — 360
 White-bellied Violet 362
Black Cap — V 119
Blongio — VII. 379
Blue Throat — V. 195
Boat-bill — VII. 426
Bonana, Lesser — III. 217
Booby, Common — VIII. 333
 Great — 337
 Lesser — 339
 Little — ib.
  Brown — ib.
 Spotted — 340
 White — 336
Boutsallick — VI. 319
Brambling — IV. 108
Brent — IX. 76
Brunet — III. 349
Brunette — VII. 472

Bucco

Bucco Capensis — VII. 91
Cayanensis — 90
Dubius — 126
Elegans — 92
Grandis — 100
Macrorynchos — 93
Niger — 97,98
Parvus — 99
Philippensis — 96
Tamatia — 88
Viridis — 101
Buceros Abyssinicus — 148
Africanus — 147
Bicornis — 150
Galeatus — 153
Hydrocorax III. 34—VII. 140
Malabaricus VII. 142
Manillensis — 137
Nasutus — 134
Panayensis — 138
Rhinoceros — 155
Bulfinch — IV. 298
Bunting, Amazon — 291
Blue — 295
Blue-faced — 153
Bourbon — 292
Brazilian — 288
Cape — 263
Cinereous — 294
Cirl — 279
Common — 284
Dominican — 138
Foolish — 282
Gray — 293
Green — 155
Hooded — 256
Lesbian — 258
Long-tailed — 140
Lorraine — 259
Louisiana — 272
Louisiane — 261
Mexican — 289
Mustachoe — 257
Olive — 290
Orange-shouldered 141
Painted — 150
Payanan — 143
Plata — 291
Psittaceous — 144
Bunting, Red-eyed — IV. 297
Reed — 253
Rice — 270
Snow — 264
Shaft-tailed — 137
Towhe — 122
Variegated — 142
Whidah — 134
Yellow — 274
Yellow-bellied Cape 262
Yellow-faced — 290
Buphaga Africana III. 154
Bustard, African — II. 44
Arabian — 42
Great — 1
Indian — 47
Little — 34
Rhaad — 52
Ruffed — 50
Thick-kneed VIII. 102
Butcher Bird — I. 239
Buzzard — 159
Ash-coloured — 177
Honey — 161
Moor — 172

Cacastol — III. 171
Cacolin — II. 430
Caica — VI. 217
Calandre — V. 47
Calao — VII. 130
Abyssinian — 148
African, or Brac — 147
Malabar — 142
Manilla — 137
Molucca — 140
of the Island of Panay 138
Philippine — 150
Rhinoceros — 155
Round helmeted 153
Calybe — III. 152
Cancroma — VII. 426
Canut — VIII. 134
Caprimulgus Acutus VI. 461
Americanus — 458
Brasilianus — 455
Carolinensis — 448
Cayanensis — 455, 459
Europeus — 436

Caprimulgus Grandis VI. 456
Griſeus — 462
Guianenſis — 463
Jamaicenſis — 452
Rufus — 464
Virginianus — 450
Caracara — II. 344
Caraya — IV. 388
Cardinal, Creſted — III. 414
Cariama — VII. 313
Carouge — III. 214
Caſſican — VII. 128
Caſſique, Yellow — III. 207
Green — 211
Creſted — 212
Caſſowary, Galeated — I. 376
New Holland — 389
Catotol — IV. 199
Caudec — 473
Caurale — VIII. 161
Ceinture de Pretre — V. 59
Cendrille — 63
Certhia Afra — 505
Braſiliana — 530
Chalybea — 494
Cruentata — 506
Cyana — 527
Cyanea — 520
Familiaris — 476
Famoſa — 512
Flaveola — 532
Gutturalis — 516
Jugularis — 501
Lotenia — 502
Mexicana — 514
Muraria — 481
Olivacea — 499
Omnicolor — 504
Philippina — 492
Pinus — 292
Pulchella — 511
Purpurea — 518
Puſilla — 490
Senegalenſis — 491
Soui-manga — 487
Sperata — 489
Spiza — 524
Variegata — 529
Violacea — 509
Certhia Zeylonica — V. 498
Chacamel — II. 346
Chaffinch — IV. 96
Charadrius, Apricarius VIII. 82
Bilobus — 99
Calidris — VII. 508
Cayanus — VIII. 100
Coronatus — 98
Gallicus — 121
Hiaticula — 88
Himantopus — 109
Melanocephalus — 101
Morinellus — 84
Oedicnemus — 102
Pileatus — 97
Pluvialis — 78
Spinoſus — 95, 96
Vociferus — 93
Charboniere — V. 394
Chatterer — III. 389
Blue-breaſted — IV. 356
Bohemian — III. 389
Carunculated — IV. 362
Pompadour — 358
Purple-throated — 355
Red — 361
Silky — 357
Variegated — 364
Cheric — V. 271
Chimer — IV. 383
Chinquis — II. 319
Churge — 47
Ciconia Nigra — VII. 261
Cincle — 524
Cochicat — 118
Cock — II. 54
of the Rock — IV. 346
Peruvian — 349
Cockatoo — VI. 80
Black — 87
Leſſer White — 83
Little fleſh-billed — 85
Red creſted — ib.
Great — ib.
White-creſted — 82
Yellow-creſted — 83
Cocotzin — II. 495
Cocquar — 306
Colemouſe — V. 401

Coleniculi

Coleniculi — II. 431
Colibri — VI. 40
Blue — 59
Carmine-throated — 54
Dotted or Zitzil — 47
Green and black — 50
Green-throated — 53
Little — 61
Rusty-bellied — ib.
Topaz — 44
Tufted — 51
Violet — 55
Violet-tailed — 52
Colin — II. 426
Great — 429
Colius, Payanensise — IV. 326
Striatus — 325
Collar, Red — VI. 57
Colma — IV. 380
Colnud, Cayenne — III. 72
Columba Canadensis — II. 494
Capensis — 490
Caroliniensis — 494
Livia — 439
Macroura — 489
Madagascariensis — 477
Marginata — 488
Palumbus — 469
Passerina — 495
Risoria — 487
Speciosa — 478
Turtur — 482
Viridis — 491
Coly — IV. 321
Colymbus Arcticus - VIII. 243
Auritus — 220
Cayenensis — 225
Cornutus — 221
Cristatus — 219
Dominicus — 231
Glacialis — 241
Grylle — IX. 301
Immer — VIII. 234
Ludovicianus — 224
Minor — 228
Obscurus — 218
Podiceps — 230
Rubricollis — 225
Stellatus — 237
Colymbus Thomensis VIII. 223
Troile — IX. 298
Urinator — VIII. 213
Commander — III. 188
Bonjour — IV. 296
Condor — I. 139
Coot, Common — VIII. 200
Crested — 209
Greater — 207
Coquillade — V. 74
Coracias, Abyssinica — III. 126
Cayanensis — 118
Garrula — ib.
Madagascariensis — 131
Orientalis — 130
Sinensis — 117
Varia — VII. 128
Cormorant — VIII. 282
Corrira, Italica — 428
Corvus Balicassius — III. 73
Calvus — 69
Canadensis — 103
Caribæus — 88
Caryocatactes — 109
Cayanus — 105
Corax — 11
Cornix — 51
Corone — 38
Cristatus — 106
Dauricus — 57
Eremita — 7
Erythorynchos — 101
Flavus — 106
Frugilegus — 46
Glandarius — 94
Graculus — 1
Hottentottus — 68
Jamaicensis — 58
Mexicanus — 91
Monedula — 59
Novæ Guineæ — 70
Nudus — 72
Papuensis — 71
Peruvianus — 102
Pica — 75
Pyrrhocorax — 65
Sibiricus — 105
Zanahoe — 93
Cotinga — IV. 351

INDEX.

Coukeels — VI. 329
Coua — 313
Coulaciffi — 148
Coulavan — III. 230
Couricaca — VII. 267
Courliri or Courlan — 425
Coyolcos — II. 430
Crab-catcher — VII. 371
Black — 377
Blue — 381
Brown-necked Blue - 382
Chalybeate — 387
Cinereous — 385
Coromandel — 376
Gray — 391
Green — 388
Iron-gray — 383
Little — 378
Mahon — 376
Purple — 385
Red-billed White — 384
Rufous — 390
Spotted Green — 389
White and Brown — 377
Cracra — 386
Crane — 277
Brown — 299
Collared — 295
Hooping — 296
Numidian — 301
White — 296
Crax, Alector — II. 327
Pauxi — 335
Creeper — V. 473
African — 505
All-green — 527
Beautiful — 511
Black and Blue — 520
Blue — 522
Black and Violet — 530
Yellow — 532
Black-headed — 524
capped — 525
Blue-headed — 526
Cayenne — 527
Ceylonefe — 498
Collared — 494
Common — 476
Famous — 512
Creeper, Green-faced V. 516
Green-gold — 504
Lotens — 502
Philippine — 492
Purple — 518
Red — 514
Red-breafted — 489
Red-fpotted — 506
Senegal — 491
Variegated — 529
Violet — 487
Violet-headed — 509
Wall — 481
Creeper-billed Brown Bird 516
Purple Bird — 518
Red Bird — 514
Crefcent — III. 456
Crefferelle — I. 226
Crick — VI. 196
Blue-faced — 195
Blue-headed — 198
Mealy — 193
Red and Blue — 194
with a Yellow Head and Throat — 190
Violet — 200
Crifpin — III. 369
Crofs-bill — 405
Crotophaga Ani — VI. 364
Major — 366
Crow, Alpine — III. 65
Bald — 69
Bare-necked — 72
Carribean — 88
Carrion — 38
Chattering — 58
Cinereous — 103
Hermit — 7
Hooded — 51
Hottentot — 68
Jamaica — 58
Mexican — 91
Leffer — 93
New Guinea — 70
Papuan — 71
Philippine — 73
Red-legged — 1
Senegal — 57
White-breafted — ib.

Cuckoo

I N D E X.

Cuckoo — VI. 262
African — 341
Black of Cayenne - 360
Little — 361
Black and White crested 309
Blue — 337
Brasilian crested — 352
Brown — 355
Brown and Yellow - 326
Brown variegated with Black — 323
Carolina — 345
Cayenne — 358
Chinese — 336
Chinese, spotted — 325
Collared — 334
Collared, crested — ib.
Coromandel, crested - 327
Egyptian — 314
Gilded — 332
Gold, Green, and White ib.
Great spotted — 308
Great Madagascar - 311
Greenish of Madagascar ib.
Honey — 338
Horned — 354
Indian spotted — 319
Laughing — 353
Little — 328
Long-bellied Rain - 347
Long-shafted — 333
Madagascar — 313
Mindanao — 320
Panayan — 326
Paradise — 333
Piaye — 358
Pisan — 309
Pointer — 338
Rain — 344
Red-cheeked — 256
Rufous spotted — 323
Rufous white — 317
Sacred — 321
St. Domingo — 357
Society — 323
Spotted — 355
Straight-heeled — 317
Variegated of Mindanao 320
White-rumped Black - 361
Cuckow, Yellow-bellied VI. 328
Cuculus Ægyptius — 314
Afer — 341
Americanus — 345
Auratus — 332
Brasiliensis — 256
Cæruleus — 337
Canorus — 262
Cayanus — 358
Cornutus — 354
Coromandus — 334
Cristatus — 313
Dominicus — 357
Flavus — 328
Glandarius — 308
Guira — 352
Honoratus — 321
Indicator — 338
Maculatus — 325
Madagascariensis — 311
Melanoleucus — 327
Mindanensis — 320
Minor — 346
Nævius — 355
Niger — 330
Orientalis — 330
Paradiseus — 333
Persa — 257
Pisanus — 309
Pluvialis — 344
Punctatus — 323
Radiatus — 326
Ridibundus — 353
Scolopaceus — 319
Senegalensis — 317
Sinensis — 336
Tahitius — 323
Tenebrosus — 361
Tranquilus — 360
Vetula — 347
Cujelier — V. 23
Cuil — VI. 321
Curassow, Crested — II. 327
Crying — 346
Cushew — 335
Curlew — VIII. 18
Bald — 30
Brown — 28
Crested — 32

Curlew,

Curlew, great, of Cayenne VIII. 46
Green — 26
Luzonian — 29
Red — 33
Red-fronted Brown — 40
White — 39
Wood — 41
Curucui, Red-bellied - VI. 246
Violet-headed — 252
Violet-hooded — ib.
Yellow-bellied — 250
Curucuckoo — 256

Darter, White-bellied VIII. 406
Daw, Alpine — III. 65
Bald — 69
Mustachio — 68
New Guinea — 70
Papuan — 71
Demi-fins — V. 325
Didus, Ineptus — I. 390
Nazarenus — 400
Solitarius — 394
Diomedea, Demersa - IX. 341
Exulans — 289
Diver, Black-throated VIII. 243
Great — 234
Great Northern — 241
Imber — 234
Little — 237
Little Northern — 243
Sea-cat — 238
Speckled — 237
Dodo, Hooded — I. 390
Nazarene — 400
Solitary — 394
Dotterel — VIII. 84
Double Spur — II. 388
Drongo — IV. 478
Dronte — I. 390
Duck — IX. 100
Beautiful, crested — 206
Black — 205
Brown — 212
Buffel-headed — 209
Collared, of Newfoundland — 210
Common tame — 100
Falcated — 232
Duck, Gray-headed - IX. 213
Harlequin — 210
King — 213
Little Brown — 242
Little Thick-headed - 20
Long-tailed from Newfoundland — 169
Mareca — 215
Mexican — 241
Muscovy — 138
Musk — 139
Red-billed Whistling 154
Red Crested — 153
Summer — 206
Tufted — 194
Velvet — 204
White-faced — 214, 236
Dunlin — VII. 524

Eagle, Bald — I. 65
Cinereous — ib.
Golden — 46
Little American — 100
Oronoco — 97
Pondicherry — ib.
Ringtail — 54
Rough-footed — 58
Sea — 76
White-tailed — 65
Egret, Demi — VII. 363
Great — 361
Little — 357
Reddish — 362
Rufous — ib.
Eider — IX. 90
Emberiza Amazona - IV 291
Bicolor — 157
Borbonica — 292
Brasiliensis — 288
Butyracea — 156
Cærulea — 295
Capensis — 262, 263
Cia — 282
Cinerea — 294
Ciris — 150
Cirlus — 279
Citrinella — 274
Cyanopsis — 153
Familiaris — 294

I N D E X.

Emberiza Flaveola — IV. 290
Granatina — 144
Grisea — 293
Hortulana — 245
Lesbia — 258
Longicauda — 141
Lotharingica — 259
Ludovicia — 261
Mexicana — 289
Miliaria — 284
Nivalis — 264
Olivacea — 290
Oryzivora — 270, 272
Panayensis — 143
Paradisea — 134
Platensis — 291
Principalis — 142
Provincialis — 257
Psittacea — 144
Regia — 137
Schæniclus — 253
Serena — 138
Vidua — 140
Viridis — 155
Erne — I. 65
Falco Æruginosus — 172
Æsalon — 232
Albicaudus — 65, 68
Albicilla — ib.
Apivorus — 161
Buteo — 159
Candidus — 195
Chrysaetos — 46
Cyaneus — 164
Formosus — 101
Fulvus — 54, 153
Furcatus — 175
Gallicus — 86
Gyrfalco — 194
Haliaetus — 70
Laniarius — 196
Leucocephalus — 65, 69
Lithofalco — 231
Nævius — 58
Nisus — 179
Ossifragus — 76
Palumbarius — 184
Piscator — 222
Ponticerianus — 97
Falco Pygargus — I. 167
Sacer — 199
Serpentarius - VII. 316
Subbuteo — I. 223
Tinnunculus — 226
Falcon, Black — 216
Cirrated — 219
Common — 202
Fisher — 222
Jer — 192
Red — 217
Red-throated — 100
Spotted — 217
Stone — 231
Fauvette — V. 110
Alpine — 146
Babbler — 128
Black Headed — 119
Blueish of St. Domingo 156
Cayenne — 155
Gray or Grisette — 125
Little — 117
Little Rufous — 137
of the Woods, or Russet 131
Reed — 134
Rufous-tailed from Cayenne — 155
Spotted — 140
Spotted from Louisiana 152
Spotted from the Cape of Good Hope - 151
Small — 152
Winter — 142
Yellow-breasted Louisiana — 154
Favourite — VIII. 197
Fieldfare — III. 265
Canada — 271
Cayenne — 270
Fig-eater — V. 140
Ash-throated Cinereous 321
Belted — 300
Black — 313
Black-cheeked — 288
Black-collared — 294
Blue — 275
Brown — 287
Brown and Yellow — 290
Cærulean — 307

Fig-

Fig-eater, Cinereous-headed — V. 286
Collared Cinereous — 298
Crested — 312
Grasset — 320
Great, of Jamaica — 322
Green and White — 284
Green and Yellow — 270
Golden-crowned — 310
Golden-winged — 309
Half-collared — 316
Olive — 314
Olive Brown — 319
Orange — 311
Orange-throated — 285
Pine — 292
Prothonotary — 315
Red-breasted — 306
Red-headed — 280
Rufous — 304
Senegal — 276
Spotted — 279
Variegated — 303
White-throated — 281
Yellow-headed — 295
Yellow-spotted — 289
Yellow-throated — 282
Finch, Amaduvade IV. 85
Bahama — 157
Black and Yellow 123
Black-faced — III. 451
Blue-bellied — IV. 81
Blue-headed — V. 329
Bonana — IV. 120
Brasilian — 144
Canary — 1
Capsa — III. 443
Chinese — IV. 125
Collared — 129
Cowpen — 120
Cuba — 92
Dusky — 75
Eustachian — 126
Frizzled — 128
Greenish — 158
Lapland — 117
Long-billed — 124
Lutean — 177
Orange — 121
Finch, Purple — IV. 317
Senegal — 87
Strasburg — 63
Variegated — 127
Yellow — 156
Fingah — I. 249
Fist of Provence — V. 184
Flamingo, Red — VIII. 437
Flavert — III. 418
Fly Bird — VI. 1
Black Long-tailed — 38
Broad-Shafted — 34
Collared — 33
Crested — 22
Eared — 31
Forked-tail Violet — 36
Least — 10
Long-tail — 37
Long-tailed Steel-Coloured — 35
Purple — 24
Racket — 23
Spotted-necked — 30
Fly-catcher — IV. 414
Active — 439
Ash-coloured — 476
Azure — 429
Black-cap — 437
Black and White — ib.
Bourbon — 426
Brown — 431
Brown-throated Senegal 428
Cat — 455
Cinereous — 438
Collared — 420
Collared Black, or Lorraine — 418
Crested — 458
Dwarfish — 447
Fork-tail — 451
King of — 446
Lemon of Louisiana 434
Little Black Aurora 441
Martinico — 436
Mexican — 200
Mutable — 459
Paradise — 452
Purple-throated — 480
Red — 418, 477

Fly-

I N D E X.

Fly-catcher, Red-eyed IV. 434
Round-crested — 442
Rufous — 433, 443
Senegal — 424
Spotted — 416
Spotted Yellow V. 289
Streaked — IV. 440
Swallow-tailed — 457
Tyrant — 472
Undulated — 423
Whiskered — 430
Yellow-bellied — 444
Yellow-crowned — 473
Foudi — III. 450
Foudi Jala — V. 108
Founingo — II. 477
Francolin — 384
Frigat — VIII. 346
Pelican — ib.
Fringilla Amandava IV. 85
Argentoratensis — 63
Benghalus — 81
Bicolor — 157
Butyracea — 156
Cælebs — 96
Canaria — 1
Capua — III. 432
Carduelis — IV. 160
Catotol — 199
Crispa — 128
Cristata — III. 451
Cyanomelas — V. 329
Domestica — III. 432
Erythrophthalma IV. 122
Eustachii — 126
Granatina — 144
Jamaica — 120
Indica — 129
Lapponica — 117
Linaria — 183
Longirostris — 124
Lutensis — 177
Maia — 92
Melba — 178
Montana — III. 445
Monticula — 455
Montifringilla IV. 108
Montium — 65
Nivalis — 118
Fringilla Pecoris IV. 120
Petronia — III. 453
Purpurea — IV. 317
Senegala — 87
Sinica — 125
Spinus — 188
Tristis — 179
Variegata — 127
Zena — 121
Fulica Aterrima — VIII. 207
Atra — 200
Cayanensis — 173
Chloropus — 163
Cristata — 209
Fistulans — 172
Flavipes — 171
Flavirostris — 197
Fusca — 168
Maderaspatana — 195
Martinica — 196
Nævia — 170
Porphyrio — 186
Purpurea — 198
Spinosa — 185
Viridis — 194
Gachet — 311
Gadwall — IX. 157
Gallinule, Brown VIII. 168
Cayenne — 173
Common — 163
Crake — 137
Crowing — 198
Favourite — 197
Green — 194
Grinetta — 170
Madras — 195
Martinico — 196
Piping — 172
Purple — 186
Spotted — 147
Yellow-legged — 171
Ganga — II. 213
Gannet — VIII. 341
Lesser — 336
Garganey — IX. 218
Garnet — VI. 46
Garrot — IX. 186
Garzette, White — VII. 355
Gelinotte — II. 204

Giarole

INDEX.

Giarole — VII. 518
Gingeon — IX. 143
Gip-gip — VII. 211
Girole — V. 45
Glareola, Nævia — VII. 518
Austriaca — 519
Glout — VIII. 172
Goat-sucker, American VI. 458
Brasilian — 455
Carolina — 448
European — 436
Grand — 456
Gray — 462
Guiana — 463
Jamaica — 452
Rufous of Cayenne 464
Sharp-tailed — 461
Spectacle — 458
Variegated of Cayenne 459
Virginian — 450
White-necked 455, 459
Godwit, Red — VII. 483
Goertan — 23
Gold Cravat — VI. 25
Goldfinch — IV. 160
American — 179
Green, or Maracacao 178
Yellow — 179
Gold Green — VI. 16
Golden Eye — IX. 186
Gonolek — I. 255
Goose — IX. 25
Armed — 64
Bastard — 58
Black-backed — 66
Canada — 71
Chinese — 61
Cravat — 71
Egyptian — 67
Esquimaux — 69
Guinea — 61
Laughing — 70
Magellanic — 57
of the Malounes — 58
Spur-winged — 64
Tame — 25
White-fronted — 70
Gorget, Green — VI. 56
Goulin — III. 380
Gracula Calva — 380
Cristatella — 324
Religiosa — 376
Grakle, Bald — 380
Minor — 376
Paradise — 383
Grayfinch — IV. 72
Grebe, Black-breasted VIII. 223
Cayenne — 225
Chesnut — 228
Circled-bill — 230
Philippine — ib.
St. Domingo — 231
Coot — 232
Crested — 219
Great — ib.
Little — 220
Dusky — 218
Eared — 220
Great — 225
Horned — 221
Little — 222
Little — 218, 228
Louisiana — 224
Pied-bill — 230
Red-necked — 225
Tippet — 213
White-winged 231
Grenadin — IV. 144
Greenfinch — 147
Varied — 158
Green-shank — VII. 481
Grigri — 120
Grinetta — VIII. 170
Grisetta — V. 76
Grivelin — III. 416
Cravated — 430
Grosbeak — 401
Abyssinian — 427
Bengal — 422
Black — IV. 316
Black-crested — 319
Blue — 315
Brown — 119
Canada — III. 418
Cardinal — 414
Cinereous — IV. 72

Grosbeak,

Grosbeak, Gray — III. 424
Java — 420
Lineated — IV. 312
Marygold — 130
Minute — 314
Molucca — III. 425
Nun — 423
Orange — IV. 310
Philippine III. 421
Pine — 412
Pin-tailed — 419
Purple — IV. 318
Red-breasted III. 416
Spotted, of the Cape of Good Hope 434
Three-toed — 429
Waxbill — IV. 89
White-billed — 311
White-headed — 94
Grous, Black — II. 184
Broad-tailed — 199
with variable Plumage — 202
Canada Hazel — 245
Hazel — 204
Scotch — 211
Long-tailed — 251
Pin-tailed — 213
Red — 221
Wood — 169
Guarona — VIII. 42
Guifette — 308
Black, or Scare-crow 309
Guignard — 84
Guillemot — IX. 298
Black — 301
Foolish — 298
Little — 301
Guinea, Pintado — II. 144
Guira-Beraba — 348
Guira-Cantara — VI. 35
Guirarou — IV. 367
Guira-Querea — VI. 452
Guitso Batito — III. 429
Guit-guit, American V. 519
Black and Blue — 520
Black and Violet — 530
Spotted, Green — 527
Variegated — 529
Gull, Arctic — VIII. 404
Black-mantled — 365
Black-toed — 408
Common — 384
Glaucous — 366, 376
Gray-mantled — 366
Gray-mantled Brown 376
and White-mantled 379
Herring — ib.
Ivory — 380
Laughing — 389
Red-legged — 386
Variegated, or Grisard 372
Wagel — ib.
Habesh of Syria — IV. 49
Hæmatopus, Ostralegus VIII. 113
Hamburgh — IV. 320
Hard-bill — III. 412
Harfang — I. 314
Harle — VIII. 248
Hawk, Gos — I. 184
Little — 190
Pigeon — 191
Sparrow — 179
Thick-billed 190
Heath Cock, Ruffed II. 246
Hedge Sparrow — V. 142
Heron, Agami — VII. 366
Black — 353
Black-capped, White 365
Black-crested, White 365
Blue — 381, 387
Brown — 365
Cayenne Night — 422
Chesnut — 373, 374
Common — 329
Crested Purple — 354
Crowned — 306
Dry — 367
Gardenian — 410
Great — 369
American — ib.
Great White — 350
Green — 388
Hudson's Bay — 370
Little White — 384
Louisiane — 390
Malacca — 377

Heron, Mexican — VII. 385
New Guinea — 377
Night — 419
Pondicherry — 392
Philippine — 378
Purple — 354
Red-legged — 374
Red-ſhouldered — 370
Rufous — 373
Scolopaceus — 425
Squacco — 375
Squaiotta — 372
Violet — 355
Yellow-crowned — 383
Hirundo Acuta — VI. 585
Ambroſiaca — 510
Americana — 581
Apus — 534
Borbonica — 578
Capenſis — 506
Cayanenſis — 558
Chalybea — 561
Cinerea — 560
Dominicenſis — 555
Eſculenta — 568
Faſciata — 509
Francica — 580
Leucoptera — 567
Melba — 548
Montana — 532
Nigra — 554
Panayana — 505
Pelaſgia — 582
Peruviana — 557
Purpurea — 563
Riparia — 526
Rufa — 505
Senegalenſis — 508
Subis — 563
Tapera — 564
Torquata — 566
Violacea — 560
Urbica — 512
Hoamy — III. 280
Hoazin — II. 337
Hobby — I. 223
Hocco — II. 327
Hociſana — III. 91
Hocti — VII. 367
Hoopoe — VI. 379
Black and White of the Cape of Good Hope 397
Madagaſcar — ib.
Hornbill, Abyſſinian - VII. 148
African — 147
Black-billed — 134
Helmet — 153
Indian III. 34. VII. 140
Manilla — 137
Panayan — 138
Philippine — 150
Pied — 142
Rhinoceros — 155
Horſeman, Common — 489
Green — 497
Striped — 492
Variegated — 494
White — 496
Hotchicat — 119
Houbara — II. 50
Houhou — VII. 368
of Egypt — VI. 314
Houtou, or Momot — 372
Humming-bird, admirable - 61
All-green — 16
Amethyſtine — 15
Black-bellied — 50
Black-breaſted — 56
Black-capped — 38
Blue-fronted — 17
Blue tailed — 48
Broad-ſhafted — 34
Carbuncle — 28
Cayenne — 29
Creſted Green — 22
Crimſon-headed — 59
Fork-tailed — 37
Fork-tailed Cayenne - 35
Leſſer — 36
Gold-throated — 25
Gray-bellied — ib.
Gray-necked — 59
Green and Blue — 27
Green-throated — 53
Leaſt — 10
Little Brown — 24

Humming Bird, Mango VI. 58
Paradise — 51
Racket-tailed — 23
Red-breasted — 54
Red-throated — 12
Ruby-necked — 19
Ruby-throated — 31
Rufous-bellied — 61
St. Domingo — 60
Sapphire — 26
and Emerald ib.
Spotted — 47
Spotted-necked — 30
Supercilious — 46
Topaz — 44
Tufted-necked — 17
Violet — 55
Violet-eared — 31
Violet-tailed — 52
White-bellied — 33
White-tailed — 57

Jabiru — VII. 270
American — ib.
Jacamar — 213
Green — 214
Long-tailed — 216
Paradise — ib.
Jacana, Black — VIII. 181
Brasilian — 183
Chesnut — 177
Green — 182
Peca — 183
Variegated — 185
Jackdaw — III. 59
Jacobine, Crested — VI. 327
Jacobine and Domino - III. 425
Jaguacati — VII. 205
Japacani — III. 183
Jay — — 94
Blue, of North America 107
Brown Canada — 103
Cayenne — 105
Peruvian — 102
Red-billed of China - 101
Siberian — 105
Yellow-bellied of Cayenne — 106
Ibijau — VI. 455
Ibis, Bald — VIII. 30
Bay — 26
Black — 17
Brown — 40
Cayenne — 41
Crested — 32
Egyptian — 13
Manilla — 28
Mexican — 43
Scarlet — 33
White — 13, 39
White-necked — 46
Wood — 267
Jendaya — VI. 225
Imbrim — VIII. 241
Italian Courier — 428
Juba, Apute — VI. 231

Kamichi — VII. 323
Katraca — II. 317
Kestrel — I. 226
Kildir — VIII. 93
Kingalite — 175
King-fisher — VII. 158
Bengal — 197
Black and White — 179
Black-capped — 183
Blue and Black of Senegal — 189
Blue and Rufous — 175
Blue-headed — 193
Brasilian spotted — 207
Cape — 178
Cayenne — 201
Cinereous — 203
Collared — 186
Crab — 176
Crab-eating — ib.
Crested — 182, 199
Gray-headed — 190
Great — 182
Great Brown — 174
Gambia — 175
Greatest — 174
Green-headed — 184
Green and Rufous 209
White — 210
Orange — 212
Little Indian — 197

I N D E X.

Kingfisher, Long-shafted VII. 192
Pied — 179
Purple — 194
Red-headed — 191
Rufous — 194
Senegal — 190
with straw-coloured head and tail — 185
Ternate — 192
Thick-billed — 178
Three-toed — 198
White-billed — 195
White-headed — 185
Yellow-fronted — 191
Kink — III. 222
Kinki-Manou of Madagascar — IV. 476
Kiolo — VIII. 155
Knot — 134
Kite — I. 153
Black — 156
Carolina — 175
Koulik — VII. 122
Labbe, or Dung-bird VIII. 400
Long-tailed — 404
Lagopede — II. 232
Lanius, Barbarus — I. 255
Cærulescens — 249
Canadensis — 256
Cayanus — 252
Collurio — 246
Curvirostris — 253
Emeria — 250
Excubitor — 239
Forficatus — IV. 478
Leucocephalus — I. 254
Leucorynchos — 251
Madagascarensis — 256
Nengeta — IV. 367
Pitangua — 471
Rufus — I. 254
Rutilus — 244
Sulphuratus — 253
Tyrannus — IV. 464
Viridis — I. 251
Lanner — 196
Lapwing — VIII. 47
Armed, of Senegal — 60
Cayenne — 65
Indies — 62
Lapwing, Armed, of Louisiana — VIII. 63
Plover — 67
Swiss — 58
Lark, African — V. 63
Brown-cheked Pennsylvanian — 55
Cape — 51
Cinereous — 62
Crested — 65
Lesser — 72
Grashopper — 36
Italian — 45
Large or Calandre — 47
Louisiana — 34
Marsh — 57
Meadow — 41
Red — 55
Rufous — 61
Rufous-backed — 21
Senegal — 76
Siberian — 59
Shore — 53
Sky — 1
Tit — 28
Undated — 74
Willow — 40
Wood — 23
Larus, Canus — VIII. 384
Catarrhactes — 368, 372
Cinerarius — 386
Crepidatus — 400
Eburneus — 380
Fuscus — 379
Glaucus — 366, 376
Hybernus — 394
Marinus — 365
Parasiticus — 404
Ridibundus — 389
Riga — 381
Linnet — IV. 51
Mountain — 65
Yellow-headed — 73
Bastard or Bimbele V. 133
Little Simon — 273
Locustelle — 40
Lohong — II. 42
Long Shank — VIII. 109
Loriot — III. 223
Chinese — 231

Loriot, Indian — III. 232
Lory, Collared — VI. 114
Crimson — 116
First Black-capped 115
Grand — 119
Great — ib.
Gueby — 118
Indian — 121
Long-tailed Scarlet 120
Molucca — 117
Paraguan — 213
Parakeet — 120
Red — ib.
and Violet — 121
Tricolor — 122
Purple-capped — 114
Red — 117
and Violet — 118
Tricolor — 115
Loxia, Abyssinia — III. 427
Astrild — IV. 89
Aurantia — 310
Bengalensis — III. 422
Bonariensis — IV. 130
Brasiliana — III. 416
Cærulea — IV. 315
Cana — 72
Canadensis — III. 418
Capensis — IV. 324
Cardinalis — III. 414
Chloris — IV. 147
Cocothraustes — III. 401
Collaria — 423
Coronata — IV. 319
Curvirostra — III. 405
Enucleator — 412
Flabellifera — 419
Fusca — IV. 119
Grisea III. 424.—IV. 73
Lineola — 312
Ludoviciana — III. 416
Maia — IV. 94
Malacca — III. 425
Minuta — IV. 314
Nigra — 316
Oryzivora — III. 420
Philippina — 421, 426
Pyrrhula — IV. 298
Torrida — 311
Loxia, Tridactyla — III. 429
Violacea — IV. 318
Luan — II. 314
Lulu — V. 72
Lumme — VIII. 243

Maccaw, Black — VI. 175
Blue and Yellow — 168
Brasilian Green — 169
Red and Yellow, from Jamaica } 158
Magpie — III. 75
Jamaica — 85
of the Antilles — 88
Senegal — 85
Magnari — VII. 265
Maia — IV. 92
Maian — 94
Maipouri — VI. 215
Mainate — III. 376
Manacus, Serena — IV. 337
Manakin — IV. 327
Black and Yellow — 334
Black-capped — 331
Black-crowned — 344
Blue-backed — 330
Gold-headed — 335
Great or Tighe — 330
Orange — 334
Papuan — 344
Purple-breasted — 353
Red and Black — 332
Red-headed — 335
Rock — 346
Variegated — 337
White-faced IV. 343—V. 339
fronted — IV. 337
headed — 335
throated — 337
Manchot, Great — IX. 338
Hopping — 346
Middle — 341
with a truncated Bill 349
Mansfeni — I. 102
Manucode — III. 144
Black — 149
Marail — II. 342
Marouette — VIII. 147

Martin — VI. 512
Sand — 526
Mascarine — VI. 105
Matuitui VII. 207 — VIII. 45
Maubeche, Common VII. 505
Gray — 507
Spotted — ib.
Meleagris, Gallopavo II. 115
Merganser — VIII. 248
Crested — 152
Little — 254
Crowned — 258
Hooded — ib.
Mantled — 255
Red-breasted — 252
Stellated — 256
Mergus, Albellus 251
Cucullatus — 258
Merganser — 248
Minutus — 256
Serrator — 252, 255
Merlin — I. 232
Merops, Angolensis VI. 428
Apiaster — 411
Badius — 420
Brasiliensis — 409
Cafer — 420
Cayanensis — 433
Chrysocephalus — 427
Cinereus — 419
Congener — 434
Erythrocephalus — 432
Erythropterus — 431
Flavicans — 418
Fusca — 402
Malaccensis — 408
Nubicus — 430
Philippinus — 429
Red and Blue — 409
Rufus — 407
Superciliosus — 422
Viridis — 424
Mew, Great Cinereous, or Blue-footed VIII. 384
Laughing — 389
Spotted — 381
White — 380
Winter — 394
Micteria, Americana VII. 270, 275
Middle Bill — V. 325
Middle-Bill Black and Blue — V. 329
Black and Rufous — 331
with a White Crest and Throat 339
Millouinan — IX. 185
Minister — IV. 76
Mittek — VIII. 174
Mocking Bird — III. 288
Moloxita — 366
Montvoyau of China VI. 463
Morillon — IX. 191
Little — 194
Motacilla, Acredula V. 290
Æquinoctialis — 314
Afra — 265
Africana — 151
Alba — 242
Albicollis — 281
Atricapilla — 119
Aurantia — 239
Auricollis — 285
Boarula — 259
Bonariensis — 331
Cærulea — 307
Cærulescens — 156
Calendula — 374
Calidris — 322
Campestris — 341
Cana — 321
Canadensis — 300, 301
Capensis — 264
Caprata — 217
Cayana — 345
Chloroleuca — 284
Chrysocephala — 311
Chrysoptera — 309
Cinerea — 252
Cristata — 312
Curruca — 128
Cyanocephala — 343
Dominica — 297
Erithacus — 171
Fervida — 216
Ficedula — 177
Flava — 256, 266
Flavescens — 277
Fulicata — 218
Fulva — 318
Fusca — 319

Motacilla Fuſcata — V. 277
Fuſcenſis — 287
Fuſcicollis — 155
Guianenſis — 176
Guira — 348
Hippolais — 110
Hottentotta — 238
Icterocephala — 295
Leucorhoa — 240
Lineata — 347
Ludoviciana — 282
Luſcinia — 78
Maculoſa — 286
Madagaſcarienſis 108, 271
Maderaſpatenſis — 267
Magna — 222
Mauritania — 275
Modularis — 142
Multicolor — 313
Nævia — 140
Noveboracenſis — 152
Oenanthe — 228
Palmarum — 333
Paſſerina — 117, 125
Pennſylvanica — 306
Penſilis — 158
Perſpicilla — 225
Petechia — 280
Philippenſis — 219
Phænicurus — 163
Pinguis — 320
Protonotarius — 315
Provincialis — 149
Regulus — 366
Rubecula — 185
Rubetra — 212
Rubicola — 203
Rufa — 137
Ruficapilla — 304
Ruficauda — 155
Salicaria — 134
Schænobænus — 131
Semitorquata — 316
Sialis — 200
Stapazina — 236
Subflava — 277
Suecica — 195
Sybilla — 220
Tigrina — 289
Tiphia — 270
Motacilla Trochilus V. 350
Troglodytes — 357
Varia — 303
Velia — 346
Vermivora — 324, 327
Virens — 294
Umbria — 153
Undata — 276
Motmot, Braſilian — VI. 372
Motteux — V. 228
Moucherolle, — IV. 450
Brown of Martinico 456
Creſted — 452
Forked-tail of Mexico 459
of the Philippines — 456
Virginian — 455
Green-creſted 458
Muſcicapa, Agilis — 439
Audax — 473
Barbata — 430
Bicolor — 437
Borbonica — 426
Cærulea — 429
Cana — 476
Carolinenſis — 455
Cayenenſis — 444
Coronata — 442
Crinita — 458
Ferox — 472
Forficata — 457
Fuliginoſa — 431
Fuſca — 437
Griſola — 416
Ludoviciana — 475
Martinica — 436, 456
Melanoptera — 428
Mutata — 459
Olivacea — 434
Paradiſi — 452
Philippenſis — 458
Rubricollis — 480
Rufeſcens — 433
Ruticilla — 441
Senegalenſis — 424
Tyrannus — 451
Variegata — 440
Virens — 438
Undulata — 423

INDEX.

Nandapca — VI. 275
Napaul — II. 315
Nightingale — V. 78
  Great — 105
Noddy — VIII. 418
Noira Lori — VI. 111
Numida, Meleagris — II. 144
Nutcracker — III. 109
Nuthatch — V. 458
  Great — 471
  Great hook-billed ib.
  Spotted — 472

Ococolin — II. 433
Old Man, or Rain Bird VI. 344
Onore — VII. 415
  Rayed — 416
  of the Woods — 417
Open Bill — 392
Orchef — III. 422
Orfraie — I. 76
Organiste — 235
Oriole, Black — III. 193
  Lesser — 194
  Crowned — 195
  Black and Yellow 207
  Crested — 212
  Golden — 223
  Icteric — 178
  Kink — 222
  Mexican — 181
  New Spain — 185
  Olive — 198, 220
  Red-winged — 188
  Ring-tailed — 182
  Striped-headed — 233
  Weaver — 199
  Whistler — 202
  White-headed — 213
  Yellow-headed — 219
Oriolus, Annulatus — 182
  Baltimore — 203
  Bonana — 214
  Capensis — 220
  Chinensis — 230
  Cinereus — 187
  Costototl — 185
  Cristatus — 211, 212
  Galbula — 223
Oriolus Japacani — III. 183
  Ictericus — 178
  Icterocephalus — 219
  Ludovicianus — 213
  Melanocephalus 231, 232
  Melancholicus — 196
  Mexicanus — 195
  Minor — 194
  Niger — 193
  Novæ Hispaniæ — 181
  Olivaceus — 198
  Persicus — 207, 209
  Phœniceus — 188
  Radiatus — 233
  Sinensis — 222
  Spurius — 205
  Textor — 199
  Viridis — 202
  Xanthornus — 217
Ortolan, Bunting — IV. 245
  Pivote — V. 184
Osprey — I. 70
  Carolina — 101
Ostrich — 323
  American — 366
Otis, Afra — II. 44
  Arabs — 42
  Bengalensis — 47
  Houbara — 50
  Rhaad — 52
  Tarda — 1
  Tetrax — 34
Ouzel, Blue — III. 312
  Ring — 299
  Rose-coloured — 306
  Solitary — 315
  Water — VIII. 126
Owl, Aluco — I. 291
  Brasilian Eared — 310
  Brown — 302
  Canada — 313, 317
  Cayenne — 316
  Great-eared — 270
  Little — 306
  Long-eared — 279
  Saint Domingo — 318
  Scops-eared — 288
  Snowy — 314
  Tawny — 294

Owl.

I N D E X.

Owl, White — I. 297
Ox-pecker — III. 154
Oyster-catcher — VIII. 113

Padda, or Rice Bird III. 420
Palamedea, Cornuta VII. 323
Cristata — 313
Palikour, or Anter — IV. 379
Palmiste — III. 361
Paradisea, Apoda — 135
Aurea — 150
Magnifica — 146
Regia — 144
Superba — 149
Viridis — 152
Paradise Bird, Blue Green ib.
Gold-breasted — 150
Greater — 135
King — 144
Magnificent — 146
Superb — 149
Paragua — VI. 213
Paroare — III. 455
Parra, Brasiliensis VIII. 183
Cayanensis — 65
Goensis — 62
Jacana — 177
Ludoviciana — 63
Nigra — 181
Senegalla — 60
Variabilis — 185
Viridis — 182
Parraka — II. 347
Parakeet, Alexandrine VI. 123
Angola, Yellow — 128
Azure-headed — 129
Black-banded, Great 139
Black-winged 151, 153
Blossom-headed — 126
Blue-faced — 131
Blue-headed — 129, 143
Blue-winged — 152
Collared — ib.
Crested — 141
Double-collared — 125
Double-ringed — ib.
Gray-breasted — 129
Gray-headed — 150
Great-collared — 123
Parakeet, Golden-winged VI. 149
Lace-winged — 132
Little — 135
Long-shafted Great 136
Lory — 127
Luzonian — 153
Malacca — 136
Mouse — 129
Mustacho — 130
Otaheitan Blue — 154
Philippine — 148
Red — 141
Red and Green — 140
Reddish-winged Great 137
Red headed — 126, 145
Guinea — ib.
Red-throated — 138
Red-winged Little — 138
Rose-ringed — 134
Rose-headed Ring — 135
Sapphire-crowned 143
Short-tailed — 142
Variegated-winged 151
Yellow — 128
Parroquet, Ara — 237
Brown-throated — 221
Cayenne — 240
Chesnut-crowned — 224
Emerald — 226
Gold-head — 243
Golden-crowned — 232
Green — 241
Least Blue and Green 242
Long-tailed Green 228
Pavouanne — 219
Red-fronted — 230
Red and Blue-headed ib.
Variegated Throat 222
Wings — 223
Yellow Guarouba — 233
Brasilian —ib.
headed — 225, 235
throated — 239
winged — 223
Parrot — 63
Agile — 196
Amboyna — 108
Red — 122

I N D E X.

Parrot, Angola Yellow — VI. 128
Ash-coloured — 88
Aurora — 186
Black or Vaza — 104
of Madagascar ib.
Bloody-billed — 107
Blue-faced — 198
Blue-fronted — 195
Blue-headed 126, 208
Brasilian Yellow — 233
Carolina — 235
Cinereous or Jaco — 88
Common Amazon's 187
Dusky — 211
Emerald — 226
Festive — 205
Gingi — 137
Gray-headed — 108
Great-bellied — 107
Great Blue headed Green 108
Green — 102
Green and Red, Chinese ib.
Hawk-headed — 103
Hooded — 217
Illinois — 231
Little Dusky — 209
Maccaw — 237
Mealy Green — 193
Noble — 182
Orange-headed — 212
Paradise — 203
Pavouane — 219
Red and Blue — 194
Red and White — 85
Red-banded — 207
Red breasted — 131
Red-headed Amazon 184
Red-throated — 210
Ruff-necked — 200
Senegal — 108
Variegated — 103
White-breasted — 215
White-fronted — 185
Yellow Amboyna — 128
Yellow-headed — 225
Yellow-winged — 190
Partridge — II. 349
African, Red — 389
Barbary, Red — 391
Partridge, Bare-necked II. 389
Damascus — 366
European, Red — 378
Gray — 352
Gray-white — 364
Greek — 369
Guernsey — 378
Guiana — IV. 412
Mountain — II. 368
New England — 394
Pearled, Chinese — 393
Red — 369
Rock or Gambra — 392
Senegal — 388
Parus, Amatorius — V, 454
Americanus — 298
Ater — 401
Atricapillus — 407
Biarmicus — 416
Bicolor — 447
Cæruleus — 412
Capensis — 439
Caudatus — 432
Ceta — 456
Cristatus — 443
Cyanus — 452
Major — 394
Narbonensis — 429
Palustris — 404
Pendulinus — 420
Sibericus — 441
Virginianus — 450
Passerinette — 117
Pavo, Bicalcaratus — II. 323
Cristatus — 253
Muticus — 320
Tibetanus — 319
Pauxi, or Stone — 335
Peacock — 253
Japan — 320
Thibet — 319
Pearl Green — VI. 60
Pelecanus, Aquilus VIII. 346
Bassanus — 337, 341
Carbo — 282
Fiber — 339
Fuscus — 278
Graculus — 290
Maculatus — 340

Pelecanus,

Pelecanus, Onocrotalus VIII. 259
Piscator — 336
Sula — 333
Thagus — 281
Pelican, Brown — 278
Indented-billed — 281
Saw-billed — ib.
White — 259
Penelope, Cristata — II. 340
Marail — 342
Satyra — 315
Penguin — IX. 330
Cape — 341
Crested — 346
Great — 333
Little — 335
Patagonian — 338
Red-footed — 349
Petrel, Antarctic — 264
Blue — 268
Blue-billed — ib.
Brown Puffin — 278
Cinereous — 256
Fulmar — 256
Giant — 271
Great Black — 278
Greatest — 271
Pintado — 258
Puffin — 273
Shear-water — ib.
Stormy — 279
White and Black — 258
White or Snowy — 266
Phœnicopterus Ruber VIII. 431
Pettychaps — V. 110
Phaeton, Æthereus VIII. 321, 322
Demersus — IX. 349
Phœnicurus — VIII. 323
Phalarope, Cinereous or Brown — VIII. 210
Gray — 212
Red — 211
with indented Festoons 212
Phasianus, Argus — II. 314
Colchicus — 286
Hybridus — 306
Cristatus — 337
Gallus — 54
Motmot — 317
Phasianus, Nycthemerus II. 311
Parraqua — 347
Pictus — 308
Pheasant, Argus — 314
Bastard — 306
Black and White China 311
Common — 286
Courier — 347
Crested — 337
Horned — 315
Iris — 323
Motmot — 317
Piauhau — IV. 480
Picus, Aurantius — VII. 25
Auratus — 39
Bengalensis — 21
Bicolor — 72
Canadensis — 67
Cardinalis — 65
Carolinus — 69, 71
Cayanensis — 30
Cinnamomeus — 34
Chlorocephalus — 37
Erythrocephalus — 55
Exalbidus — 32
Flavipes — 54
Goensis — 20
Goertan — 23
Lineatus — 51
Major — 57
Manillensis — 19
Martius — 41
Melanochloros — 29
Minor — 61
Moluccensis — 66
Multicolor — 35
Nubicus — 64
Passerinus — 28
Philipparum — 18
Pileatus — 48
Principalis — 46
Pubescens — 73
Rubricollis — 53
Rufus — 36
Senegalensis — 24
Striatus — 26
Tricolor — 68
Tridactylus — 75
Varius — 74

I N D E X.

Picus, Villosus — VII. 72
  Viridis — 6
Pigeon — II. 435
  Great Crowned — 480
  Ring — 469
  Scollop-necked — 478
  Wild — 439
Pimatol — III. 172
Pimtail — IX. 166
Pipra — IV. 327
  Albifrons IV. 343 — V. 339
  Atricapilla — IV. 341
  Aureola — 332,334
  Erythrocephala — 335
  Gutturalis — 337
  Manacus — 331
  Papuensis — 344
  Pareola — 330
  Rubetra — 339
  Rupicola — 346
Pitpit, blue — V. 344
  Blue-capped — 347
  Green — 343
  Varegated — 346
Plastron, Black — VI. 59
  White — 58
Platelea Leucorodia VII. 431
Plotus, Anhinga - VIII. 406
Plover — 71
  Alwargrim — 82
  Armed of Cayenne - 100
  Black-headed — 101
  Crean-coloured — 121
  Crested — 95
  Crowned — 98
  Golden — 78
  Great — 102
  Hooded — 97
  Long-legged — 109
  Noisy — 93
  Ring — 88
  Spur-wing — 96
  Wattled — 99
  Wreathed — 98
Plume, White — IV. 343
Pluvian — VIII. 101
Pochard — IX. 181
Polochian — VI. 408
Popinjay, Aurora-headed 212
Popinjay, Brown — VI. 211
  Mailed — 204
  Paradise — 203
  Purple-bellied — 207
  Red-banded — ib.
  Violet — 209
  with a Blue Head and Throat — 208
Porzana — VIII. 169
Pratincole, Austrian - VII. 517
  Senegal — 518
  Spotted — ib.
Procellaria, Æquinoctialis IX. 278
  Antarctica — 264
  Cærulea — 268
  Capensis — 258
  Gigantea — 271
  Glacialis — 256
  Nivea — 266
  Pelagica — 279
  Puffinus — 273
  Vittata — 268
Promerops, Blue-winged VI. 400
  Brown, with spotted Belly — 401
  Cape — ib.
  Crested — 399
  Grand — 403
  Great — ib.
  Mexican — 401
  New Guinea, Brown - 402
  Orange — 405
  Striped-bellied Brown 402
Promerup — 399
Psittacus Accipitrinus - 103
  Æruginosus — 221
  Æstivus — 187
  Agilis — 196
  Alexandri 123,125,129,134
  Amboinensis — 122
  Ana — 224
  Aracanga — 158
  Ararauna — 168
  Ater — 175
  Aterrimus — 87
  Atricapillus — 139
  Aureus — 232
  Aurora — 186
  Autumnalis — 198

INDEX.

Psittacus Borneus — VI. 120
Cæruleocephalus — 194
Canus — 150
Capensis — 152
Carnicularis — 230
Carolinensis — 235
Chrysopterus — 149
Collarius — 210
Cristatus — 82
Cyanocephalus — 126
Domicella — 114
Dominicensis — 207
Erithacus — 88
Erythrocephalus 126, 135, 136
Eupatria — 137
Festivus — 205
Galgulus — 143, 148
Garrulus — 111
Gramineus — 108
Grandis — 119
Guarouba — 233
Guebiensis — 118
Guianensis — 219
Hæmatodus — 131
Havanensis — 195
Jandaya — 225
Javanicus — 141
Incarnatus — 138
Indicus — 121
Leucocephalus - 185, 207
Lory — 115
Ludovicianus — 212
Macao — 158
Macrorhyncos — 107
Makawuanna — 237
Mascarinus — 105
Melanocephalus — 215
Melanopterus — 151
Menstruus — 208
Minor — 153
Murinus — 129
Niger — 104
Nobilis — 182
Ochrapterus — 190
Olivaceus — 132
Ornatus — 127
Paradisi — 203
Psittacus Paraguanus - VI. 213
Passerinus — 242
Pertinax — 231
Pileatus — 217
Pullarius — 145
Pulverulentus — 193
Pondicerianus — 130
Puniceus — 116
Purpureus — 209
Ruber — 117
Rufirostris — 228
Senegalus — 108
Severus — 169
Sinensis — 102
Smaragdinus — 226
Solstitialis — 128
Sordidus — 211
Sosove — 240
Sulphureus — 83
Taitianus — 154
Taraba — 184
Tirica — 241
Torquatus — VI. 152, VII. 118
Tovi — VI. 239
Tui — 243
Violaceus — 200
Virescens — 223
Viridis — 250
Psophia, Crepitans — IV. 390
Ptarmigan — II. 232
Hudson's Bay — 242
Puffin — IX. 304
of Kamtschatka — 312
Purre, or Stint — VII. 521
Pygargue — I. 65

Quadricolor — III. 424
Quail — II. 396
Chinese — 422
Great, Polish — 419
Madagascar — 423
Malouine — 421
Noisy — 423
White — 420
Quapactol, or the Laugher VI. 353

Rail,

I N D E X.

Rail, Banded — VIII. 153
Barbary — 176
Brown — 151
Cayenne — 155
Little — 159
Spotted — 156
Jamaica — 158
Land — 137
Little — 159
Long-billed — 154
Virginian — 157
Water — 144
Rallus, Aquaticus — 144
Bengalensis — VII. 497
Carolinus — VIII. 157
Cayanensis — 155
Crex — 137
Fuscus — 151
Jamaicensis — 158
Longirostris — 154
Minutus — 159
Philippensis — 150
Porzana — 147
Striatus — 152
Torquatus — 153
Variegatus — 156
Ramphastos, Aricari — VII. 120
Cæruleus — 124
Dicolorus — 113
Luteus — 124
Momota — VI. 372
Pavoninus — VII. 119
Picatus — 116
Piperivorus — 122
Piscivorus — 115
Toco — 112
Tucanus — 113
Viridis — 120
Raven — III. 11
Indian, of Bontius — 34
Recurvirostra, Alba VII. 486
Avosetta — VIII. 422
Red Black — III. 417
Red-breast — V. 185
Blue, of North America — 200
Redpoll, Lesser — IV. 183
Red-shank, or Pool Snipe VII. 490
Red-shank, White — VII. 496
Redstart — V. 163
Red Tail — 171
Guiana — 176
Red-wing — III. 273
Rhaad — II. 52
Rhynchops, Nigra - VIII. 412
Riband, Blue — IV. 353
Ringtail — I. 167
Rochier — 231
Roller — III. 115
Abyssinian — 126
Angola — 127
Cayenne — 118
Chinese — 117
Garrulous — 118
Madagascar — 131
Mexican — 132
Paradise — 133
of the Indies — 130
Oriental — ib.
Pied — VII. 128
Rook — III. 46
Rose-throat — 416
Royal Bird — VII. 306
Ruff and Reeve — 498
Runner — VIII. 428

Sacre — I. 199
Sanderling, or Curwillet VII. 508
Sandpiper, Cayenne VIII. 65
Common — VII. 514
Dusky — 505
Freckled — 507
Goa — VIII. 62
Gray — 67
Green — VII. 509
Grisled — 507
Louisiane — VIII. 63
Senegal — 60
Shore — VII. 494
Spotted — VIII. 132
Striated — VII. 492
Swiss — VIII. 58
San-hia — VI. 336
Sarcelle, Brown and White IX. 243
Carolina — 242
Chinese — 233

Sarcelle,

I N D E X.

Sarcelle, Common — IX. 218
Coromandel — 231
Egyptian — 229
Feroe — 235
Java — 232
Little — 222
Long-tailed Rufous — 239
Madagascar — 230
Mexican — 241
Soucrourette — 237
Soucrourou — 236
Spinous-tailed — 238
Summer — 225
White and Black — 240
Sassebe — VI. 210
Savacou — VII. 426
Savana — IV. 451
Schet of Madagascar — 459
Schet-be — I. 254
Schomburger — III. 196
Scopus Umbretta — VII. 423
Scoter — IX. 196
Broad-billed — 205
Double — 204
Screamer, Crested — VII. 313
Horned — 323
Scolopax, Ægocephala — 483
Alba — VIII. 39
Arquata — 18
Calidris — VII. 490
Candida — 496
Capensis - 472, 474, 475
Fedoa — 484
Fusca — { VII. 485 / VIII. 40 }
Gallinago — VII. 463
Gallinula — 470
Glottis — 481
Guarauna — VIII. 42
Lapponica — VII. 482
Limosa — 479
Luzoniensis — VIII. 29
Paludosa — VII. 460
Phœopus — VIII. 24
Rusticola — VII. 442
Totanus — 480
Scops — I. 288
Sea Lark — VII. 521
Sea Partridge, Brown — 518
Sea Partridge, Collared VII. 519
Gray — 517
Sea Swallow — VIII. 297
Great — 302
Great Alar Extent - 313
Great of Cayenne - 315
Lesser — 307
of the Philippines - 312
Secretary — VII. 316
Vulture — ib.
Senegal — IV. 87
Radiated — 89
Serevan — 91
Shaft, Blue — VI. 48
White — 46
Shag — VIII. 290
Shear-bill — 412
Sheldrake — IX. 171
Short Tail — III. 373
Shoveler — IX. 160
Shrike — I. 237
Barbary — 255
Bengal — 250
Brasilian — IV. 471
Cayenne — I. 252
Crested — 256
Gray — IV. 367
Fork-tailed — { I. 249 / IV. 478 }
Great Cinereous - I. 239
Hook-billed — 253
Madagascar — 256
Red-backed — 246
Rufous — 254
Tyrant — IV. 465
White-headed — I. 254
Yellow-bellied — 253
Sifilet — III. 150
Sincialo — VI. 228
Siskin — IV. 188
Black Mexican — 199
Mexican — 200
Sitta, Europæa - V. 458, 468
Jamaicensis — 466, 468
Major — 471
Nævia — 472
Sitelle — 458
Sirli — 63

Skimmer,

Skimmer, Black — VIII. 412
Smew — 254
Smiring — 171
Snipe — VII. 463
Cape — 473
China — 475
Dusky — 485
Jack — 470
Jadreka — 479
Madagascar — 474
Madras — 475
Spotted — 481
Snow Finch — IV. 118
Soco — VII. 364
Sosove — VI. 240
Soui Manga — V. 487
Bourbon — 507
Collared — 494
Iris — 504
Long-tailed — 508
Violet-hooded - 509
Glossy Gold Green 511
Great Green - 512
Purple — 493
Purple-breasted Olive 498
Red-breasted Green - 505
Red-breasted Purple-chesnut — 489
Violet — 492
Sparrow, Beautiful marked III. 452
Black — 441
Blue — 450
Crested Tree — 451
Date — 443
Green — 449
House — 432
Little Senegal — IV. 92
Ring — III. 453
Little — 455
Tree — 445
Spicifere — II. 320
Spipolette — V. 40
Spoonbill, White — VII. 431
Stare, Brown-headed - III. 170
Cape, or Pied — 167
Common — 155
Louisiana — 169
Stare, Magellanic - III. 173
Mexican — 171
Sterna, Cayanensis VIII. 315
Fissipes — 309
Fuliginosa — 313
Hirundo — 302
Minuta — 307
Nævia — 308
Nigra — 311
Panayensis — 312
Stolida — 418
Stone Chat — V. 203
Great — 222
Luzonian — 217
Madagascar, or Fitert 220
of the Cape of Good Hope — 223
of the Philippines - 218
Great — 219
Senegal — 216
Stork — VII. 243
American — 265
Black — 261
White — 243
Sturnus, Capensis — III. 167
Cinclus — VIII. 126
Ludovicianus — III. 168
Mexicanus — 171
Milibaris — 173
Vulgaris — 155
Strix, Aluco — I. 291
Bubo — 270
Flammea — 297
Funerea — 313, 317
Otus — 279
Passerina — 306
Scops — 288
Stridula — 294
Ulula — 302
Struthio Camelus — 323
Cassuarius — 376
Cuculatus — 399
Rhea — 366
Sugar Bird — V. 532
Sultana Hen, or Porphyrion VIII. 186
Brown — 195
Green — 194

INDEX.

Sultana Hen, Little - VIII. 196
Swan — IX. 1
Swallow — VI. 466
Ambergris — 510
Ash-bellied — 560
Black — 554
Blue of Louisiana — 560
Brasilian — 564
Brown-collared — 566
Chimney, or Domestic 493
Crag — 532
Esculent — 568
Gray-rumped — 580
Great Rufous-bellied of Senegal — 588
Peruvian — 557
Rufous-rumped — 581
St. Domingo — 555
Senegal — 508
Sharp-tailed Black of Martinico — 585
Brown of Louisiana — 582
Wheat — 578
White-bellied — 509
Cayenne — 567
cinctured — 509
winged — 567
Swift — 534
White-bellied - 548, 555
collared — 558
Swift Runner — VIII. 121

Tacco — VI. 347
Tait-sou — 337
Tamatia — VII. 88
Beautiful — 92
Black and White — 93
Collared — 91
with the Head and Throat Red — 90
Tanagra, Atra — IV. 211
Atricapilla — 212
Bonariensis — 204
Brasiliensis — 214
Cayana — 225, 240
Cristata — 203
Cyanea — 76
Dominica — 218
Episcopus — 219
Tanagra, Grisea — IV. 228
Guinensis — 224
Gularis — 220
Gyrola — 233
Jacapa — 215
Jacarina — 237
Magna — 203
Mexicana — 230
Mississippensis — 210
Pileata — 232
Rubra — 205, 209
Striata — 213
Tatao — 228
Tricolor — 227
Violacea — 238
Virens — 221
Tanagre, Bishop — 219
Black and Blue — 223
Black and Rufous — 214
faced — 211
headed — 212
throated — 231
Blue — 230
Canada — 209
Crested — 203
Furrow-clawed — 213
Golden — 238
Grand — 203
Gray — 228
Gray-headed — 224
Green — 221
Green-headed — 227
Hooded — 232
Jacarini — 237
Mississippi — 210
Negro — 240
Olive — 222
Paradise — 228
Red — 209
Red-breasted — 215
Red-headed — 220, 233
Rufous-headed — 225
Saint Domingo — 218
Scarlet — 205
Small — 233
Syacu — 234
Turquoise — 214
Violet — 204

Tanombe

I N D E X.

Tanombe — III. 345
Tantalus, Albicollis VIII. 46
Calvus — 30
Cayanenſis — 41
Criſtatus — 32
Falcinellus — 26
Griſeus — 45
Ibis — 13
Loculator — VII. 267
Manillenſis — VIII. 28
Mexicanus — 43
Niger — 17
Ruber — 33
Taparara — VII. 201
Tarier — V. 212
Tavoua — VI. 205
Tcha-chert-bé — I. 254
Teal, African — IX. 229
Blue-winged — 237
Chineſe — 233
Common — 222
Coromandel — 231
Madagaſcar — 230
St. Domingo — 239
Spinous tailed — 238
Summer — 225
Tern, Black — VIII. 309
Cayenne — 315
Common — 302
Kamtſchatkan — 308
Leſſer — 307
Panayan — 312
Sooty — 313
Tetéma — IV. 380
Tetrao Albus — II. 242
Alchata — 213
Bicalcaratus — 388
Bonaſia — 204
Canadenſis — 245
Chinenſis — 422
Chrokiel — 419
Cinereus — IV. 408
Coturnix — II. 396
Coyolcos — 430
Criſtatus — 428
Damaſcenus — 366
Falklandicus — 421
Francolinus — 384
Guianenſis — IV. 412
Tetrao Lagopus II. 221,232
Major — IV. 406
Marilandus — II. 394
Mexicanus — 431
Montanus — 368
Novæ Hiſpaniæ 429
Nudicollis — 389
Perdix — 352
Perlatus — 393
Petroſus — 392
Phaſianellus — 251
Rufus — 369
Soui — IV. 410
Striatus — II. 423
Suſcitator — ib.
Tetrix — 184
Togatus — 246
Variegatus — IV. 408
Urogallus — II. 169
Throſtle — III. 246
Thruſh — 234
Abyſſinian — 368
African — 323
Alarum — IV. 376
Amboina — III. 354
Ant — IV. 379
Aſh-coloured III. 343
Aſh-rumped — 357
Barbary — 277
Barred-tail — IV. 388
Black-breaſted III. 352
Black-cheeked — 359
Black-chinned — 370
Black-creſted IV. 381
Black-headed — 388
Blue-tailed — 376
Black-throated III. 341
Black-winged IV. 384
Blue — III. 312
Bourbon — 355
Brunet — 349
Cape — 353
Cayenne — 270
Ceylon — 332
Chiming — IV. 383
Chineſe — III. 280
Creſted, of China — 324
Crying — 337
Dominican — 356

Thruſh

INDEX.

Thrush Ethiopian — III. 367
Ferruginous — 286
Gilded — 331
Glossy — 327
Guiana — 254
Hispaniola — 364
Indian — 343
King — IV. 374
Little — III. 255,281
Madagascar — 345
Mauritius — 347
Mimic — 288
Mindanao — 346
Missel — 260
Musician — IV. 385
Nun — III. 366
Olive — 340
Orange-bellied — 335
Palm — 361
Philippine — 280
Pigeon — 339
Reed — 257
Red-breasted — 271
Red-legged — 278
Rufous — 363
Rufous-naped IV. 380
Rufous-winged III. 325
Senegal — 344
Small — 280
Solitary — 315
Speckled — IV. 378
Spectacle — III. 326
Surinam — 360
Water — VIII. 132
Whidah — III. 362
White-chinned — 351
White-eared IV. 382
White-rumped III. 336
Yellow-breasted — 364
Tiklin — VIII. 150
Brown — 151
Collared — 153
Striped — 152
Tinamous, Cinereous IV. 408
Great — 406
Little — 410
Variegated — 408
Tirica — VI. 241
Titiri, or Pipiri — IV. 464
Titmice — V. 379
Titmouse Amorous 454
Bearded — 416
Black — 456
Blue — 412
Cape — 439
Collared — 449
Crested — 443
of Carolina — 447
Great — 394
Great Blue — 452
Languedoc — 429
Long-tailed — 432
Penduline — 420
Siberian — 441
Toupet — 447
Yellow Rump — 450
Throated Gray 451
Tock — VII. 134
Toco — 112
Tocolin — III. 187
Tocro — IV. 412
Todus Cæruleus VII. 222
Cinereus — 221
Varius — 223
Tody, Blue — 222
Cinereus — 221
Green — 219
North American — ib.
Orange-bellied Blue 222
South American, or Tic-tic — 221
Variegated — 223
Tolcana — III. 170
Toucan — VII. 103
Aracari — 120
Black-billed — 124
Blue — ib.
Collared — 118
Green — 120
Pavouine — 119
Piperine — 122
Preacher — 116
Red-bellied — ib.
Yellow-throated — 113
Toucnam Courvi — III. 421
Toui, Yellow-throated VI. 239

I N D E X.

Touraco — VI. 257
Tourocco — II. 489
Tcurte — 494
Touyon — I. 366
Traquet — V. 203
Tringa Alpina VII. 472, 524
Calidris — 505
Canutus — VIII. 134
Cinclus — VII. 521
Equeſtris — 489
Fulicaria VIII. 211
Fuſca — VII. 518
Griſea — 507
Helvetica VIII. 58
Hyperborea — 210
Hypoleucos VII. 514
Interpres — VIII. 123
Lobata — 212
Maculata — 132
Nævia — VII. 507
Ochropus — 494, 509
Pugnax — 498
Squatarola — VIII. 67
Striata — VII. 492
Vanellus — VIII. 47
Trochilus Albus — VI. 52
Amethyſtinus — 15
Auratus — 17, 46
Auritus — 31
Bicolor — 26
Campylopterus — 34
Carbunculus — 28
Colubris — 12
Criſtatus — 22
Cyaneus — 59
Cyanurus — 48
Dominicus — 60
Elatus — 28
Fimbriatus — 30
Forficatus — 37
Furcatus — 36
Gramineus — 56
Hirſutus — 61
Holoſericeus — 50
Glaucopus — 17
Jugularis — 54
Leucogaſter — 25
Leucurus — 57
Longicaudus — 23
Trochilus Macrourus VI. 35
Maculatus — 53
Margaritaceus — 59
Melliſugus — 29
Mellivorus — 33
Minimus — 10
Moſchitus — 19
Mungo — 58
Ouriſſia — 27
Paradiſeus — 51
Pegaſus — 25
Pella — 44
Polytmus — 38
Punctulatus — 47
Ruber — 24
Rubineus — 31
Sapphirinus — 26
Superciliosſus — 46
Thaumantius — 61
Violaceus — 55
Viridiſſimus — 16
Troglodyte — V. 357
Trogon Curucui — VI. 246
Violaceus — 252
Tropic Bird — VIII. 316
Great — 321
Little — 322
Red-ſhafted — 323
Troupiale — III. 175
Black — 193
-capped — 195
Little — 194
Olive — 198
Spotted — 196
Whiſtler — 202
Trumpeter Gold-breaſted — IV. 390
Tufted-neck — VI. 17
Turdus Abyſſinicus III. 368
Æneus — 327
Æthiopicus — 367
Alapi — IV. 388
Amboinenſis — III. 354
Ater — 341
Arundinaceus — 257
Aurantius — 351
Auritus — IV. 382
Bambla — 384

Turdus

I N D E X.

Turdus Barbaricus — III. 277
Bicolor — 336
Borbonica — 355
Cafer — 353
Canorus — 337
Cantans — IV. 385
Capenſis — III. 349
Cayanenſis — 270
Chryſogaſter — 335
Cinereus — 343
Cinnamomeus — 352
Cirrhatus — IV. 381
Cochinchinenſis III. 370
Colma — IV. 380
Columbinus — III. 339
Coraya — IV. 338
Cyanurus — 376
Cyanus — III. 312
Dominicanus — 356
Eremita — 321
Erythropterus — 325
Formicivorus - IV. 379
Guianenſis — III. 254
Hiſpaniolenſis — 364
Indicus — 343
Iliacus — 273
Leucogaſter — 362
Lineatus — IV. 378
Madagaſcarienſis III. 345
Manillenſis — 320
Mauritianus — 347
Merula — 292
Migratorius — 271
Mindanenſis — 346
Minor — 255
Monacha — 366
Morio — 323
Muſicus — 246
Nigerrimus — 359
Olivaceus — 340
Orientalis — 357
Orpheus — 291
Palmarum — 361
Pectoralis — 364
Perſpicillatus — 326
Philippenſis — 280
Pilaris — 265
Plumbeus — 278

Turdus Polyglottus III. 288
Rex — IV. 374
Roſeus — III. 306
Rufus — 286
Rufifrons — 363
Saxatilis — 309
Senegalenſis — 344
Sinenſis — 280
Solitarius — 315
Surinamus — 360
Tinniens — IV. 376
Tintinnabulatus — 383
Torquatus — III. 299
Trichas — V. 288
Viſcivorus — III. 260
Urovang — 338
Zeylonus — 332
Turkey — II. 115
Turnix — 423
Turn-Stone — VIII. 123
Turtle, Collared — II. 487
Common — 482
Turtlette — 490
Turvert — 491
Twite — IV. 66
Tyrant — 463
of Carolina — 469
Cayenne — 472
Louiſiana — 475

Vanga — I. 253
Vardiole — III. 92
Variole — V. 61
Vengoline — IV. 70
Verdin of Cochin China III. 370
Vintſi — VII. 199
Ultra-marine — IV. 48
Umbre, Tufted — VII. 423
Vouroudriou — VI. 341
Upupa Aurantia — 405
Capenſis — 397
Epops — 379
Magna — 403
Mexicana — 400
Paradiſea — 399
Promerops — 401
Vultur, Cinereous — I. 116
Criſtatus — 117

I N D E X.

Vultur Fulvus — I. 110
Gryphus — 139
Leucocephalus — 122
Papa — 126
Percnopterus — 108
Vulture — 104
Alpine — 108
Ash-coloured — 122
Carrion — 130
Cinereous — 116
Fulvous — 110
Hare — 117
King of — 126

Wagtail, African — V. 265
Cape — 264
Cinereous — 252
Gray — 259
Pied — 267
Timor — 266
White — 242
Yellow — 256
Warbler, Æquinoctial — 311
African — 151
Babbling — 128
Banana — 336
Belted — 300
Black-throated — 301
Bloody-side — 304
Blue — 200
Blue-gray — 156
Blue-headed — 343
Blue-striped — 347
Blue-throated — 195
Bourbon — 273
Cærulean — 307
Cayenne — 344
Citron-bellied — 277
Crested — 312
Dark — 222
Dartford — 149
Dusky — 277
Epicurean — 177
Flaxen — 277
Gold-winged — 309
Gray-throated — 321
Grasset — 320
Green — 294
Green Indian — 270

Warbler, Green and White — V. 284
Guira — 348
Half-collared — 316
Hang-nest — 323
Jamaica — 297
Louisiane — 282
Luzonian — 217
Madagascar — 108
Maurice — 275
New York — 152
Olive Brown — 319
Orange-bellied — 318
Orange-headed — 311
Orange-throated — 285
Palm — 333
Passerine — 117, 125
Philippine — 219
Pine — 292
Prothonotary — 315
Quebec — 295
Red-bellied — 346
Red-headed — 280
Red-throated — 306
Reed — 131
Rufous — 137
Rufous and Black — 313
Rufous-tailed — 153
Saint Domingo — 281
Sedge — 134
Simple — 341
Sooty — 218
Spectacle — 225
Sultry — 216
Sybil — 220
Undated — 276
White-chinned — 331
White-eyed — 271
White-poll — 303
Yellow-bellied — 155
Yellow-breasted — 288
Yellow-rumped — 286
Water Hen — VIII. 163
Great — 169
Great of Cayenne — 173
Little — 168
Wheat Ear — V. 288
Greenish Brown — 239

INDEX.

Wheat Ear, Great or Cape V. 238
  Orange-breasted — 239
  Senegal or Rufous — 240
Whimbrel — VIII. 24
  Brasilian — 42
Whin Chat — V. 212
Whip poor Will — VI. 450
Whistler, Black-billed IX. 156
  Crested — 153
  with red Bill and yellow Nostrils 154
White John — I. 86
Wigeon — IX. 143
Widow — IV. 132
  Dominican — 138
  Extinct — 144
  Fire-coloured — 143
  Gold-collared — 134
  Great — 140
  Orange-shouldered — 141
  Shaft-tailed — 137
  Speckled — 142
Wood-chat — I. 244
Woodcock — VII. 442
  Savanna — 460
Woodpecker — 1
  Bengal — 21
  Black — 41
  Black-breasted — 35
  Brown — 66
  Cardinal — 65
  Carolina — 69
  Cayenne — 30
  Crimson-rumped — 23
  Ferruginous — 34
  Goa — 20
  Gold-backed — 24
  Gold-crested — 29
  Gold-winged — 39
  Gray-headed of the Cape of Good Hope — 25
  Great of the Philippines 18
  Great Striped of Cayenne — 29
  Great Variegated — 65
  Greater Spotted — 57
  Green — 6
Woodpecker, Green of Senegal — VII. 23
  Hairy — 72
  Larger Red-crested 48
  Least of Cayenne — 38
  Lesser Black — 54
  Lesser Spotted — 61
  Lineated — 51
  Little — 73
  Little Brown Spotted 66
  Little Olive of St. Domingo — 28
  Little Striped of Cayenne — 30
  Little Striped of Senegal — 21
  Little Variegated of Virginia — 73
  Little Yellow-throated 37
  Manilla — 19
  Minute — 38
  Nubian — 64
  Orange — 25
  Passerine — 28
  Pileated — 48
  Rayed — 26
  Red-headed — 55
  Red-necked — 53
  Rufous — 36
  Southern Three-toed 75
  Spotted Indian — 21
  Spotted of Canada — 67
  Spotted of the Philippines — 19
  Striped of Louisiana — 71
  Striped of St. Domingo 26
  Varied — 68
  Variegated Jamaica 69
  Variegated of Carolina 74
  Variegated of Encenada 72
  Variegated Undated 75
  White-billed — 46
  Yellow of Cayenne 32
  Yellow-bellied — 74
Woodpecker Creeper — 77
Worabee — IV. 46
Worm Eater — V. 327
Wren, Common — 357

Wren,

I N D E X.

Wren, Gold-crested — V. 366
Scotch — 290
Titmouſe — 377
Yellow — 350
Great — 356
Wryneck — VII. 79

Xochitol — III. 185

Yacou — II. 340
Yellow Neck — V. 158
Yunx Minutiſſimus — VII. 38
Torquilla — 79

Zanoe — III. 93
Zilatat — VII. 390
Zonecolin — II. 428

F I N I S.

DIRECTIONS to the BINDER.

Plate	Vol.	Page	Plate	Vol.	Page	Plate	Vol.	Page
1	I.	46	46	II.	444	91	IV.	96
2	—	70	47	—	445	92	—	137
3	—	76	48	—	450	93	—	140
4	—	86	49	—	452	94	—	144
5	—	104	50	—	467	95	—	147
6	—	126	51	—	451	96	—	150
7	—	151	52	—	453	97	—	160
8	—	159	53	—	469	98	—	203
9	—	167	54	—	482	99	—	219
10	—	172	55	—	487	100	Should be Blue Tanagre	230
11	—	179	56	—	487			
12	—	184	57	III.	1			
13	—	192	58	—	11	101	Should be Ortolan	245
14	—	199	59	—	38			
15	—	202	60	—	51	102	—	274
16	—	205	61	—	59	103	—	284
17	—	223	62	—	65	104	—	298
18	—	226	63	—	75	105	—	321
19	—	232	64	—	94	106	—	327
20	—	239	65	—	109	107	—	346
21	—	246	66	—	118	108	—	353
22	—	270	67	—	131	109	—	376
23	—	279	68	—	135	110	—	390
24	—	288	69	—	144	111	—	406
25	—	294	70	—	154	112	—	416
26	—	297	71	—	155	113	—	451
27	—	302	72	—	178	114	—	464
28	—	306	73	—	223	115	V.	1
29	—	323	74	—	257	116	—	23
30	II.	1	75	—	260	117	—	28
31	—	54	76	—	292	118	—	36
32	—	115	77	—	299	119	—	65
33	—	144	78	—	306	120	—	78
34	—	169	79	—	309	121	—	110
35	—	184	80	—	312	122	—	119
36	—	204	81	—	376	123	—	142
37	—	213	82	—	389	124	—	146
38	—	232	83	—	401	125	—	185
39	—	253	84	—	414	126	—	195
40	—	286	85	—	432	127	—	203
41	—	306	86	—	453	128	—	242
42	—	327	87	—	455	129	—	269
43	—	331	88	IV.	51	130	—	357
44	—	378	89	—	85	131	—	401
45	—	396	90	—	92	132	—	416

Plate	Vol.	Page	Plate	Vol.	Page	Plate	Vol.	Page
133	V.	432	174	VII.	270	219	VIII.	302
134	—	458	175	—	277	220	—	316
135	Fig. 1. ſhould be Small Creeper of France	480	176	—	301	221	—	333
			177	—	306	222	—	346
			178	—	316	223	—	365
			179	—	323	224	—	366
			180	—	329	225	—	372
136	—	481	181	—	357	226	—	400
137	VI.	1	182	—	394	227	—	406
138	—	40	183	—	419	228	—	412
139	—	82	184	—	426	229	—	418
140	—	104	185	—	431	230	—	422
141	—	105	186	—	442	231	—	431
142	—	117	187	—	463	232	IX	1
143	—	145	188	—	479	233	—	25
144	—	169	189	—	490	234	—	61
145	—	185	190	—	498	235	—	67
146	—	186	191	—	502	236	—	81
147	—	202	192	—	514	237	—	90
148	—	204	193	VIII.	13	238	—	100
149	—	231	194	—	18	239	—	100
150	—	245	195	—	24	240	—	138
151	—	257	196	—	47	241	—	144
152	—	313	197	—	78	242	—	144
153	—	317	198	—	88	243	—	157
154	—	337	199	—	102	244	—	166
155	—	366	200	—	109	245	—	171
156	—	372	201	—	113	246	—	191
157	—	379	202	—	123	247	—	196
158	—	402	203	—	126	248	—	218
159	—	411	204	—	137	249	—	218
160	—	436	205	—	144	250	—	233
161	—	504	206	—	161	251	—	256
162	VII.	6	207	—	163	252	—	258
163	—	41	208	—	177	253	—	277
164	—	79	209	—	186	254	—	279
165	—	88	210	—	200	255	—	289
166	—	96	211	—	219	256	—	298
167	—	113	212	—	228	257	—	304
168	—	128	213	—	234	258	—	330
169	—	143	214	—	241	259	—	330
170	—	158	215	—	248	260	—	333
171	—	214	216	—	254	261	—	338
172	—	219	217	—	259	262	—	349
173	—	243	218	—	282			